Formal Methods Pacific '97

Springer
Singapore
Berlin
Heidelberg
New York
Barcelona
Budapest
Hong Kong
London
Milan
Paris
Santa Clara
Tokyo

Formal Methods Pacific '97

Proceedings of FMP '97
9-11 July 1997
Wellington, New Zealand

Lindsay Groves
Steve Reeves
(Editors)

Springer

Lindsay Groves
Department of Computer Science
Victoria Universityof Wellington
P O Box 600
Wellington
New Zealand

Steve Reeves
Department of Computer Science
University of Waikato
Hamilton
New Zealand

ISBN 981-3083-31-X

Typesetting: Camera-ready by authors
5 4 3 2 1 0

Preface

Formal Methods Pacific '97 (FMP'97) is a combined conference incorporating the 6th Australasian Refinement Workshop and the 3rd New Zealand Formal Program Development Colloquium. Both series of meetings were initiated to provide a forum for discussion of current research on mathematically-based techniques for design and development of computer systems. Both have previously been kept very informal, so as to be easy to organise and cheap to run.

The Australian Refinement Workshop (ARW) series began in 1990 as a way of bringing together various researchers in Australia working on program refinement, typically based on formalisms such as Z, VDM and the refinement calculus. Over the seven years since the first workshop, the focus has broadened to encompass a wider range of formalisms and techniques. In particular, there has been a growth in interest in tools to support formal software development. The name was also changed from Australian to Australasian Refinement Workshops, to reflect the growing participation from New Zealand. The New Zealand Formal Program Development Colloquium (NZFPDC) series began in 1994 with similar aims, but brought together a smaller and more varied group, with interests in functional languages, type theory and process algebras, as well as the model-based techniques that provided the focus of the ARW meetings.

Formal Methods Pacific came about because those involved in organising ARW and NZFPDC both suggested it was Wellington's turn to host the next meeting; so it was decided to run the two together. The name "Formal Methods Pacific" was chosen to reflect the wider range of interests and geographical spread of the expected participants. The choice of name was the subject of some debate, since many of the techniques we discuss would be better described as "rigorous" or "mathematically based", rather than "formal", and the term "method" is sometimes taken to suggest mechanical procedures — and we are certainly not presuming that we represent the entire Pacific region in these matters. This name was ultimately adopted because it was shorter and more readily understood than any of the alternatives we considered, but these concerns raise important issues for the "Formal Methods" community to consider as the discipline matures and evolves.

As well as deciding to combine with NZFPDC for their 1997 meeting, the ARW organisers felt that the quality of papers presented at recent ARWs was such that they deserved to be published in the form of proceedings, and that producing refereed proceedings was necessary if we were to continue to attract such papers. In order to still retain the essential flavour of informal interaction, and a forum for discussion of current work, we decided to solicit papers in two streams: *Completed Work* and *Work in Progress*. Over 20 papers were submitted in the *Completed Work* category, of which 13 were accepted after refereeing and revision, and are included in these proceedings. Of course, the term *Completed Work* should not be taken as meaning that the work is completely finished, but rather that it has reached the stage of maturity required for refereed publication. *Work in Progress* papers were vetted only for relevance to the conference, and a later deadline was imposed. At the time of writing, 16 papers had been accepted in this category, and extended abstracts have been included in these proceedings; full versions of these papers will be made available at the FMP'97 Web page (`http://www.mcs.vuw.ac.nz/comp/FMP97/`).

We have been fortunate in being able to bring Carroll Morgan to FMP'97 as our invited speaker, to talk about his recent work on probabilistic predicate transformers. Carroll will be presenting three papers, but (very much in keeping with his subject!) exactly what they will be is as yet undetermined. We include in the proceedings an extended abstract, an abstract and one complete paper, as a representative sample on this work.

Carroll Morgan's attendance at FMP'97 has been made possible by the generous sponsorship of the Centre for Discrete Mathematics and Theoretical Computer Science (CDMTCS) at Auckland and Waikato Universities, the University of Waikato Computer Science Department, and the Victoria University of Wellington School of Mathematical and Computing Sciences.

The FMP'97 Organising Committee comprised Lindsay Groves (Convenor), David Carrington, Ian Hayes, Neil Leslie, Steve Reeves, Ken Robinson and Trevor Vickers. The Programme Committee consisted of the members of the Organising Committee, along with Peter Bancroft, Jeremy Gibbons, Jim Grundy, Paddy Krishnan, Brendon Mahony, Ray Nickson and Mark Utting. The *Completed Work* papers were refereed by the Programme Committee, with assistance from Michael Butler, David Edmond, John Hosking, Andrew Martin and Martin Purvis.

We wish to thank our sponsors for their support; the Programme Committee and other referees for their assistance; the CDMTCS directors, Douglas Bridges and Cris Calude, for agreeing to publish these proceedings in the Springer-Verlag DMTCS series; and Ian Shelley and Gillian Chee at Springer Singapore for their help and advice.

These proceedings were formatted with LaTeX using Springer's LLNCS style. We wish to thank Jeremy Gibbons and Jim Grundy for technical assistance with the LaTeX 2.09 and LaTeX2e versions of the LLNCS style.

Finally, we wish to thanks the authors, without whom none of this would have been possible, for their promptness (usually) in producing initial and revised versions of their papers and abstracts, and their co-operation in producing final versions for the proceedings.

Lindsay Groves, Wellington
Steve Reeves, Hamilton
April 1997

Table of contents

Invited Papers

Completed Work: refereed papers

Work in Progress: extended abstracts

Probabilistic Imperative Programming:
A Rigorous Approach*
(Extended Abstract)

Karen Seidel, Carroll Morgan and Annabelle McIver**

Programming Research Group, Oxford University
{carroll,anabel,karen}@comlab.ox.ac.uk

Kozen's operational [8] and predicate-transformer [9] approaches to probabilistic semantics were restricted to deterministic programs; yet during the same period and subsequently the trend in standard (non-probabilistic) program development was to include demonic nondeterminism as a natural part of the process [3]. In the decades since then, the importance of demonic nondeterminism as 'underspecification' has been increasingly recognised [7,17,1], and it it is crucial to the notion of imperative program 'refinement' [2,13–15].

Jifeng He (with CAR Hoare) [5] and we (with JW Sanders) [12] have extended Kozen's work to include nondeterminism naturally, in principle bringing within reach for probabilistic programs all the well-developed techniques for rigorous specification and development of standard imperative programs.

The contribution of this report is to develop the probabilistic/demonic models' definitions and properties in a more practical direction, and to show by counter-example that the 'obvious' probabilistic logic cannot deal with nondeterminism:

- The operational interaction between demonic and probabilistic choice is given in terms of a simple gambling game involving demons and dice, facilitating informal but rigorous reasoning;
- The link with predicate transformers is explained in terms of the 'expected winnings' when gambling as above;
- The probabilistic healthiness conditions governing the new predicate transformer algebra are related directly to elementary probability theory;
- The flexibility of the framework is illustrated by calculating not only expected correctness but also expected efficiency ('Mr Bean's socks'); and
- The obvious alternative — and simpler — approach in terms of probabilities and standard postconditions is shown not to be compositional when nondeterminism is present.

We summarise the syntax and semantics of the probabilistic language, using the 'Monty Hall problem' as an example.

---

* Submitted to *Science of Computer Programming*; an earlier version is available at
http://www.comlab.ox.ac.uk/oucl/publications/tr/TR-6-96.html.
** McIver is, and Seidel was, supported by the EPSRC.

References

1. J.-R. Abrial. *The B Book: Assigning Programs to Meanings*. Cambridge University Press, 1996.
2. R.-J. R. Back. A calculus of refinements for program derivations. *Acta Informatica*, 25:593–624, 1988.
3. E.W. Dijkstra. *A Discipline of Programming*. Prentice Hall International, Englewood Cliffs, N.J., 1976.
4. W. Feller. *An Introduction to Probability Theory and its Applications*, volume 2. Wiley, 2 edition, 1971.
5. Jifeng He, A. K. McIver, and K. Seidel. Probabilistic models for the guarded command language. To appear in FMTA'95 Special Issue of Science of Computer Programming. Also available via *http* [16], 1996.
6. C. Jones. Probabilistic nondeterminism. Monograph ECS-LFCS-90-105, Edinburgh Univ. Edinburgh, U.K., 1990. (PhD Thesis).
7. C.B. Jones. *Systematic Software Development using VDM*. Prentice-Hall, 1986.
8. D. Kozen. Semantics of probabilistic programs. *Journal of Computer and System Sciences*, 22:328–350, 1981.
9. D. Kozen. A probabilistic PDL. In *Proceedings of the 15th ACM Symposium on Theory of Computing*, New York, 1983. ACM.
10. A. K. McIver and C. C. Morgan. Probabilistic predicate transformers: part 2. Technical Report PRG-TR-5-96, Programming Research Group, March 1996. Also available via *http* [16].
11. C. C. Morgan. Proof rules for probabilistic loops. In He Jifeng, John Cooke, and Peter Wallis, editors, *Proceedings of the BCS-FACS 7th Refinement Workshop*, Workshops in Computing. Springer Verlag, July 1996. *http://www.springer.co.uk/ewic/workshops/7RW*.
12. C. C. Morgan, A. K. McIver, and K. Seidel. Probabilistic predicate transformers. *ACM Transactions on Programming Languages and Systems*, 18(3):325–353, May 1996.
13. C. C. Morgan and T. N. Vickers, editors. *On the Refinement Calculus*. FACIT Series in Computer Science. Springer-Verlag, Berlin, 1994.
14. J. M. Morris. A theoretical basis for stepwise refinement and the programming calculus. *Science of Computer Programming*, 9(3):287–306, December 1987.
15. G. Nelson. A generalization of Dijkstra's calculus. *ACM Transactions on Programming Languages and Systems*, 11(4):517–561, October 1989.
16. PSG. Probabilistic Systems Group: Collected reports. *http://www.comlab.ox.ac.uk/oucl/groups/probs/bibliography.html*.
17. J.M. Spivey. *Understanding Z: a Specification Language and its Formal Semantics*. Cambridge University Press, 1988.

Probabilistic Data Refinement
(Abstract)

Annabelle McIver and Carroll Morgan

Programming Research Group, Oxford University
{anabel,carroll}@comlab.ox.ac.uk

A module is said to be refined by another if the second can replace the first in any program without detection. Standard techniques for proving refinements are well established, and involve setting up a correspondence between the (abstract) states in the specification module and the (concrete) states in the implementation module. The correspondence induces a relation of data refinement between the components of the modules.

Such techniques are inadequate when the modules contain probabilistic choice however — for that we need a correspondence between probability distributions over abstract states and probability distributions over concrete states, rather than the usual individual state-to-state relations. We show how the standard methods can be extended in that way, to allow for probabilistic data refinements, and we illustrate the techniques by applying them to the Steam Boiler Example — widely known as as a test case for formal system specification.

One nice way to view data refinement in general is to think of it determining for the customer what can be achieved, given the constraints imposed by the components available for use in the implementation, or alternatively determining the requirements forced on the probabilistic behaviour of the components once the customer has stated his limitations on the global behaviour. In the probabilistic setting that generates quantitative relationships between abstract and concrete reliabilities, expressed as probabilities of success or failure, and we can ask: "given the reliability of the available components, what is the most reliable system we can build from them?"; or "given the contracted reliability of the required system, what are the cheapest components we can use to build it?".

Of some theoretical interest is the extent to which the standard techniques do *not* replay in the probabilistic setting, due to unexpected cross-couplings between probability and nondeterminism. That has implications for modularisation and information hiding, and may lead to techniques for quantifying security.

A Probabilistic Temporal Calculus Based on Expectations

Carroll Morgan and Annabelle McIver*

Programming Research Group, Oxford University.
{carroll,anabel}@comlab.ox.ac.uk.

Abstract. Generalising Boolean-valued predicates to *expectations* — functions from the state space into $[0, 1]$ — allows the definition of probabilistic temporal operators that treat explicit probabilities as well as demonic nondeterminism and divergence.

The conventional operational interpretation of the temporal operators does not generalise so easily: although one may speak of "satisfying a predicate" in the standard case, it is not meaningful to "satisfy an expectation". That difficulty is avoided by giving the operational interpretation of the operators for the probabilistic case in terms of various kinds of gambling game.

Keywords: Temporal logic, probability, formal semantics, program correctness, weakest precondition.

1 Introduction

Temporal logic specialises modal logic to the case in which the "worlds" are states of a computation and the relation between worlds is determined by the state-to-state evolution of some computation; the temporal formulae express properties of (future) states which the computation might or might not reach [2].

Probability can be added to temporal logic in a variety of ways. At one extreme the computation is probabilistic but the temporal formulae do not refer to probabilities in any way; the notion of validity is altered from "true" to "holds with probablility 1" [8]. At the other extreme are logics which allow virtually complete access to explicit probabilities, introducing for example a real-valued term $\Pr \phi$ giving for any formula ϕ the probability of its holding [4, for dynamic logic]. Further variation is provided in both cases by the extent to which demonic non-determinism is allowed and, if it is, to what fairness constraints it must conform.

This report takes a middle course, generalising $\{0, 1\}$-valued predicates to $[0, 1]$-valued expectations (following Kozen [7]) and building a logic[2] based on formulae over those: both demonic non-determinism and divergence (catastrophic

---

* McIver is supported by the EPSRC.

[2] At least we hope to do so: at present we have no "favourite" axioms, and are exploring the logic's suitability in calculations based on laws proved direct from the semantics.

non-termination, or **abort**) are included naturally. The use of $[0, 1]$ means that numbers do occur explicitly in the calculations (in contrast to [8]); however the use of expectations rather than explicit probabilities may mean that the expressiveness is not as great as [4].

To recover explicit probabilities, when desired, we rely on this observation:

> If a random variable $\mathcal{A}$ is $\{0, 1\}$-valued over some state space S, selecting as a characteristic function some sub-space S' of it, then the expected value of $\mathcal{A}$ over a probability distribution $\Pr$ on S is in fact the probability $\Pr(S')$ that $\Pr$ assigns to S' directly.

For example, the probability that a computation will establish some standard predicate ϕ is the expected value of the random variable *1 if ϕ else 0* over the distribution of final states realised by executing the program.

Following [13] we first base standard temporal logic on predicate transformers rather than explicit relations and then, by generalising those to *probabilistic* predicate transformers [12], we show how a probabilistic temporal logic arises naturally; the simplicity of the definitions in [13] is retained.

A satisfactory operational interpretation of the probabilistic logic cannot however be based on the notion of "achieving a predicate", as it is in the standard case, since that is not meaningful for formulae whose value lies strictly between 0 and 1. For the probabilistic case we regard the underlying computation as a gambling game: the player(s) seek to maximise or minimise the expectations using strategies that determine whether or not to "continue playing".[3]

2   Standard temporal logic

Standard temporal logic [2] extends classical predicate logic with modal operators based on distinguished "instants of time" called "states"; the instants are determined by an underlying computation that "steps" from one state to the next.

Conventional models for standard branching-time temporal logic extend models for standard predicate logic so that classical (non-temporal) formulae are interpreted within the states separately, and the "step" relation between states is used to interpret the temporal operators.

The conventional approach thus sees the underlying computation as a relation between initial and (possibly several, with nondeterminism) final states. In this section we review an alternative but equivalent model in which the computation is a predicate transformer rather than a relation [13]. Predicate transformers map sets of final states to sets of initial states, and we show how the standard temporal operators are interpreted in that case.

---

[3] This interpretation is related to Back and von Wright's use of games to motivate the interplay of angelic and demonic nondeterminism expressed by predicate transformers [1]; Stirling suggests a similar game-based interpretation of validity in the modal μ-calculus [17]. In both of those cases only standard predicates are used.

2.1 Syntax of formulae

For classical predicate logic, to which the temporal operators are added, we use the conventional syntax over variables $a, \cdots, z$, with constant, function and relation symbols taken directly from mathematics as needed.

The propositional connectives are false, true, $\neg$ (highest precedence), $\wedge$, $\vee$, $\Rightarrow$ and $\Leftrightarrow$ (lowest precedence); the quantifiers are $\forall, \exists$ with their scope always indicated by explicit parentheses, as in $(\forall x \bullet \mathcal{P})$ for formula $\mathcal{P}$.

The syntax for temporal operators is $\circ\mathcal{P}$ (next), $\Diamond\mathcal{P}$ (eventually), $\Box\mathcal{P}$ (always) and $\mathcal{P} \rhd \mathcal{Q}$ (until); they have highest precedence of all and associate to the right.

2.2 Classical semantics

To interpret classical formulae we fix a universe $\mathbb{V}$ of values and a meaning over $\mathbb{V}$ for any constant, function and relation symbols used. The *state space*

$$S \ := \ Var \to \mathbb{V} \, ,$$

is then the set of functions from the set of variables Var to the values $\mathbb{V}$. (The symbol $:=$ means "is defined to be".)

Formulae denote *predicates*, which are functions from S to the Booleans, or equivalently are subsets of S; we then say that a classical formula is *true* (or *false*) in a state — similarly *holds* (or *does not hold*) in a state, or that a state *satisfies* (or *does not satisfy*) a formula — if the corresponding predicate gives the Boolean *true* (or *false*) when applied to that state.

2.3 Temporal operators informally

For the temporal operators we add to the interpretation above an *underlying computation*, represented by a *predicate transformer* over the state space — the one tranformer gives for each state the possible "single steps" that can take that state to each of its successor states. Usually we call the transformer *step*, and so we have

$$step\colon \ \mathbb{P}S \leftarrow \mathbb{P}S \, ,$$

in which we write the functional arrow backwards to emphasise that the transformer takes final predicates to initial ones. If *step* is deterministic then for each state there is just one successor; where it is nondeterministic there may be more than one.

When convenient, we assume *step* is the meaning of some (syntactic) program *Step* written in the language of guarded commands: thus we have $step = \mathrm{wp}.Step$, where "wp" is the weakest-precondition semantic function [3].

We assume that *step* satisfies the usual *healthiness conditions* for predicate transformers [3], in particular that it is "feasible" and "positively conjunctive":

$$\begin{aligned} &step.false \equiv false & &\text{feasible} \\ &step.(\mathcal{P} \wedge \mathcal{Q}) \equiv step.\mathcal{P} \wedge step.\mathcal{Q} & &\text{positively conjunctive.} \end{aligned}$$

In this section we give an operational intuition for the temporal operators by referring to the execution of various program fragments.

Next **informally** The formula $\circ\mathcal{P}$, pronounced "next $\mathcal{P}$", holds in the current state if one execution of the underlying computation *step* is guaranteed to establish $\mathcal{P}$.

Remembering that $\mathcal{P}$ is (interpreted as) a predicate over the state space, and that $step = \text{wp}.Step$, we have

$$\circ\mathcal{P} \quad\equiv\quad \text{wp}.Step.\mathcal{P} \ . \tag{1}$$

We note the following:

1. If nondeterminism is present in *Step* it is understood *demonically*, so that $\circ\mathcal{P}$ holds only if $\mathcal{P}$ holds in *all* of the next states that *Step* can reach from the current state.
2. Divergence — or "aborting" behaviour — in *Step* is also interpreted demonically: if *Step* can diverge in the current state, then $\circ\mathcal{P}$ does not hold for any $\mathcal{P}$. (Even $\circ\textbf{true}$ does not hold when *Step* diverges.)

Eventually **informally** The formula $\Diamond\mathcal{P}$, pronounced "eventually $\mathcal{P}$", holds if repeated execution of *Step* (including possibly no executions at all) is guaranteed to establish $\mathcal{P}$.

Using programming idioms[4] we have

$$\Diamond\mathcal{P} \quad\equiv\quad \text{wp}.(\textbf{do}\ \neg\mathcal{P}\ \rightarrow\ Step\ \textbf{od}).\textsf{true} \ , \tag{2}$$

so that $\Diamond\mathcal{P}$ holds if execution of the above loop is guaranteed to terminate (in a state satisfying $\mathcal{P}$).

Thus if Program (2) can loop forever then $\Diamond\mathcal{P}$ is false; similarly $\Diamond\mathcal{P}$ is false if *Step* can diverge before $\mathcal{P}$ is established.

Always **informally** The formula $\Box\mathcal{P}$, pronounced "always $\mathcal{P}$", holds if repeated execution of *Step* (including no executions at all) is guaranteed to preserve the truth of $\mathcal{P}$.

We have

$$\Box\mathcal{P} \quad\equiv\quad \text{wnp}.(\textbf{do}\ \mathcal{P}\ \rightarrow\ Step\ \textbf{od}).\textsf{false} \ , \tag{3}$$

where wnp (weakest *nonterminating* precondition) differs from wp only in interpreting looping as success rather than failure,[5] so that for example[6]

$$\text{wnp}.(\textbf{do}\ \textsf{true}\ \rightarrow\ \textbf{skip}\ \textbf{od}).\textsf{false} \quad\equiv\quad \textsf{true} \ .$$

---

[4] This is why the syntactic form *Step* is convenient — though of course it is not necessary, since the semantics of the loop can be given in terms of *step* directly.

[5] We see in Sec. 2.4 that wnp simply takes the greatest- rather than least fixed point for loops, acting like wp otherwise; it differs from the weakest *liberal* precondition (wlp) in that wnp applies wp (rather than wnp again "hereditarily") to the loop body. Thus divergence in the body is treated as failure.

[6] We note in passing that $\Box$, itself a predicate transformer, is therefore not feasible even though *step* is.

Thus $\Box\mathcal{P}$ holds if execution of the above loop, from the current program state, is guaranteed *not* to terminate.

Nondeterminism and divergence continue to be interpreted demonically, so that $\Box\mathcal{P}$ is false if any of the possible executions diverges, or reaches a state not satisfying $\mathcal{P}$.

Unless **informally** The formula $\mathcal{P} \rhd \mathcal{Q}$, pronounced "$\mathcal{P}$ unless $\mathcal{Q}$", holds if repeated execution of *Step* is guaranteed to preserve $\mathcal{P}$ unless $\mathcal{Q}$ is established. We have

$$\mathcal{P} \rhd \mathcal{Q} \quad \equiv \quad \mathrm{wnp.}(\mathbf{do}\ (\mathcal{P} \wedge \neg\mathcal{Q}) \rightarrow \textit{Step}\ \mathbf{od}).\mathcal{Q}\ , \tag{4}$$

so that $\mathcal{P} \rhd \mathcal{Q}$ holds if execution of the above loop is guaranteed either to loop forever or to terminate establishing $\mathcal{Q}$.

Note that if termination can occur when $\mathcal{Q}$ does not hold then $\mathcal{P} \rhd \mathcal{Q}$ is false. Comparing the programs (3) and (4) shows that (as usual) we have the equivalence

$$\Box\mathcal{P} \quad \equiv \quad \mathcal{P} \rhd \mathsf{false}\ ,$$

so that *always* is a special case of *unless*.

2.4   Temporal semantics

Using the informal program-based descriptions of the previous section, we now give the precise meaning of the temporal operators. Once the underlying computation is given they, like classical formulae, denote predicates over the state space.

In each case the conventional wp-semantics gives the definition directly (perhaps after simplification) from Programs (1)–(4) above, except that as noted we use greatest- (ν) rather than least- (μ) fixed points for loops under wnp.

We fix the underlying computation *step*, and to avoid proliferation of "semantic brackets" we use $\mathcal{A}$, $\mathcal{B}$ on the left to stand for formulae while on the right they stand for the corresponding meanings (predicates).[7] The Boolean operators are understood to act pointwise over predicates, and the ordering with respect to which the least- and greatest fixed points are taken is the pointwise extension of *false* $\sqsubseteq$ *true*.

Definition 2.1 *next*: $\circ\mathcal{A} := step.\mathcal{A}$.   ∎

Definition 2.2 *eventually*: $\Diamond\mathcal{A} := (\mu\mathcal{X} \bullet \mathcal{A} \vee \circ\mathcal{X})$.   ∎

Definition 2.3 *always*: $\Box\mathcal{A} := (\nu\mathcal{X} \bullet \mathcal{A} \wedge \circ\mathcal{X})$.   ∎

Definition 2.4 *unless*: $\mathcal{A} \rhd \mathcal{B} := (\nu\mathcal{X} \bullet \mathcal{B} \vee \mathcal{A} \wedge \circ\mathcal{X})$.   ∎

The definitions of this section are based on those of Morris [13], and are related to similar definitions later given by Lukkien and van de Snepscheut [9, 6]. General properties of the temporal operators will be examined when we treat their probabilistic versions.

---

[7] We reserve metavariables $\mathcal{P}$, $\mathcal{Q}$ for classical formulae.

3 Probabilistic temporal logic

Our principal contribution is to generalise the previous section by using probabilistic [12] rather than standard predicate transformers, and to explore the algebra and interpretation of the resulting probabilistic temporal calculus. We generalise in three stages:

1. (Secs. 3.1, 3.2) Whereas standard formulae are expressions of type Boolean over the state space, probabilistic formulae are expressions of type real. We say that probabilistic formulae denote *expectations* (rather than predicates). In this presentation we restrict expectations to lie non-strictly between 0 and 1.[8] Predicates embed into expectations by taking *false* to (the constant function) 0 and *true* to 1.
2. (Sec. 3.3) The underlying computation is extended to include probabilistic behaviour. Syntactically we add a new operator $_p\oplus$ that selects its left (or right) operand with probability p (or $1-p$); semantically the program becomes an expectation transformer, generalising predicate transformers.
3. (Sec. 3.4) The definitions Defs. 2.1–2.4 of the temporal operators are adjusted slightly, so that they operate over expectations rather than predicates.

The above are dealt with in three successive sections further below; first, however, we review the role of expectations as a basis for probabilistic programming logic in general.

3.1 Expectations of probabilistic programs

In probability theory a *random variable* is some real-valued function of the *sample space*; *selections* may be made from the sample space according to a given *probability distribution*. The *expectation* of the random variable is then its "average" value over many selections.

For us the state space is the sample space; executing a probabilistic program determines a distribution over the (final) state space; and the postcondition is a random variable whose expectation is taken over that distribution.

To regard a (standard) postcondition as a random variable we let it take the value 1 when satisfied by a state and 0 when it is not — then the probability of the program's making the postcondition true (regarding it as a predicate) is the same as the postcondition's expected value (regarding it as a random variable over the distribution of final states).

Consider for example a coin-flipping game represented by the program

$$c := H \; {}_{\frac{1}{2}}\oplus \; c := T \, ,$$

and a "payoff" postcondition represented by the expectation

$$1 \; \underline{\text{if}} \; c{=}H \; \underline{\text{else}} \; 0$$

---

[8] We use the $[0, 1]$ range here for compatibility with [11], given our interest in loops. Other possibilities are $[0, \infty)$ and bounded above as in [12], or even $[0, \infty]$.

so that the player wins £1 (say) if he flips heads, and nothing if he flips tails. The expected win *per flip* is then 50p, exactly the probability (in £'s) of establishing the standard postcondition $c = H$.

In fact expectations are more general than explicit probabilities, and in a necessary way — probabilities alone do not distinguish the two programs

$$x := A \sqcap (x := B {}_{\frac{1}{2}}\!\oplus x := C) \tag{5}$$

and

$$(x := A \sqcap x := B) {}_{\frac{1}{2}}\!\oplus (x := A \sqcap x := C). \tag{6}$$

Since nondeterministic choices as above are always understood demonically — they take the alternative whose chance of establishing the postcondition is the lesser — the two programs (5) and (6) agree on their probability of establishing each of the eight possible standard postconditions locating x in the various subsets of $\{A, B, C\}$. On the basis of questions "what is the probability of establishing the standard $\mathcal{P}$?" those programs are indistinguishable, no matter which $\mathcal{P}$ is chosen.

For example, the probability of establishing postcondition $x \neq C$ is

$$1 \sqcap (\tfrac{1}{2} \cdot 1 + \tfrac{1}{2} \cdot 0) = \tfrac{1}{2} \text{ for Program (5)}$$
$$\text{and } \tfrac{1}{2}(1 \sqcap 1) + \tfrac{1}{2}(1 \sqcap 0) = \tfrac{1}{2} \text{ for Program (6)}.$$

Yet the programs *should* be distinguished, since Program (5) treats B and C equally (if the result is not A, then it is equally likely to be B or C), but Program (6) does not. (See [15] for further discussion of the example.)

If we use a probabilistic postcondition — an expectation — such as

$$\begin{array}{l} \tfrac{1}{2} \text{ if } x = A \\ 1 \text{ if } x = B \\ 0 \text{ if } x = C, \end{array} \tag{7}$$

then the programs are distinguished. For Program (5) we have pre-expectation

$$\frac{1}{2} \sqcap (\frac{1}{2} \cdot 1 + \frac{1}{2} \cdot 0) \;=\; \frac{1}{2},$$

but for Program (6) we have

$$\frac{1}{2}(\frac{1}{2} \sqcap 1) + \frac{1}{2}(\frac{1}{2} \sqcap 0) \;=\; \frac{1}{4}.$$

In [12] it is shown that it is the interaction of demonic and probabilistic choice that requires proper expectations, rather than predicates and probabilities, and that expectations are exactly "general enough".

3.2   Syntax and semantics of expectations

Probabilistic formulae are written as expressions of type $\mathbb{R}$ over the variables of the program, and are restricted to the range $[0, 1]$. Standard (Boolean-valued) formulae are embedded using the brackets $[\cdot]$, whose effect semantically is to convert *false* to 0 and *true* to 1. Thus for example expectation (7) could be written

$$[x = A]/2 + [x = B] \ .$$

Where convenient we use the operators $\sqcup$ (maximum) and $\sqcap$ (minimum) instead of $\wedge$ and $\vee$, since for example

$$[P \wedge Q] \quad \equiv \quad [P] \sqcap [Q] \ ;$$

and for quantifiers we can use

$$(\sqcup x \bullet A) \quad \text{distributed maximum}$$
$$(\sqcap x \bullet A) \quad \text{distributed minimum}$$

to generalise $\exists$ and $\forall$ over formulae in which x may be free.

Finally, by analogy with $\equiv$ (equal in all states) for comparing standard formulae, we generalise semantic entailment by defining

$$A \Rrightarrow B := \text{in all states } A \text{ is no more than } B, \text{ and}$$
$$A \Lleftarrow B := \text{in all states } A \text{ is no less than } B.$$

Thus $\equiv$, $\Rrightarrow$ and $\Lleftarrow$ are relations between probabilistic formulae, while $=$, $\leq$ and $\geq$ retain their usual (Boolean-valued) meanings *within* formulae.

This table illustrates our terminology for formulae:

	classical	temporal
standard	$x = 1$	$\Diamond(x = 1)$
probabilistic	$[x = 1]/2$	$\Diamond[x = 1]/2$

3.3   Expectation transformers

The probabilistic semantics of a simple imperative language is given in Fig. 1, including explicit operators for nondeterministic ($\sqcap$) and probabilistic ($_p\oplus$) choice: such programs denote *expectation transformers*.

Standard semantics — using predicate transformers — is preserved by the extension in the sense that, for standard program *Prog* and standard formula P, we have

$$[\text{wp}_s.Prog.P] \quad \equiv \quad \text{wp}_p.Prog.[P] \ ,$$

where (here only) we write wp_s, wp_p explicitly for the standard, probabilistic weakest precondition functions respectively.

The healthiness conditions for probabilistic programs — extending the feasibility and positive conjunctivity conditions for standard programs — are given in [12], and will be introduced as required below. We note here for well-definedness that wp.*Prog*.A is an expectation (lies in $[0, 1]$) if A is.

$$\text{wp.}\mathbf{abort}.\mathcal{P} := 0$$
$$\text{wp.}\mathbf{skip}.\mathcal{P} := \mathcal{P}$$
$$\text{wp.}(x := E).\mathcal{P} := \mathcal{P}^x_E$$
$$\text{wp.}(\mathit{Prog, Prog'}).\mathcal{P} := \text{wp.}\mathit{Prog}.(\text{wp.}\mathit{Prog'}.\mathcal{P})$$
$$\text{wp.}(\textbf{if } B \textbf{ then } \mathit{Prog} \textbf{ else } \mathit{Prog'} \textbf{ fi}).\mathcal{P} := \text{wp.}\mathit{Prog}.\mathcal{P} \textbf{ if } B \textbf{ else } \text{wp.}\mathit{Prog'}.\mathcal{P}$$
$$\text{wp.}(\mathit{Prog} \sqcap \mathit{Prog'}).\mathcal{P} := \text{wp.}\mathit{Prog}.\mathcal{P} \wedge \text{wp.}\mathit{Prog'}.\mathcal{P}$$
$$\text{wp.}(\mathit{Prog} \,{}_p{\oplus}\, \mathit{Prog'}).\mathcal{P} := p \cdot \text{wp.}\mathit{Prog}.\mathcal{P} + (1-p) \cdot \text{wp.}\mathit{Prog'}.\mathcal{P}$$
$$(\mu X \bullet C) := \text{least fixed point of } \mathit{cntx}, \text{ defined}$$
$$\text{wp.}C = \mathit{cntx}.(\text{wp.}X) \text{ for any } X.$$

Fig. 1. Probabilistic wp-semantics as expectation transformers

Lemma 3.1 *(probabilistic) feasibility*   For any probabilistic program *Prog* and expectation $\mathcal{A}$ we have

$$0 \quad \Rrightarrow \quad \text{wp.}\mathit{Prog}.\mathcal{A} \quad \Rrightarrow \quad \sqcup \mathcal{A} \ .$$

Proof: Use either structural induction over the semantics in Fig. 1 or direct reference to the healthiness conditions for expectation transformers [12].   ∎

We conclude this section with three examples of probabilistic reasoning, based on simple coin-flipping programs. We write $c := (H \,{}_p{\oplus}\, T)$ to abbreviate

$$c := H \,{}_p{\oplus}\, c := T \ ,$$

and write $c := (H \sqcap T)$ similarly; we use $\overline{p}$ for $1-p$.

Example: two successive probabilistic choices   Two coins c, d are flipped in succession, the first giving heads H with probability p (giving tails T with probability $\overline{p}$) and the second giving heads with probability q. The probability that after both flips the coins will show the same face is the probability that the program

$$c := (H \,{}_p{\oplus}\, T);\ d := (H \,{}_q{\oplus}\, T)$$

will establish the standard postcondition $c = d$; using expectations, we calculate from Fig. 1

$$
\begin{aligned}
&\text{wp.}(c := (H \,{}_p{\oplus}\, T));\ d := (H \,{}_q{\oplus}\, T)).[c = d] \\
\equiv\ &\text{wp.}(c := (H \,{}_p{\oplus}\, T)). && \text{sequential composition} \\
&\quad (\text{wp.}(d := (H \,{}_q{\oplus}\, T)).[c = d]) \\
\equiv\ &\text{wp.}(c := (H \,{}_p{\oplus}\, T)). && \text{probabilistic choice} \\
&\quad (q \cdot \text{wp.}(d := H).[c = d] + \overline{q} \cdot \text{wp.}(d := T).[c = d]) \\
\equiv\ &\text{wp.}(c := (H \,{}_p{\oplus}\, T)).(q[c = d]^d_H + \overline{q}[c = d]^d_T) && \text{assignments} \\
\equiv\ &\text{wp.}(c := (H \,{}_p{\oplus}\, T)).(q[c = H] + \overline{q}[c = T]) && \text{substitutions} \\
\equiv\ &p(q[H = H] + \overline{q}[H = T]) + \overline{p}(q[T = H] + \overline{q}[T = T]) && \text{as above} \\
\equiv\ &p(q \cdot 1 + \overline{q} \cdot 0) + \overline{p}(q \cdot 0 + \overline{q} \cdot 1) && [\mathit{false}] \equiv 0;\ [\mathit{true}] \equiv 1 \\
\equiv\ &pq + \overline{p}\,\overline{q} \ .
\end{aligned}
$$

Since the "post-expectation" $[c = d]$ is standard[9], we recognise the "pre-expectation" $pq + \overline{p}\,\overline{q}$ as indeed the probability of the program's establishing the standard postcondition $c = d$. Note however that the intermediate expectation (between the two statements)

$$q[c = H] + \overline{q}[c = T]$$

is *not* standard.

Example: nondeterministic then probabilistic choice In this case the first coin is "placed" by a demon striving to make the probability of establishing $c = d$ as low as possible; the second coin is flipped, as before. We have therefore the program

$$c := (H \sqcap T); \ d := (H \,_q\!\oplus T) \,,$$

and calculate

$$
\begin{aligned}
&\text{wp}.(c := (H \sqcap T); \ d := (H \,_q\!\oplus T)).[c = d] \\
\equiv\ &\text{wp}.(c := (H \sqcap T)).(q[c = H] + \overline{q}[c = T]) && \text{as above} \\
\equiv\ &\quad (q[H = H] + \overline{q}[H = T]) && \text{nondeterministic choice} \\
&\quad \sqcap (q[T = H] + \overline{q}[T = T]) \\
\equiv\ &\quad q \sqcap \overline{q} \,.
\end{aligned}
$$

Here the demon — given first choice — selects for c the face that is *least* likely to result from the subsequent flip of d, thus minimising the probability that $c = d$ will be established by the second flip.

Example: probabilistic then nondeterministic choice In this case the first coin is flipped and the second is placed. The program is

$$c := (H \,_p\!\oplus T); \ d := (H \sqcap T) \,,$$

and we calculate

$$
\begin{aligned}
&\text{wp}.(c := (H \,_p\!\oplus T); \ d := (H \sqcap T)).[c = d] \\
\equiv\ &\text{wp}.(c := (H \,_p\!\oplus T)). && \text{nondeterministic choice; assignment} \\
&\quad ([c = H] \sqcap [c = T]) \\
\equiv\ &\text{wp}.(c := (H \,_p\!\oplus T)).[c = H \wedge c = T] && \sqcap\text{-}\wedge \text{ correspondence} \\
\equiv\ &\text{wp}.(c := (H \,_p\!\oplus T)).[\textit{false}] \\
\equiv\ &\text{wp}.(c := (H \,_p\!\oplus T)).0 \\
\equiv\ &\quad 0 \,.
\end{aligned}
$$

In this case the demon — given second choice — is able on every occasion to make d differ from c, no matter how the flip of c resolved. Thus the least *guaranteed* chance of the program's establishing $c = d$ is just 0 — there is no guarantee at all, since we must always assume the demon might act against us.

---

[9] It is the embedding of a standard predicate.

3.4   Probabilistic temporal operators

The final step for defining probabilistic versions of the temporal operators is to adjust Defs. 2.1–2.4 from Sec. 2.4 so that they act over expectations rather than predicates. We replace $\wedge,\vee$ by $\sqcap,\sqcup$ respectively and take fixed points of real- rather than Boolean-valued functions:

Definition 3.2 probabilistic *next*: $\circ\mathcal{A} := step.\mathcal{A}$.   ∎

Definition 3.3 probabilistic *eventually*: $\Diamond\mathcal{A} := (\mu\mathcal{X} \bullet \mathcal{A} \sqcup \circ\mathcal{X})$.   ∎

Definition 3.4 probabilistic *always*: $\Box\mathcal{A} := (\nu\mathcal{X} \bullet \mathcal{A} \sqcap \circ\mathcal{X})$.   ∎

Definition 3.5 probabilistic *unless*: $\mathcal{A} \rhd \mathcal{B} := (\nu\mathcal{X} \bullet \mathcal{B} \sqcup \mathcal{A} \sqcap \circ\mathcal{X})$.   ∎

To establish the credibility of the definitions, in subsequent sections we give both an operational interpretation of the probabilistic temporal operators and an investigation of the laws they satisfy.

As a start, however, we recall here the programs from Sec. 2.3, allowing the computation *Step* now to be probabilistic but stipulating for simplicity that the formula $\mathcal{P}$ remain standard. Then in particular Program (2) becomes

$$\mathrm{wp.}(\mathbf{do}\ \mathcal{P}{=}0\ \rightarrow\ Step\ \mathbf{od}).\mathcal{P}\ , \tag{8}$$

with the equality $\mathcal{P}{=}0$ "converting" the expectation $\mathcal{P}$ to a Boolean, as required syntactically for the guard, and the explicit postcondition $\mathcal{P}$ replacing **true**, simply to emphasise that the "aim" of the loop is to establish $\mathcal{P}$.[10]

When $\mathcal{P}$ is standard, the expression (8) gives $\Diamond\mathcal{P}$ in agreement with Definition 3.3 — and so we can as usual interpret (8) as a probability: in a given state the value of $\Diamond\mathcal{P}$ is indeed the probability that, starting from that state, repeated execution of *Step* will establish $\mathcal{P}$ eventually.

Similar reasoning applies to $\circ P$, $\Box\mathcal{P}$ and $\mathcal{P} \rhd \mathcal{Q}$, based on Programs (1), (3) and (4) when as above $\mathcal{P}$ and $\mathcal{Q}$ are standard (and classical), showing that standard temporal formulae over probabilistic computations may be interpreted as probabilities directly, as one would hope. That interpretation is sufficient for programs containing no demonic nondeterminism.

As noted at (5) and (6) however, standard formulae have insufficient resolution when probabilistic and demonic nondeterminism interact: one must use expectations. And the above analogies fail — for all four programs — in the more general case when the formulae are not standard, because it does not make sense to speak of "establishing" a proper (not standard) expectation.

To interpret probabilistic temporal formulae, therefore, we rely on the shift of viewpoint described in the next section.

---

[10] By analogy with predicates, we say that a standard expectation is *established* in a state where its value is 1.

4 Temporal operators as games

If one considers a program's establishing a postcondition to be "winning a game", then our generalisation to expectations simply adds the extra information of how much is won: the standard case embeds as "win 1 if the postcondition is established; win 0 if it is not".

For an operational interpretation of probabilistic temporal formulae we consider *poker machines*[11]; they have a window in which various symbols are visible; pulling a handle changes the symbols probabilistically; and each configuration of visible symbols is associated with a certain reward (possibly 0). In computational terms, each configuration of symbols shown in the window is a state; the underlying computation is the effect of one pull of the handle; and the post-expectation is the "pay function", giving the worth to the player of each symbol configuration should he press *pay* at that point.

Each of the temporal operators corresponds to a different way of gambling with the machine, and we consider them in turn.

4.1 The *next* game

The simplest game is *next* where, direct from Definition 3.2, we can see that the game consists of pulling the handle exactly once and then pressing *pay*: the expectation $\circ A$ evaluated over the initial configuration (showing *before* the handle is pulled) gives the win determined jointly by the expectation A and the probabilistic distribution of final configurations (that could show *after* the handle is pulled).

Suppose for example the configurations are just natural numbers $\mathbb{N}$, the effect of a handle pull is the computation $n := (n + 1 \ {}_p\!\oplus 0)$, and the pay function is $n/(n + 1)$. Then the expected win from a single handle pull is

$$
\begin{aligned}
&\circ(n/(n + 1)) \\
\equiv\quad &\text{wp.}(n := (n + 1 \ {}_p\!\oplus 0)).(n/(n + 1)) && \text{Definition 3.2} \\
\equiv\quad &p \cdot (n + 1)/(n + 2) + \overline{p} \cdot 0 && \text{probabilistic choice; assignment} \\
\equiv\quad &p(n + 1)/(n + 2) \, ,
\end{aligned}
$$

from which it is evident that the expected win does depend on the initial configuration: a gambler starting at 0 every time would expect an average win per game of $p/2$ in the long run; but the larger the initial configuration, the larger the expected win from a single handle pull would be (approaching p in the limit).

4.2 The *eventually* game

For *eventually* the player pulls the handle repeatedly, and decides on the basis of the current configuration whether to stop and press *pay*, or to pull again. Thus for $\Diamond A$ we consider an expectation

$$
\text{wp.}(\textbf{do } G \to Step \textbf{ od}).A \, , \tag{9}
$$

---

[11] *slot* machines (US); *fruit* machines (UK).

in which the guard G — a predicate — represents the gambler's *strategy*: if it is true (in a configuration) then he plays again; if it is false, he presses *pay* and stops playing.[12]

For the gambler some strategies G are clearly worse than others. The strategy $G := \mathsf{true}$ continues the game forever, which we interpret as losing (as winning 0 no matter what $\mathcal{A}$ is) since *pay* is never pressed; the strategy false represents pressing *pay* without pulling the handle at all, in which case the amount won is determined by the current configuration.[13]

For standard pay function $\mathcal{P}$, which pays 1 for some configurations and 0 for all others, an optimal strategy is given by the predicate $\mathcal{P}{=}0$ that continues play as long as the current configuration pays nothing — since things can only get better. Whenever $\mathcal{P}{=}1$, however, the gambler should stop immediately — he can't win more than 1, and could well win less if he continued.

Indeed (recall Program (8)) the expected win above is $\Diamond\mathcal{P}$ and remarkably, as the following lemma shows, even for proper expectations $\mathcal{A}$ the expectation $\Diamond\mathcal{A}$, applied to the initial state, gives the best that can be obtained for *any* strategy.

Lemma 4.1 For underlying computation *Step*, expectation $\mathcal{A}$ and strategy G we have

$$\mathrm{wp}.(\mathbf{do}\ G \to \mathit{Step}\ \mathbf{od}).\mathcal{A} \quad \Rrightarrow \quad \Diamond\mathcal{A}\ ,$$

showing that $\Diamond\mathcal{A}$ is an upper bound over all strategies G for the eventually game.

Proof: Compare Definition 3.3 with the usual least-fixed-point definition of $\mathbf{do}\cdots\mathbf{od}$, and use monotonicity of μ.   ■

More significant is the complementary result, that for any computation *Step* and expectation $\mathcal{A}$ there is a strategy G whose expected win approaches $\Diamond\mathcal{A}$ arbitrarily closely — in fact if the state space is finite, there is a G that realises $\Diamond\mathcal{A}$ exactly.

We sketch a proof of the above, relying on informal operational arguments rather that the wp-definitions, for the case in which *Step* is deterministic and non-divergent; we assume at first that the state space is finite.[14]

The optimal strategy will be the predicate $\mathcal{A} < \Diamond\mathcal{A}$, and our first lemmas establish that it guarantees termination of the game.

Lemma 4.2 For any (probabilistic) loop with deterministic and non-divergent body, the (standard) predicate "has probability 0 of termination" is invariant.

---

[12] Note we consider only strategies based on the configuration: the predicate G cannot express strategies like "play 5 times then stop".

[13] On a real poker machine the handle must be pulled at least once after inserting your money: the normal game would thus be $\circ\Diamond\mathcal{A}$.

[14] The full proof, using wp for rigour and allowing both nondeterminism and divergence, has the same structure as the following but depends substantially on technical results from [12, 11].

None of the technical results elsewhere in the paper depend on this section.

Proof: Let $T.s$ give the probability of termination from state s; and consider a single execution of the loop body that takes some initial state s to a final state s'. We show the contrapositive, that $T.s' > 0$ implies $T.s > 0$ also.

Since the body is deterministic and non-divergent, the transition from s to s' has some non-zero probability p — otherwise it could never be taken.[15] That gives

$$T.s \ \geq \ p \cdot T.s' \ > \ 0$$

as required. ∎

Lemma 4.3 If a non-empty (standard) predicate I is preserved by the computation *Step*, and the state space is finite, then $\mathcal{A}.s = (\Diamond\mathcal{A}).s$ holds for some state s satisfying I.

Proof: Suppose first that I is **true**. From Definition 3.3 we have $\mathcal{A} \Rrightarrow \Diamond\mathcal{A}$ for any $\mathcal{A}$ and, with Lemma 3.1 as well, we have also that $\Diamond\mathcal{A} \Rrightarrow \sqcup\mathcal{A}$. Thus because the state space is finite we can choose state s with $\mathcal{A}.s = \sqcup\mathcal{A}$, and have then

$$\sqcup\mathcal{A} \ = \ \mathcal{A}.s \ \leq \ (\Diamond\mathcal{A}).s \ \leq \ \sqcup\mathcal{A} \ ,$$

so that $\mathcal{A}.s = (\Diamond\mathcal{A}).s$ — and any s satisfies I when I is **true**.

Now if I is not **true**, but still is non-empty and invariant, we simply restrict the whole system to I; the above argument repeated "within I", in that sense, gives the s required as before. ∎

Lemma 4.4 For deterministic and non-divergent *Step*, the loop

$$\textbf{do} \ \mathcal{A}{<}\Diamond\mathcal{A} \rightarrow \textit{Step} \ \textbf{od} \tag{10}$$

terminates with probability 1 from every initial state.

Proof: Let Z be the predicate denoting the states at which the probability of termination is 0. From Lemma 4.2 we have invariance of Z; from Lemma 4.3 we have that $\mathcal{A} = \Diamond\mathcal{A}$ is attained at some s in Z if Z is non-empty. But that would be a contradiction, since when $\mathcal{A}.s = \Diamond\mathcal{A}.s$ termination from s is immediate, and so s could not lie in Z.

Hence Z is empty: the probability of (10)'s termination is everywhere non-zero.

Since the state space is finite, we conclude from the *0-1 law*[16] [5, 11] that the probability of termination is in fact everywhere 1. ∎

We have shown that the strategy $\mathcal{A} < \Diamond\mathcal{A}$, the very reasonable "keep going as long as taking $\mathcal{A}$ now is strictly worse than waiting for $\mathcal{A}$ later", is terminating at least. We finish off as follows.

---

[15] This is where determinism is used. If nondeterminism were present, a transition could be taken even though its minimum *guaranteed* probability was zero.

[16] The *0-1 law* states that if the probability of a process' eventual escape from a system is bounded away from 0, then in fact that probability is 1.

Lemma 4.5 For deterministic and non-divergent *Step*, and finite state space, we have

$$\Diamond\mathcal{A} \;\Rrightarrow\; \text{wp.}(\textbf{do }\mathcal{A}{<}\Diamond\mathcal{A} \to Step\textbf{ od}).\mathcal{A} \tag{11}$$

Proof: We use the probabilistic analogue of "loop invariants": for general loop

$$\textbf{do }G \to Body\textbf{ od} \;,$$

if expectation $\mathcal{B}$ is *invariant* — if $\mathcal{B}\sqcap[G] \Rrightarrow \text{wp.}Body.\mathcal{B}$ holds — and the loop terminates with probability 1, then

$$\mathcal{B} \;\Rrightarrow\; \text{wp.}(\textbf{do }G \to Body\textbf{ od}).(\mathcal{B}\sqcap[\neg G]) \;. \tag{12}$$

(See [16], [11, Theorem A.3].)

To use that result, first we show that $\Diamond\mathcal{A}$ is invariant:

$$
\begin{aligned}
&\quad \Diamond\mathcal{A}\sqcap[\mathcal{A} < \Diamond\mathcal{A}] \\
\Rrightarrow&\quad \Diamond\mathcal{A}\sqcap[\Diamond\mathcal{A} = \text{wp.}Step.\Diamond\mathcal{A}] &&\text{Definition 3.3}\\
\Rrightarrow&\quad \text{wp.}Step.\Diamond\mathcal{A} \;.
\end{aligned}
$$

But we know from Lemma 4.4 that the loop terminates, so we use (12) to conclude with

$$
\begin{aligned}
&\quad \Diamond\mathcal{A} \\
\Rrightarrow&\quad \text{wp.}(\textbf{do }\mathcal{A} < \Diamond\mathcal{A} \to Step\textbf{ od}). &&\text{termination; loop rule above}\\
&\quad (\Diamond\mathcal{A}\sqcap[\mathcal{A} \geq \Diamond\mathcal{A}]) \\
\Rrightarrow&\quad \text{wp.}(\textbf{do }\mathcal{A}{<}\Diamond\mathcal{A} \to Step\textbf{ od}).\mathcal{A} &&[\mathcal{A} \geq \Diamond\mathcal{A}]\sqcap\Diamond\mathcal{A} \Rrightarrow \mathcal{A}
\end{aligned}
$$

as required.   ∎

Putting Lemmas 4.1 and 4.5 together gives

Lemma 4.6 For deterministic and non-divergent *Step*, and finite state space, we have

$$\Diamond\mathcal{A} \;\equiv\; \text{wp.}(\textbf{do }\mathcal{A}{<}\Diamond\mathcal{A} \to Step\textbf{ od}).\mathcal{A} \;,$$

so showing that $\mathcal{A} < \Diamond\mathcal{A}$ is the optimal strategy.   ∎

Before moving to infinite state spaces, we give an example showing that finiteness is indeed necessary for Lemma 4.5. Let *Step* be $n := n + 1$ over the infinite state space $\mathbb{N}$, and define $\mathcal{A} := n/(n + 1)$. From Definition 3.3 we have $\Diamond\mathcal{A} \equiv (\sqcup N \bullet (\Diamond\mathcal{A})_N)$, where

$$
\begin{aligned}
(\Diamond\mathcal{A})_0 &\equiv 0 \\
(\Diamond\mathcal{A})_1 &\equiv n/(n+1) \sqcup \text{wp.}(n := n + 1).0 \\
&\equiv n/(n+1) \\
(\Diamond\mathcal{A})_2 &\equiv n/(n+1) \sqcup \text{wp.}(n := n + 1).(n/(n+1)) \\
&\equiv (n+1)/(n+2) \\
&\;\;\vdots \\
(\Diamond\mathcal{A})_N &\equiv n/(n+1) \sqcup \text{wp.}(n := n + 1).(\Diamond\mathcal{A})_{N-1} \\
&\equiv (n + N - 1)/(n + N) \;,
\end{aligned}
$$

and hence — taking the limit — we see that $\Diamond\mathcal{A} \equiv 1$. But then $\mathcal{A} < \Diamond\mathcal{A}$ is identically true, and with that strategy the expectation

$$\text{wp.}(\textbf{do true} \to n:=\ n+1\ \textbf{od}).(n/(n+1))$$

is 0 everywhere, due to non-termination.

Thus for infinite state spaces we must reconsider Lemmas 4.3 and 4.4, where the finiteness assumption is used. Let some predicate F over the state space be true for only finitely many states, and define

$$Step_F := \textbf{if } F \textbf{ then } Step \textbf{ fi}$$
$$\Diamond_F\mathcal{A} := \text{``}\Diamond\mathcal{A} \text{ interpreted over computation } Step_F\text{''}$$

Then it is immediate from Definition 3.3 that $(\Diamond_F\mathcal{A}).s = \mathcal{A}.s$ for any state s not satisfying F, and that suffices to recover both lemmas: for Lemma 4.3 note that if $I \not\subseteq F$ then $(\Diamond_F\mathcal{A}).s = \mathcal{A}.s$ is attained for $s \in I-F$, and that otherwise I is finite; for Lemma 4.4 we have that the probability of termination is bounded away from 0, since it is 1 except in finitely many states and non-zero for the rest.

Thus from Lemma 4.6 we have

$$\begin{aligned}
&\Diamond_F\mathcal{A}\\
\equiv\ &\text{wp.}(\textbf{do } \mathcal{A}{<}\Diamond_F\mathcal{A} \to Step_F \textbf{ od}).\mathcal{A} \qquad\qquad (13)\\
\equiv\ &\text{wp.}(\textbf{do } \mathcal{A}{<}\Diamond_F\mathcal{A} \to Step \textbf{ od}).\mathcal{A}\ ,
\end{aligned}$$

where we can replace $Step_F$ in the body by $Step$ because the body is executed only within F, where they do not differ. That gives our principal theorem for *eventually*:

Theorem 4.7 For deterministic, non-divergent and continuous computation $Step$, and expectation $\mathcal{A}$, the expectation $\Diamond\mathcal{A}$ is the supremum over all standard strategies G of

$$\text{wp.}(\textbf{do } G \to Step \textbf{ od}).\mathcal{A}\ . \qquad\qquad (14)$$

Proof: From (13) we have an explicit set of strategies $G_F := \mathcal{A}{<}\Diamond_F\mathcal{A}$ whose limit in (14) attains $(\sqcup F \bullet \Diamond_F\mathcal{A})$; from Definition 3.3 and the assumed continuity of $Step$, however, it is routine to show that

$$(\sqcup F \bullet \Diamond_F\mathcal{A})\ \equiv\ \Diamond\mathcal{A}\ .$$

$\blacksquare$

It is instructive to specialise Lemma 4.6 to the standard case: we then have

$$\begin{aligned}
&\Diamond\mathcal{P}\\
\equiv\ &\text{wp.}(\textbf{do } \mathcal{P}{<}\Diamond\mathcal{P} \to Step \textbf{ od}).\mathcal{P}\\
\equiv\ &\text{wp.}(\textbf{do } (\mathcal{P}{=}0) \wedge (\Diamond\mathcal{P}{=}1) \to Step \textbf{ od}).\mathcal{P}\ . \qquad \text{\small $\mathcal{P}$ is standard}
\end{aligned}$$

Thus for standard $\mathcal{P}$ the "recommended" strategy is to continue playing as long as $\mathcal{P}$ does not hold ($\mathcal{P}{=}0$) but only *provided* it will hold eventually ($\Diamond\mathcal{P}{=}1$) — since otherwise there is no point in continuing, and one might as well give up now. The earlier operational interpretation of $\Diamond\mathcal{P}$ in Program (8) remains valid, but its strategy is missing the second conjunct: if that gambler is ignorant of $\Diamond\mathcal{P}{\neq}1$ he continues to play, but can never win.

4.3   The *always* game

It can be shown [10] that for deterministic and non-diverging *step* we have

$$step.(1 - \mathcal{A}) \;\;\equiv\;\; 1 - step.\mathcal{A} \;,$$

where the subtraction $1-$ acts pointwise over $\mathcal{A}$; then straightforward comparison of Definitions 3.3 and 3.4 gives us the following lemma.

Lemma 4.8 If the underlying computation is deterministic and non-diverging, then

$$\square\mathcal{A} \;\;\equiv\;\; 1 - \Diamond(1 - \mathcal{A})$$

for any expectation $\mathcal{A}$.   ∎

With that lemma and Theorem 4.7 we can calculate the strategy for an *always* game, as follows: writing $\overline{\mathcal{A}}$ for $1 - \mathcal{A}$ when convenient, we have

$$
\begin{aligned}
&\square\mathcal{A} \\
\equiv\;& 1 - \Diamond\overline{\mathcal{A}} &&\text{Lemma 4.8}\\
\equiv\;& 1 - (\sqcup F \bullet \text{wp.}(\textbf{do } \overline{\mathcal{A}}{<}\Diamond_F\overline{\mathcal{A}} \to \textit{Step } \textbf{od}).\overline{\mathcal{A}}) &&\text{Theorem 4.7}\\
\equiv\;& (\sqcap F \bullet 1 - \text{wp.}(\textbf{do } \overline{\mathcal{A}}{<}\Diamond_F\overline{\mathcal{A}} \to \textit{Step } \textbf{od}).\overline{\mathcal{A}}) &&\\[2ex]
\equiv\;& &&\text{\textit{Step} deterministic and non-divergent: see below}\\
&(\sqcap F \bullet \text{wnp.}(\textbf{do } \overline{\mathcal{A}}{<}\Diamond_F\overline{\mathcal{A}} \to \textit{Step } \textbf{od}).\mathcal{A}) &&\\[2ex]
\equiv\;& (\sqcap F \bullet \text{wnp.}(\textbf{do } \mathcal{A}{>}\square_F\mathcal{A} \to \textit{Step } \textbf{od}).\mathcal{A}) \;, &&\text{Lemma 4.8}
\end{aligned}
$$

where by analogy with $\Diamond_F$ we define $\square_F\mathcal{A}$ to be $\square\mathcal{A}$ interpreted over $\textit{Step}_F$ instead of *Step*. For the deferred justification note that the $(1-)$ converts the least fixed point to a greatest fixed point, while at the same time replacing $step.\mathcal{X}$ by $1 - step.\overline{\mathcal{X}}$ in the fixed-point equation: but since *step* is deterministic and non-divergent those two terms are equal.

Thus $\square\mathcal{A}$ is given as an *infimum* of strategies strategies of the gambling game. The analogue of Lemma 4.1 shows that it is in fact the infimum over *all* strategies D of

$$\text{wnp.}(\textbf{do } D \to \textit{Step } \textbf{od}).\mathcal{A} \;,$$

in which the decision to stop or continue is taken by a *demon* with the gambler's worst interests at heart. The demon will not force the gambler to play forever, however, since wnp interprets non-termination as success and the demon strives for (the gambler's) failure.

4.4   The *unless* game

The game for $\mathcal{A} \rhd \mathcal{B}$ combines the eventually and always games: the gambler tries to maximise the expectation, while the demon tries to minimise it. Let the gambler's strategy be G and the demon's D; the game is then

$$
\begin{aligned}
(\nu X \bullet \quad &\textbf{if } \neg P \textbf{ then resultis } \mathcal{B}\\
&\textbf{elsif } \neg D \textbf{ then resultis } \mathcal{A}\\
&\textbf{else } \textit{Step}; X) \;.
\end{aligned}
\tag{15}
$$

On each step the gambler decides whether to stop and accept $\mathcal{B}$ in the current state; if the gambler did not stop then the demon decides whether to force him to stop, and accept $\mathcal{A}$ instead; if neither stop, one step is taken and the game continues (with non-termination interpreted as success for the gambler).

For strategies G, D and expectations $\mathcal{A}, \mathcal{B}$ denote by $U_{G,D}.(\mathcal{A}, \mathcal{B})$ the result of playing the game (15) above. It can be shown that

$$(\sqcap D \bullet U_{G,D}.(\mathcal{A}, \mathcal{B})) \quad \Rrightarrow \quad \mathcal{A} \rhd \mathcal{B} \quad \Rrightarrow \quad (\sqcup G \bullet U_{G,D}.(\mathcal{A}, \mathcal{B})) \,,$$

which states (on the left) that for all gambler's strategies G the demon can ensure that he wins no more than $\mathcal{A} \rhd \mathcal{B}$, and (on the right) that for all demon's strategies D the gambler can win at least $\mathcal{A} \rhd \mathcal{B}$. In the limit we have

$$(\sqcup P \bullet (\sqcap D \bullet U_{G,D}.(\mathcal{A}, \mathcal{B}))) \quad \equiv \quad \mathcal{A} \rhd \mathcal{B} \quad \equiv \quad (\sqcap D \bullet (\sqcup P \bullet U_{G,D}.(\mathcal{A}, \mathcal{B}))) \,,$$

showing that if both players use their best strategy, the result is $\mathcal{A} \rhd \mathcal{B}$.

5   Conclusion

The generalisation from predicates to expectations is suggested by Kozen for imperative (but deterministic) programs [7]; in [12] we extend that to include nondeterminism. Morris' formulation of temporal properties as predicate transformers [13] then gives access to a probabilistic temporal logic.

Our principal technical contributions are two. The first is demostrating the agreement between the simple fixed-point definitions in Sec. 3.4 and the operational interpretations as games in Sec. 4 — that for example $\Diamond \mathcal{A}$ is the least upper bound of all "seek to maximise" game strategies based on $\mathcal{A}$. Such an agreement is needed to justify our fixed-point definitions, in spite of their simplicity: we must be sure they correspond to some actual behaviour. Back and von Wright [1] and Stirling [17] suggest similar uses for games, but over predicates (Boolean-valued) rather than the more general expectations (real-valued).

Our second substantial contribution (not reported here) is the identification of *sub-linearity* as the key property of the *next* operator. It is based on the healthiness conditions for probabilistic predicate transformers [12]: the property is that

$$\circ(a\mathcal{A} + b\mathcal{B} \ominus c) \quad \Lleftarrow \quad a(\circ\mathcal{A}) + b(\circ\mathcal{B}) \ominus c \,,$$

for arbitrary expectations $\mathcal{A}, \mathcal{B}$ and non-negative reals a, b and c with $a\mathcal{A}$ denoting the pointwise multiplication of expectation $\mathcal{A}$ by scalar a, and $x \ominus y$ denoting $(x - y) \sqcup 0$.

When $\mathcal{A}$ and $\mathcal{B}$ are standard the property specialises to distributivity of conjunction, as in a modal algebra.

Using elementary fixed-point properties of the remaining operators *eventually, always* and *unless*, we are able with sub-linearity to establish probabilistic analogues for many if not most of the axioms for standard branching-time temporal logic.

References

1. R.-J.R. Back and J. von Wright. Interpreting nondeterminism in the refinement calculus. In He Jifeng, John Cooke, and Peter Wallis, editors, *Proceedings of the BCS-FACS 7th Refinement Workshop*, Workshops in Computing. Springer Verlag, July 1996. Invited lecture. See *http://www.springer.co.uk/ewic/workshops/7RW*.

2. M. Ben-Ari, A. Pnueli, and Z. Manna. The temporal logic of branching time. *Acta Informatica*, 20:207–226, 1983.

3. E.W. Dijkstra. *A Discipline of Programming*. Prentice Hall International, Englewood Cliffs, N.J., 1976.

4. Yishai A. Feldman and David Harel. A probabilistic dynamic logic. *J. Computing and System Sciences*, 28:193–215, 1984.

5. S. Hart, M. Sharir, and A. Pnueli. Termination of probabilistic concurrent programs. *ACM Transactions on Programming Languages and Systems*, 5:356–380, 1983.

6. Wim Hesselink. Safety and progress of recursive procedures. *Formal Aspects of Computing*, 7:389–411, 1995.

7. D. Kozen. A probabilistic PDL. In *Proceedings of the 15th ACM Symposium on Theory of Computing*, New York, 1983. ACM.

8. Daniel Lehmann and Saharon Shelah. Reasoning with time and chance. *Information and Control*, 53:165–198, 1982.

9. J.J. Lukkien and J.L.A. van de Snepscheut. Weakest preconditions for progress. *Formal Aspects of Computing*, 4:195–236, 1992.

10. A. K. McIver and C. C. Morgan. Probabilistic predicate transformers: part 2. Technical Report PRG-TR-5-96, Programming Research Group, March 1996. Also available via *http* [14].

11. C. C. Morgan. Proof rules for probabilistic loops. In He Jifeng, John Cooke, and Peter Wallis, editors, *Proceedings of the BCS-FACS 7th Refinement Workshop*, Workshops in Computing. Springer Verlag, July 1996. *http://www.springer.co.uk/ewic/workshops/7RW*.

12. C. C. Morgan, A. K. McIver, and K. Seidel. Probabilistic predicate transformers. *ACM Transactions on Programming Languages and Systems*, 18(3):325–353, May 1996.

13. J. M. Morris. Temporal predicate transformers and fair termination. *Acta Informatica*, 27:287–313, 1990.

14. PSG. Probabilistic Systems Group: Collected reports. *http://www.comlab.ox.ac.uk/oucl/groups/probs/bibliography.html*.

15. K. Seidel, C. C. Morgan, and A. K. McIver. An introduction to probabilistic predicate transformers. Technical Report PRG-TR-6-96, Programming Research Group, February 1996. Also available via *http* [14].

16. M. Sharir, A. Pnueli, and S. Hart. Verification of probabilistic programs. *SIAM Journal on Computing*, 13(2):292–314, May 1984.

17. Colin Stirling. Local model checking games. In *CONCUR 95*, number 962 in LNCS, pages 1–11. Springer Verlag, 1995. Extended abstract.

Type Extension and Refinement

Peter Bancroft
School of Computing Science
Queensland University of Technology
Brisbane 4001, Australia
e-mail: bancroft@fit.qut.edu.au

Ian Hayes
School of Information Technology
The University of Queensland
Brisbane 4072, Australia
e-mail: ianh@it.uq.edu.au

Abstract. This paper extends the methods of the refinement calculus to allow the derivation of certain kinds of Oberon-like programs. The use of records, pointers, opaque types and type extension distinguishes the work from previous examples. A case study for a queue abstract data type illustrates a method for the derivation of a generic, linked implementation. Firstly, a sequence of integers is refined to a sequence of integer records with link pointers. Then the sequence of records is refined to a generic linked list that is close to Oberon code. Pointers are facilitated by declaring an explicit local data store for each queue variable. The advantage of Oberon over Modula-2 is the ability to separate the final code into two modules — a generic queue that maintains the linked data structure invariant and an integer instantiation of the queue, using type extension. This separation of concerns (isolation of the linked data structure properties) is the main benefit of the approach.

1   Introduction

The refinement calculus [1, 15, 12] provides formal methods for the derivation of code from abstract specifications. Fundamental to the approach is the inclusion of a wide-spectrum language which allows the simultaneous use of specification statements and code. In the process of refinement, all specification statements are replaced by code. The target language is usually an extended version of the guarded command language [7] which may be easily transliterated into Modula-2. The case study in [2] demonstrated the use of data refinement techniques to produce pointer implementations which are also close to Modula-2.

In this paper, the earlier work on pointers is combined with type extension [21], a language feature which supports object-orientation. The aim is to derive an Oberon implementation for an abstract data type module which is "generic" in the sense used in [17] — a module containing a linked stucture is imported into an abstract data type module which instantiates the generic data type to provide a particular application. This is an example of inheritance being used to simulate genericity [10]. In the derived program, the target language uses the refinement calculus notation of [12] with type declarations, records and pointers added to produce code that is close to Oberon. The resulting code has the advantage that the complexity of maintaining the linked data structure is separated from any concern over the actual contents of its nodes.

The inclusion of pointer variables raises the usual problem of aliasing. We take several decisions to deal with this.

- To make reasoning about pointers manageable, a single abstract data structure is represented by an explicit local store [19], into which pointers may reference.
- In the interface to a module, we eliminate aliasing through pointers, in the manner of the languages Euclid [9] and Turing [8].
- To avoid direct pointer manipulation, we make strict use of opaque types in client programs, as in Modula-2 and the limited private types of Ada [6].

The model for local data stores takes into account the requirements of an Oberon-like implementation. It is necessary to be able to dereference pointers to obtain the dynamic value of objects, but more importantly the model must provide access to the dynamic type of objects so that pointers can reference values that are extensions of their base type. Type extension is explicit and *name* based and the dynamic store is represented as a partial function from locations to type-labelled values as in [5].

Calculation of an implementation module from an abstract specification for an opaque type is facilitated using the technique described in [3, 4]. This is a modification of that in [13], to deal with modules in which there is no local state and the coupling invariant is parameterized for each occurrence of the opaque type in the formal parameters of a procedure. The method also allows for the data refinement of a module with several different opaque types as demonstrated in the case study below.

Section 2 contains a brief description of type extension. For a detailed formal semantics refer to [5]. A case study is introduced in Section 3. In Section 4, a context is defined for subsequent refinement — types and their values, assignment compatibility and a model for the explicit data store are described. In Section 5 a *Store* abstract data type is introduced, while in Section 6 a generic linked list is defined. These modules, together with the type extension context are used to derive a generic implementation for the given example, in Section 7. The example presented in the case study serves to illustrate useful methods for deriving a class of Oberon programs.

2   Type Extension

Type extension is the creation of new record types from existing record types by the addition of extra fields. In the object-oriented languages Oberon [17, 20] and Oberon-2 [16], type extension and a relaxation of the usual assignment compatibility rules provide the mechanism by which single inheritance is facilitated. This approach, as suggested in [18], has also been adopted for Ada 95 [6].

A record extension is considered to be compatible with the original record type in that the values of the extension can be assigned to variables of the original type. The extra fields cannot be stored in such a variable and are removed. This is known as the *projection* of the value into the original type. Other than the creation of a hierarchy, record extension is of little interest on its own, given a projection semantics for assignment.

More importantly, pointer types may also be extended. A pointer variable is declared with reference to a particular base record type (giving its *static* type), but can be assigned a value that refers to extensions of that type. The difference between a pointer assignment and the corresponding record assignment is that the extra fields of the extended base type remain accessible through the pointer (providing a possibly extended *dynamic* type).

The following is a typical declaration block in an Oberon-like language:

> **type** *rec*1 = **record** *f*1 : **integer**; *f*2 : **boolean end**;
> **var** *r*1 : *rec*1;
> **type** *ptr*1 = **pointer to** *rec*1;
> **var** *p*1 : *ptr*1;
> **type** *rec*2 = **record**(*rec*1) *f*3 : **integer end**;
> **var** *r*2 : *rec*2;
> **type** *ptr*2 = **pointer to** *rec*2;
> **var** *p*2 : *ptr*2;

Variable *r2* has type *rec2* which is an extension of type *rec1*. Variable *r1* can be assigned *r2* but the field *f3* will be dropped off. Pointer variable *p1* is of type *ptr1* but can be dynamically assigned a value of type *ptr2*. The extra field *f3* remains accessible through *p1* which now has static type *ptr1* but dynamic type *ptr2*.

Access to the extra fields of an extended pointer variable is subject to a *type guard*, which is an assertion that the dynamic type of the pointer is an extension of the type named in the guard. In Oberon, the expression,

```
p1(ptr2)^.f3
```

requires that variable *p1* has been extended so that it has dynamic type (at least) *ptr2*, a pointer to a record including a field *f3*.

The use of guarded expressions gives rise to two new kinds of program statements involving pointer variables. Firstly, there is *guarded assignment* such as,

```
p2 := p1(ptr2)
```

Here, *p1* is asserted to have dynamic type at least *ptr2* which must be an extension of the static type of *p2* (extension of a type includes the type itself). Its value may therefore be assigned to *p2*. The guard is necessary only when the static type of *p1* is not an extension of the static type of *p2*.

Secondly, Boolean expressions, such as

```
p1 IS ptr2
```

may be used, to check whether the dynamic type of pointer variable *p1* is at least *ptr2*. It is usual to express such a constraint in the guards of an *if* or *while* statement whose body relies on the dynamic type of extended variable *p1*. The dynamic check may avoid a runtime error in the evaluation of a guarded expression.

3   A Case Study

The specification in Figure 1 includes the type $IntQ$, with an abstract representation, which is an unbounded sequence of integers, indexed from zero. $IntQ$ is

$$
\begin{aligned}
&\textbf{module } IntFIFO \triangleq \\
&\textbf{opaque type } IntQ \triangleq seq_0 \textbf{ integer} \\
&\qquad \textbf{initially}(q : IntQ)\ q = \langle\,\rangle; \\
&\textbf{procedure } Join(\textbf{value result } q : IntQ;\ \textbf{value } n : \textbf{integer}) \triangleq \\
&\qquad q : [\,\textbf{true},\ q = q_0 \frown \langle n \rangle\,]; \\
&\textbf{procedure } Leave(\textbf{value result } q : IntQ;\ \textbf{result } m : \textbf{integer}) \triangleq \\
&\qquad q, m : [\,q \neq \langle\,\rangle,\ q_0 = \langle m \rangle \frown q\,]; \\
&\textbf{procedure } Concat(\textbf{value result } q : IntQ;\ \textbf{value result } s : IntQ) \triangleq \\
&\qquad q, s : [\,\textbf{true},\ q = q_0 \frown s_0 \wedge s = \langle\,\rangle\,] \\
&\textbf{end}
\end{aligned}
$$

Fig. 1. Integer Queues Specification

exported as an opaque type from the module, allowing the declaration of a number of variables of that type in a client program. The client program can only operate on such variables by calls to procedures in the abstract data type module — no knowledge of the implementation of the opaque type is available. Each instance of the $IntQ$ opaque type is initialised to an empty sequence on entry to the enclosing scope, according to the **initially** clause. The abstract specification corresponds to a definition module in Modula-2 or a package specification in Ada.

Each procedure in the $IntFIFO$ module is defined by a *specification statement* [11]. In the specification $w : [pre, post]$, w is a variable vector called the *frame* and denotes which variables may be changed by the operation. Informally, if the program fragment starts in a state where pre is true, it will terminate in a state where $post$ is true and the only variables that may have changed are those in the vector w.

The $IntFIFO$ module in Figure 1 is to be refined to a pointer implementation, using two data refinement steps. The initial data refinement replaces the abstract sequence by a sequence of records, consisting of an integer and a pointer field. The pointer values are constrained in the data type invariant to ensure the linked list property, giving a module *IntFIFO*1 with the following data type declarations.

module *IntFIFO*1 $\triangleq$
local type *IntPtr* $\triangleq$ **pointer to** *IntRec*;
local type *IntRec* $\triangleq$ **record**
$\qquad\qquad\qquad next$: *IntPtr*;
$\qquad\qquad\qquad item$: **integer**
$\qquad\qquad$ **end**;
local type *Store* $\triangleq$ *IntPtr* $\nrightarrow$ *IntRec*;

opaque type *IntQ* $\triangleq$ **record**
$\qquad\qquad\qquad nodes$: seq_0 *IntRec*;
$\qquad\qquad\qquad mem$: *Store*
$\qquad\qquad$ **where**
$\qquad\qquad\qquad (\forall\, i : 0..\#nodes - 2 \bullet$
$\qquad\qquad\qquad nodes(i).next \in \text{dom}\ mem\ \wedge$
$\qquad\qquad\qquad mem(nodes(i).next) = nodes(i + 1)) \wedge$
$\qquad\qquad\qquad (last\ nodes).next = \textbf{nil}$
$\qquad\qquad$ **end**

$$\vdots$$

A suitable coupling invariant relating the abstract and concrete types is given by,

$$q = (\lambda\, i : \text{dom}\ w.nodes \bullet (w.nodes(i)).item))$$

We define the following auxiliary function,

$$\begin{array}{|l}
qval : \text{seq}_0\ IntRec \rightarrow IntQ \\
\hline
(\forall\, w : \text{seq}_0\ IntRec \bullet \\
\qquad\qquad qval(w) = (\lambda\, i : \text{dom}\ w.nodes \bullet (w.nodes(i)).item))
\end{array}$$

which retrieves the corresponding *IntQ* and allows us to express the coupling invariant more concisely.

$$CI(q : IntFIFO.IntQ, w : IntFIFO1.IntQ) \triangleq q = qval(w)$$

Data refinement follows the methods of [3, 4]. The initialisation calculator is

$$initC \triangleq (\exists\, s \bullet CI(s, t) \wedge initA)$$

where s is an abstract state vector, t is a concrete state vector, *initC* is the initialisation for the concrete module and *initA* is the abstract initialisation. In this example,

$$\begin{aligned}
initC &\equiv (\exists\, q \bullet CI(q, w) \wedge q = \langle\,\rangle) \\
&\equiv w.nodes = \langle\,\rangle \\
&\Leftarrow w.nodes = \langle\,\rangle \wedge w.mem = \{\,\}
\end{aligned}$$

The calculator for a downward simulation of a specification statement is

$$s, x : [preA, postA] \sqsubseteq$$
$$t, x : [(\exists\, s \bullet preA \wedge CI(s, t)),$$
$$(\forall\, s_0 \bullet preA_0 \wedge CI(s_0, t_0) \Rightarrow (\exists\, s \bullet postA \wedge CI(s, t)))]$$

where s is an abstract state vector, x is an output vector, t is a concrete state vector and CI is the coupling invariant. A zero subscript in the postcondition denotes the initial value of a variable, while $preA_0$ is an abbreviation for $preA[s \backslash s_0]$ (all occurrences of s replaced by s_0). The calculator must be extended in the $Concat$ operation to include both parameters of type $IntQ$. By applying this calculator to each of the abstract procedures, we obtain the concrete module, $IntFIFO1$.

module $IntFIFO1 \triangleq$

$$\vdots$$

declarations as above

$$\vdots$$

$$\textbf{initially}(w : IntQ)\; w.nodes = \langle\,\rangle \wedge w.mem = \{\,\}$$
procedure $Join(\textbf{value result}\; w : IntQ;\; \textbf{value}\; n : \textbf{integer}) \triangleq$
$$w : [\textbf{true}, qval(w) = qval(w_0) \frown \langle n \rangle];$$

procedure $Leave(\textbf{value result}\; w : IntQ;\; \textbf{result}\; m : \textbf{integer}) \triangleq$
$$w, m : [qval(w) \neq \langle\,\rangle, qval(w_0) = \langle m \rangle \frown qval(w)];$$

procedure $Concat(\textbf{value result}\; w : IntQ;\; \textbf{value result}\; t : IntQ) \triangleq$
$$w : [\textbf{true}, qval(w) = qval(w_0) \frown qval(t_0) \wedge t = \langle\,\rangle]$$
end

The $IntFIFO1$ module is a concrete specification for a pointer implementation. (The constraints on the pointers are not evident in the procedure specifications as they are in the *where* clause of the opaque type declaration.) The definition for a local store is appropriate for a type-strict language such as Modula-2 but we shall later need to refine it to accommodate type extension. The opaque type $IntQ$ is a sequence of records in a local store. The effect is to duplicate the original integer sequence but to introduce links from one element of the sequence to the next. A further data refinement to introduce start and end pointers and to remove the *nodes* sequence, together with the application of the laws of algorithm refinement as in [12], takes the specification to code. This code may be transliterated into Modula-2 by replacing the local stores with a single global store (see [2]); however, the aim of this paper is to derive Oberon-like code. In the next section, a context for the subsequent development of this specification is described, taking into consideration the effects of type extension.

4   The Programming Semantics Context

This section contains a brief description of a type extension context — definitions that are needed to derive an Oberon-like implementation. Full details of an abstract syntax for a language with type extension, as well as a static and dynamic semantics in a denotational style are given in [5]. This may be extended to include a formal semantics for the use of variable references as local stores (*collections* in [8]). Only detail relevant to this paper is discussed here.

A basic type *Id* provides a set of identifiers for the abstract language. The possible types of the language are given by a disjoint union.

$$Type ::= \ldots \mid record \langle\!\langle Id \twoheadrightarrow Type \rangle\!\rangle \mid ptr \langle\!\langle Type \rangle\!\rangle \mid$$
$$extend \langle\!\langle Id \times (Id \twoheadrightarrow Type) \rangle\!\rangle \mid collection \mid \ldots$$

Included in this disjoint union are record, pointer and extension types. The function *record* takes a partial function from field names to their types and constructs a record type. The function *ptr* is used to construct a pointer type to a record or extension base type. An extension type consists of an *Id*, which is constrained to be the name of a record or another extension type, together with some extra fields. The type *collection* is used to declare local store variables.

The dynamic store is modelled as separate local stores (collections) for each instance of an opaque type. A stored value consists of a type and a value.

<pre>
┌─ StoredVal ──────────────────────────────────
│   dyntype : Type
│   dynval : Value
└──
</pre>

Assuming a set of available locations, *Loc*, the values that may be attributed to variables and expressions of the allowed types are given by

$$Value ::= \ldots \mid recval \langle\!\langle Id \twoheadrightarrow Value \rangle\!\rangle \mid refval \langle\!\langle Loc \rangle\!\rangle \mid coll \langle\!\langle Store \rangle\!\rangle \mid \ldots$$

where, the set of all pointer values is

$$PtrVal == refval (\!\mid Loc \mid\!)$$

and *Store* is defined as a partial function from pointer values to stored values,

$$Store == PtrVal \twoheadrightarrow StoredVal$$

This means that each location in the store is "tagged" with the type that it holds.

For a store to be *well-formed* there are several constraints.

- The *StoredVal* in each location must contain a *dynval* that is an element of its *dyntype*.
- The dynamic type of any pointer variable must be an extension of its static type.
- A value in a store may be a collection and each collection consists of a store. The stores within different collections must have no locations in common, and furthermore they must have no locations in common with the top-level store.

The semantic function for *Type* in a given static environment is

$$\mid \quad Val : Type \to \mathbb{P}\ Value$$

Given any type t, $Val(t)$ returns the allowable values of the type.

Type extension allows the creation of a type hierarchy through single inheritance. In a given static environment, the relation, *ext* is specified as

$$\mid \quad ext : Type \leftrightarrow Type$$

so that $t1 \mapsto t2 \in ext$ is interpretted as "t1 is extended by t2".

An important feature of a language that allows type extension is the relaxation of assignment compatibility rules. In a given static environment,

$$\begin{array}{|l}
TypeCompat : Type \leftrightarrow Type \\
\hline
TypeCompat = ext \cup \mathrm{id}\ Type
\end{array}$$

This means that types are compatible if they are the same or if the second is an extension of the first. When this is applied to assignment, it is clear that static type checking allows greater generality than type strict languages such as Modula-2.

5   A Stores Module

An abstract data type module for local stores is defined so that we can introduce a new *Store* type with appropriate operations. This module is similar to that suggested in [2] except that it is modified to be able to cope with pointer variables having possibly different dynamic types. The *Stores* module we introduce is more general because it allows a pointer variable of any base type to be passed as an actual parameter to the included procedures. The other important feature is that in any given local store, pointer values are mapped to a *StoredVal* — the definition of a well-formed store (Section 4) in this context allows pointers to refer to extended values.

A client module that uses the *Stores* module can declare any number of instances of type *Store*. Variables of this type are the *collections* of Section 4 and, in our case study, are local to a particular record type, in this instance *IntQ*. Note that *Store* is an opaque type, as the only operations to be allowed are those given as procedures in the module; whereas *StoredVal* is a local type, with the module exporting a function *dtype*.

module *Stores* $\triangleq$
local type *StoredVal* $\triangleq$ **record**
$$\begin{array}{l}
dyntype : Type; \\
dynval : Value
\end{array}$$
                     end;
opaque type *Store* $\triangleq$ *PtrVal* $\twoheadrightarrow$ *StoredVal*
    initially$(s : Store)\ s = \{\ \}$;

function $dtype(\textbf{value }s : Store; \textbf{value }p : PtrVal)\ :\ Type \triangleq$
    $\{p \in \text{dom }s\}$
    return $(s(p)).dyntype;$

procedure $New(\textbf{value result }s : Store;$
                  $\textbf{value }t : Type;\ \textbf{result }p : PtrVal) \triangleq$
    $s, p : [\text{dom }s \neq PtrVal,$
         $(\exists\, r : Val(t) \bullet p \notin \text{dom }s_0 \wedge$
               $s = s_0 \cup \{p \mapsto StoredVal(t, r)\})];$

procedure $Dispose(\textbf{value result }s : Store;\ \textbf{value }p : PtrVal) \triangleq$
    $s : [p \in \text{dom }s, s = \{p\} \lhd s_0];$

procedure $Transfer(\textbf{value result }s1 : Store;$
                  $\textbf{value result }s2 : Store;\ \textbf{value }ptrs : \mathbb{F}\ PtrVal) \triangleq$
    $s1, s2 : [ptrs \subseteq \text{dom }s1,$
        $s1 = ptrs \lhd s1_0 \wedge s2 = s2_0 \cup (ptrs \lhd s1_0) \wedge$
                  $ptrs \cap \text{dom }s2_0 = \{\,\}]$

        $\vdots$

end

In the derivation to follow, pointer dereferencing and dynamic type are used frequently. These rely on the context suggested by the definitions in the abstract language, as implemented in *Stores* abstract data type module. The abstract syntax used to define the language is unwieldy so we adopt a more concrete syntax for the rest of the paper. As an example, consider the following declarations.

local type $QPtr \triangleq$ **pointer to** $QRec;$
local type $QRec \triangleq$ **record**
                    $next : QPtr;$
                    $item : \textbf{integer}$
           end;
opaque type $Queue \triangleq$ **record**
                    $start, end : QPtr;$
                    $mem : Store$
           end;

  var $q : Queue;$

Using concrete syntax, we express the dereferenced value and dynamic type of $q.start$ as,

  $q.mem[q.start]$ and
  $dtype(q.mem, q.start)$

The use of square brackets denotes selection of the *dynval* field from the *StoredVal* pair.

The stores module includes the procedure *Transfer*, which allows pointers to be transferred from one store to another. It also indirectly ensures that stores remain pairwise disjoint on their domains. This specification for *Transfer* is found in [19] and is infeasible unless we can be sure that different stores do not overlap because of the condition, $ptrs \cap \operatorname{dom} s2_0 = \{\,\}$. The consequence of this is that any implementation for *New* must return pointer values from a single global store, and this is what happens in practice.

6   A Generic Module

To facilitate a generic implementation for our integer queue, we introduce a module which declares a generic list data type and provides suitable operations. It is generic in that, although it consists of nodes that contain only *next* pointers, by using type extension, these nodes may include objects of any type. The importance of the *GenFIFO* module is its availability for reuse by import, into a client program. The procedures in the *GenFIFO* module handle all the pointer manipulation required to maintain a linked list. Any operations on the nodes themselves are dealt with by the client module. It is necessary to import the *Stores* module to perform certain pointer operations with respect to the appropriate local store. *GenFIFO* acts as a template for linked lists of any type, with the usual *start* and *end* pointer implementation.

```
module GenFIFO ≜
import Stores;
opaque type GenPtr ≜ pointer to GenRec;
opaque type GenRec ≜ record next : GenPtr end;

type GenQ ≜ record
                  start, end : GenPtr;
                  mem : Store
              where (start, end ∈ dom mem) ∨
                    (start = nil ∧ end = nil)
              end
      initially(z : GenQ) (z.start = nil ∧ z.end = nil);

procedure Join(value result z : GenQ; value p : GenPtr) ≜
    if z.start = nil → z.start, z.end := p, p
    [] z.start ≠ nil → z.mem[z.end].next, z.end := p, p
    fi;

procedure Leave(value result z : GenQ; result p : GenPtr) ≜
    p := z.start;
    if z.start = z.end → z.start, z.end := nil, nil
    [] z.start ≠ z.end → z.start := z.mem[z.start].next
    fi;
```

> **procedure** *Concat*(**value result** z : $GenQ$; **value result** t : $GenQ$) $\triangleq$
>   **if** $z.start$ = **nil** $\rightarrow$ $z.start, z.end$:= $t.start, t.end$
>   [] $z.start$ $\neq$ **nil** $\rightarrow$ $z.mem[z.end].next, z.end$:= $t.start, t.end$
>   **fi**;
> $Stores.Transfer(t.mem, z.mem, \mathrm{dom}(t.mem))$;
> $t.start, t.end$:= **nil**, **nil**
> **end**

It is important to note that, in the *Join* procedure, even though p is a *GenPtr*, it may in fact refer to an *IntRec*, as described in Section 4. The intention is that p already refers to a new node suitably initialised in another module. The *Concat* operation is interesting because of the call to *Transfer* which removes all of the nodes from $t.mem$ and inserts them into $z.mem$.

A client module (in this case a module for integer queues) needs to be aware of the local store associated with the type *GenQ*. For this reason, *GenQ* cannot be opaque in *GenFIFO*. There is a danger in visibly exporting a type like this but we justify it by the fact that it is subject to no further data refinement and that it is not available to the main program but only to the module that is to extend it. In this respect, Oberon uses an export mark to indicate which descriptors are visible and in this convention we would need only to mark *mem*. The *GenQ* data type is used to redeclare an opaque type in the integer queue module, below. Note also that there is no need to initialise *mem* to the empty set here as that is done in the *Stores* module.

7   An Oberon-Like Implementation

With the context described in Section 4, we calculate a second data refinement using the module *IntFIFO*1 (Section 3) to obtain an Oberon-like implementation. The new definition for *Store* (Section 5) is introduced to allow easy access to the dereferenced value and dynamic type of each pointer variable. At the same time concrete data types for an integer queue are declared using the generic linked list module of Section 6. As in Section 3, opaque type data refinement calculators are used, but this time two opaque types are refined simultaneously.

The *GenFIFO* and *Stores* modules provide the mechanism for the new concrete data types in the refinement. With the data type declarations in these modules, integer-node pointer and record types are introduced, using type extension. An *IntRec* is a *GenRec* extended with an *item* field of type **integer**. This allows an *IntPtr* type which is an extension of *GenPtr* to be declared.

> **local type** *IntPtr* $\triangleq$ **pointer to** *IntRec*;
> **local type** *IntRec* $\triangleq$ **record**(*GenRec*)
>   *item* : **integer**
>   **end**;

The opaque type declaration for an integer queue type is the same as *GenQ* — *start* and *end* pointers and a local store which is a refinement of the original store; however the nodes of a generic queue are not guaranteed to contain an **integer** value. For this reason, we conjoin a suitable type invariant.

> **opaque type** $IntQ \triangleq GenFIFO.GenQ$
>   **invariant** $(x : IntQ)\ (\forall sv : \operatorname{ran} x.mem \bullet sv.dyntype = IntRec)$;

This constraint is essential if the original specification is to be implemented. The generic queue used in the application must contain only nodes of type $IntRec$.

The coupling invariant relates the old and new declarations of $IntQ$ and $Store$, as both of these types are simultaneously data refined (in a new module known as $IntFIFO2$).

$$CI_{QUEUE}(w : IntFIFO1.IntQ, z : IntFIFO2.IntQ) \triangleq$$
$$w.start = z.start \land w.end = z.end \land$$
$$CI_{Store}(w.mem, z.mem)$$

where,

$$CI_{Store}(s : IntFIFO1.Store, \sigma : IntFIFO2.Store) \triangleq$$
$$s = \{pv : \operatorname{dom} \sigma \bullet pv \mapsto \sigma(pv).dynval\}$$

The data refinement removes the sequence altogether, leaving a linked list of nodes with *start* and *end* pointers. Algorithm refinement is calculated, with the *Stores* module now being imported into the $IntFIFO2$ implementation module. Variables of type *Store* are automatically initialised to the empty set, in the *Stores* module. With the introduction of type extension, we find it necessary to introduce a convenient notation for type-guarded assignment which also considers the use of collections. The expression, $p(mem, ty)$ indicates that pointer variable p is extended to type ty in collection mem.

> **module** $IntFIFO2 \triangleq$
> **import** $Stores, GenFIFO$;
>
> **local type** $IntPtr \triangleq$ **pointer to** $IntRec$;
> **local type** $IntRec \triangleq$ **record**$(GenRec)$
> $item$: **integer**
>   **end**;
>
> **opaque type** $IntQ \triangleq GenFIFO.GenQ$
>   **invariant** $(x : IntQ) \triangleq (\forall sv : \operatorname{ran} x.mem \bullet sv.dyntype = IntRec)$;
>
> **procedure** $Join($**value result** $z : IntQ;$ **value** $n :$ **integer**$) \triangleq$
> **var** $p : IntPtr \bullet$
> $Stores.New(z.mem, IntPtr, p)$;
> $z.mem[p] := IntRec(n, \textbf{nil})$;
>   **if** $z.start = \textbf{nil} \rightarrow z.start, z.end := p, p$ $\mathcal{G}$
> $[]\ \ z.start \neq \textbf{nil} \rightarrow z.mem[z.end].next, z.end := p, p$ $\mathcal{G}$
>   **fi**;

```
procedure Leave(value result z : IntQ; result m : integer) ≜
var p : GenPtr •
    p := z.start;                                                        𝒢
    if z.start = z.end → z.start, z.end := nil, nil                      𝒢
    [] z.start ≠ z.end → z.start := z.mem[z.start].next                  𝒢
    fi;
    {dtype(z.mem, p) = IntPtr}
    m := z.mem[p(z.mem, IntPtr)].item
    (* Dispose(z.mem, p) *)

procedure Concat(value result z : IntQ; value result t : IntQ) ≜
    if z.start = nil → z.start, z.end := t.start, t.end                  𝒢
    [] z.start ≠ nil → z.mem[z.end].next, z.end := t.start, t.end        𝒢
    fi;
    Stores.Transfer(t.mem, z.mem, dom(t.mem));                           𝒢
    t.start, t.end := nil, nil                                           𝒢

end
```

The context described in Section 4 is assumed here to give a more familiar notation to our program. In particular, we use dereferenced variable references which have the advantage of being able to be conveniently used as r-values or l-values in assignment statements. An alternative to this is to define a *Dereference* procedure (as in [19]) which makes the notation awkward and quite different from Oberon. The use of the imported function *dtype* similarly improves the readability of the program.

The implementation in this module is close to Oberon and uses the generic data types for type extension definitions. Several statements in the program are marked $\mathcal{G}$. These commands are all related to the maintenance of the linked data structure — in this case a simple linked list. They are unaffected by the actual *item* field in a record. These lines of code are now replaced by appropriate procedure calls to the imported *GenFIFO* module.

In the final module, Figure 2, it is assured that all nodes are of type *IntRec* by the procedures — the type invariant is distributed in the algorithm refinement. The initialisation of *IntQ* is implicit from the definition of *GenFIFO.GenQ* and *Stores.Store*. Note that the creation of a new integer node must occur in the *IntFIFO2* module and not *GenFIFO* as the call to *Stores.New* relies on a pointer of type *IntPtr*. The other interesting point is the type guard in the *Leave* procedure, $p(z.mem, IntPtr)$, which asserts that p is of type *IntPtr* even though the node returned is from a *GenQ*.

7.1 Type Guards and Weakest Preconditions

As we have seen in the case study, type-guarded assignment has a richer semantics than simple assignment (see [5]). There are conditions implicit in assignment that are usually resolved statically or which result in a run-time error. In this

```
module IntFIFO3 ≜
import Stores, GenFIFO;
local type IntPtr ≜ pointer to IntRec;
local type IntRec ≜ record(GenRec)
                          item : integer
                   end;
opaque type IntQ ≜ GenFIFO.GenQ;

procedure Join(value result z : IntQ;  value n : integer) ≜
var p : IntPtr •
    Stores.New(z.mem, IntPtr, p);
    z.mem[p] := IntRec(n, nil);
    GenFIFO.Join(z, p)

procedure Leave(value result z : IntQ;  result m : integer) ≜
var p : GenFIFO.GenPtr •
    GenFIFO.Leave(z, p);
    {dtype(z.mem, p) = IntPtr}
    m := z.mem[p(z.mem, IntPtr)].item
    (* Dispose(z.mem, p) *)

procedure Concat(value result z : IntQ;  value result t : IntQ) ≜
    GenFIFO.Concat(z, t)

end
```

Fig. 2. Generic Code for An Integer Queue

section, we make these conditions explicit for examples where type guards are used. We define the meaning of an assignment as,

$$wp(X := E, pred) \triangleq ValidVarref(X) \land ValidExp(E) \land$$
$$TypeCompat(X, E) \land pred[E/X]$$

$ValidVarref(X)$ and $ValidExp(E)$ may be, for example, checks to ensure that X and E are well-defined. $TypeCompat$ is usually dealt with by the compiler and is discussed in [14]. If these three conditions are omitted, we obtain the familiar definition for assignment. When type guards are present, there is of course a need to satisfy some of the checks, dynamically.

For the case of an expression which includes a type-guarded pointer used as an r-value in an assignment, we can provide a weakest precondition semantics. For example, consider $E(q(mem, t))$, where E is an expression that includes the type-guarded pointer q.

$$wp(X := E(q(mem, t)), pred) \triangleq (t, dtype(mem, q)) \in ext \land$$
$$wp(X := E(q), pred)$$

In this example, *ValidExp* is defined as $(t, dtype(mem, q)) \in ext$. Implicit in this is the condition that $q \in \text{dom } mem$ which is the assumption in the definition of *dtype* (Section 4).

In the assignment of a dereferenced pointer, $mem[p(mem, t)].x := E$, where p is a pointer, t is a pointer type, x is a field in a record and E is an expression, the weakest precondition semantics is

$$wp(mem[p(mem, t)].x := E, pred) \triangleq (t, dtype(mem, p)) \in ext \wedge$$
$$wp(mem[p].x := E, pred)$$

ValidVarref$(mem[p(mem, t)].x)$ is defined as $(t, dtype(mem, p)) \in ext$ (and there may also need to be a check on the validity of E, depending on whether E is a dereferenced pointer or type-guarded expression). In both of these examples, we have omitted the type-compatibility check as this can be performed statically even where type guards are used.

7.2 Leave and Type Guards

Perhaps the most interesting feature of this development is the inclusion of type guards in the refinement calculus notation, in the modules *IntFIFO2* and *IntFIFO3*. In the *Leave* procedure of *IntFIFO2*, we have introduced the following statement,

$$m := z.mem[p(z.mem, IntPtr)].item$$

This is an instance of type-guarded assignment as described in Section 2, but with a concrete syntax that incorporates the collections. The derivation of this code is considered here.

After the second data refinement step (from *IntFIFO1*), routine algorithm refinement of the concrete specification for *Leave* gives us

$$m := z.mem[p].item$$

This statement is not yet code nor is it immediately obvious that the final type-guarded assignment is a refinement of the statement. That is because the type invariant for *IntQ*, which states that all pointers are of type *IntPtr*, must be applied to each procedure. In this case we obtain

$$\sqsubseteq \{dtype(z.mem, p) = IntPtr\}$$
$$m := z.mem[p].item$$

Weakening the assertion to allow p to be any extension of *IntPtr* and assuming an implicit static environment gives,

$$\sqsubseteq \{(IntPtr, dtype(z.mem, p)) \in ext\}$$
$$m := z.mem[p].item$$

Using the weakest precondition of Section 7.1, we obtain the desired result.

$$m := z.mem[p(z.mem, IntPtr)].item$$

The corresponding Oberon command omits the use of local stores and allows implicit dereferencing of pointers to records giving, $m := p(IntPtr).item$.

7.3   Concat

The *Concat* procedure is interesting because it involves two parameters of type *IntQ*. This means that its derivation cannot be dealt with using the data refinement techniques of [12] and relies on the method of [3]. As well as this, the effect of *Concat* on the local stores of z and t reveals the importance of this approach to pointers. In the absence of parameter aliasing, we can be sure that the operation proceeds exactly as we expect — there are no problems due pointer aliasing. The operation actually transfers all the nodes of $t.mem$ to $z.mem$. If a single global data store is used for the *Concat* operation, there is the possibility that the two lists may overlap for some suffix. The operation then creates a data structure with cycles — certainly not a linked list. We can be sure that our local stores do not overlap because of the specification for the *Stores* module in Section 5.

8   Review of Results

In this paper, we have introduced refinement methods suitable for the derivation of Oberon-like code. Our methods rely on the assumption of a context such as that described in Section 4 which includes the use of type extension and local stores. In Oberon, type extension is implemented using pointers which are dealt with here by declaring an explicit local store associated with each instance of an opaque type. Several data refinement calculations are employed in the derivation. The most important step introduces a more general model for the store which allows each local store to contain values belonging to any extension of a base type.

In the case study for a generic queue, we derive a concrete module, *IntFIFO3*, which may be easily transliterated into an Oberon implementation of the specification in Figure 1. Apart from minor syntactic variations, the main difference between our code and Oberon code involves our use of local stores rather than a single monolithic data store. A further data refinement can be calculated to remove local stores altogether as they are non-intersecting. Each location-store pair can be refined to a unique location in a global store as in [19]. As well as the removal of all references to local stores in the code of Figure 2, the *Transfer* operation is refined to **skip**.

The main advantage of the implementation obtained is in the separation of concerns — the generic queue module maintains the invariant for the linked data structure (a queue), while a separate module implements the required application (a queue of integers). The result of this paper is that we have provided data refinement techniques for the derivation of an Oberon program which uses inheritance to simulate genericity.

References

1. R. J. R. Back. A calculus of refinements for program derivations. *Acta Informatica*, 25(7):593–624, 1988.

2. P. G. Bancroft. Pointers in refinement calculus: a case study. In *Proceedings of the 7th Australian Software Engineering Conference*, pages 11–19. IREE, July 1993.

3. P. G. Bancroft and I. J. Hayes. Refinement of opaque types. Technical Report 267, University of Queensland, Key Centre for Software Technology, Department of Computer Science, April 1993. An earlier draft was presented at: 2nd Australian Refinement Workshop, 1992.

4. P. G. Bancroft and I. J. Hayes. Refining a module with opaque types. In Gopal Gupta, George Mohay, and Rodney Topor, editors, *Proceedings of the 16th Australian Computer Science Conference*, pages 615–624. Australian Computer Society, February 1993.

5. P. G. Bancroft and I. J. Hayes. A formal semantics for a language with type extension. In Jonathan P. Bowen and Michael Hinchey, editors, *ZUM'95: The Z Formal Specification Notation*, volume 967 of *Lecture Notes in Computer Science*, pages 299–314. Springer-Verlag, 1995.

6. J. G. P. Barnes. *Programming in Ada 95*. Addison-Wesley, 1995.

7. E. W. Dijkstra. *A Discipline of Programming*. Prentice Hall, 1976.

8. R. C. Holt and J. N. P. Hume. *Introduction to Computer Science using the Turing Programming Language*. Reston Pub. Co., 1984.

9. J. J. Horning. A case study in language design: Euclid. In F. L. Bauer and M. Broy, editors, *Program Construction*, volume 69 of *Lecture Notes in Computer Science*, pages 125–132. Springer-Verlag, 1979.

10. B. Meyer. Genericity versus inheritance. In N. Meyrowitz, editor, *OOPSLA'86 Proceedings*, number 11 in SIGPLAN NOTICES, pages 391–405. ACM, November 1986.

11. C. C. Morgan. The specification statement. *ACM Transactions on Programming Languages and Systems*, 10(3), July 1988.

12. C. C. Morgan. *Programming from Specifications*. Prentice-Hall, 2nd edition, 1994.

13. C. C. Morgan and P.H.B. Gardiner. Data refinement by calculation. *Acta Informatica*, 27:481–503, 1990.

14. C. C. Morgan and T. Vickers. Types and invariants in the refinement calculus. *Science of Computer Programming*, 14:281–304, 1990.

15. J. M. Morris. Laws of data refinement. *Acta Informatica*, 26:287–308, 1989.

16. H. Mössenböck and N. Wirth. The programming language Oberon-2. *Structured Programming*, 12(4), 1991.

17. M. Reiser and N. Wirth. *Programming in Oberon: Steps beyond Pascal and Modula*. Addison-Wesley, 1992.

18. E. Seidwitz. Object-oriented programming through type extension in Ada9X. *Ada Letters*, 11(2):86–97, March 1991.

19. M. Utting. Reasoning about aliasing. In *Proceedings of the Fourth Australian Refinement Workshop (ARW-95)*, pages 195–211. School of Computer Science and Engineering, The University of New South Wales, April 1995.

20. N. Wirth. The programming language Oberon. *Software Practice and Experience*, 18(7), July 1988.

21. N. Wirth. Type extensions. *ACM Transactions on Programming Languages and Systems*, 10(2):204–214, April 1988.

The Refinement Calculator:
Proof Support for Program Refinement

Michael Butler,[1] Jim Grundy,[2] Thomas Långbacka,[3,4]
Rimvydas Rukšėnas,[4,5] and Joakim von Wright[5,4]

[1] Dept. of Electronics & Computer Science, University of Southampton
[2] Dept. of Computer Science, The Australian National University
[3] Dept. of Computer Science, University of Helsinki
[4] Turku Centre for Computer Science (TUCS)
[5] Dept. of Computer Science, Åbo Akademi University

Abstract. We describe the Refinement Calculator, a tool which supports the application of the refinement calculus to program development. The tool uses a general mechanism for transformational reasoning with HOL as the underlying proof system. A graphical user interface provides the user with menus of transformations and the ability to select and focus on subcomponents of a specification using simple mouse operations. The refinement-oriented transformations are illustrated with a case study.

1 Introduction

The *refinement calculus* [2, 20, 22] is a formalisation of the stepwise refinement method of program construction. The required behaviour of a program is specified as an abstract, possibly executable, program which is then refined by a series of correctness-preserving transformations into an efficient, executable program.

When using formalisms like the refinement calculus, the derivations that transform an initial specification to an executable program are usually long and error-prone. Even small refinement steps typically generate large, and therefore difficult to manage, proof conditions. In order to make the task manageable in practice, tools are needed. In this paper, we present a tool called the *Refinement Calculator* which we have developed to support the application of refinement transformations and the proof of verification conditions introduced by individual refinement steps.

Our aim has been to develop a tool that is both user-friendly and reliable. User-friendliness is hard to measure, but it typically means providing a suitable graphic user interface (GUI). GUIs often work in a fashion where the user selects (using the mouse) the data to manipulate and then chooses the desired operation from a menu (or the other way around). Typically, applications from different domains are similar in appearance. This is desirable, since it makes moving between application domains easy. In our tool, proofs are carried out primarily in the way described above, i.e. by selecting information using the mouse, and then applying menu commands to operate on the selected data. Little typing is required of the user, contrary to current practice with most proof tools.

An important aspect of the Refinement Calculator is its reliance on *transformational* reasoning. In transformational reasoning a proof is constructed by transforming an initial expression, via a series of relation preserving steps, to a final expression that has some desired property. Program refinement is a typical example of such a proof activity: refinement being the relation preserved, and executability or efficiency being the property desired. In effect, this paper will describe a mechanisation of a general method of transformational reasoning, including a graphical user interface, and its subsequent specialisation for use with program refinement. An overview of the Refinement Calculator has been presented in an earlier paper [5]. Here we discuss more of the development history of the tool and the rationale for various design decisions as well. We also give a more detailed view of the tool from a user's perspective.

The remainder of the paper is organised as follows: After a general introduction to the refinement calculus in Sect. 2, we discuss some of the design considerations in the development of the tool in Sect. 3. We then give an overview of HOL and window inference in Sect. 4. Sect. 5 describes the graphical user interface and the general support it gives for transformational reasoning, while Sect. 6 describes the refinement specific features of the Refinement Calculator. Sect. 7 illustrates the use of the Refinement Calculator with an example program derivation.

2 Refinement Calculus

Stepwise refinement is a method for developing programs from high-level specifications into efficient implementations. In this approach, the development of a program includes the proof of its correctness. This can be compared with program verification, where the program is first developed, using informal methods, and then checked for desired correctness properties. The refinement calculus is based on the predicate transformer (weakest precondition) calculus of Dijkstra [8]. The calculus was originally developed for the refinement of sequential programs, but was later extended to deal with parallel and distributed programs through the action system formalism [3]. It is a calculus of program transformations that preserve total correctness. If S and S' are statements (program fragments), then the refinement $S \sqsubseteq S'$ holds if and only if S' satisfies every total correctness assertion that S satisfies. Thus, if we have proved the refinement formally, then S' is guaranteed to be a correct implementation of S. The refinement relation $\sqsubseteq$ is a preorder (reflexive and transitive). Furthermore, the ordinary control structures of programming (sequential composition, conditional composition and iteration) are monotonic with respect to refinement. Transitivity implies that we can do refinements stepwise, and monotonicity implies that we can focus on substatements when we do refinement (if the refinement $S \sqsubseteq S'$ holds, then $C[S] \sqsubseteq C[S']$ holds, i.e. we can replace S by S' in any program context $C[\cdot]$). The refinement calculus supports a transformational style of program development; starting from an initial specification S_0, we make a series of refinement steps,

$$S_0 \sqsubseteq S_1 \sqsubseteq \ldots \sqsubseteq S_n$$

Each individual refinement step $S_i \sqsubseteq S_{i+1}$ is justified by reference to a *refinement rule*. The application of a refinement rule typically involves three activities:

- selecting the substatement that is to be transformed;
- matching the selected statement to the rule;
- verifying conditions associated with the rule.

Some rules operate on large program components (e.g., high-level rules that replace local variables in a block) while others make only small changes (e.g., rewriting the right hand side of an assignment). Rules are often very general, so parameters must be instantiated before they can be applied. Finally, many rules have side conditions (e.g., restricting free occurrences of a variable or requiring equivalence between two expressions) that need to be checked.

Program development in the refinement calculus is creative work but it also involves tedious proof details. To make program refinement more practical, we need tools that take care of the details. One such kind of tool is a *refinement editor*, which keeps track of applicability and side conditions [11, 28]. A more advanced tool is a *refinement calculator*, where the application of a refinement rule leads to a formal proof of the refinement step in a mechanised logic (i.e. the logic of a theorem proving system). By representing the semantics of the refinement calculus in the logic and proving soundness of the refinement rules, we can get a complete guarantee that the tool permits only valid refinement steps.

3   Design Considerations

In this section we discuss the overall architecture of the Refinement Calculator tool. We also discuss some of the major design decisions that were made while developing the tool.

3.1   A Layered Design

The Refinement Calculator has been built in a number of layers. Fig. 1 illustrates the different layers in the design. The bottom layer is the HOL system, described further in Sect. 4.1. The HOL window Library, described in more detail in Sect. 4.2, forms the next layer.

The Refcalc theory is a formalisation of the refinement calculus in HOL. The development of this theory was originally based only on the 'pure' HOL system (as indicated in Fig. 1). Quite soon it became evident the theory was too difficult to use as intended. Later, when the HOL window Library implemented for the window inference system of transformational reasoning in HOL, the Refcalc theory was quickly adapted for use with it.

Another layer, called TkWinHOL, is a GUI to the HOL window Library. TkWinHOL is a tool in its own right and can be used to develop HOL proofs under window inference. It was, however, designed with the intention of being used as a basis for the Refinement Calculator. Although use of the HOL window

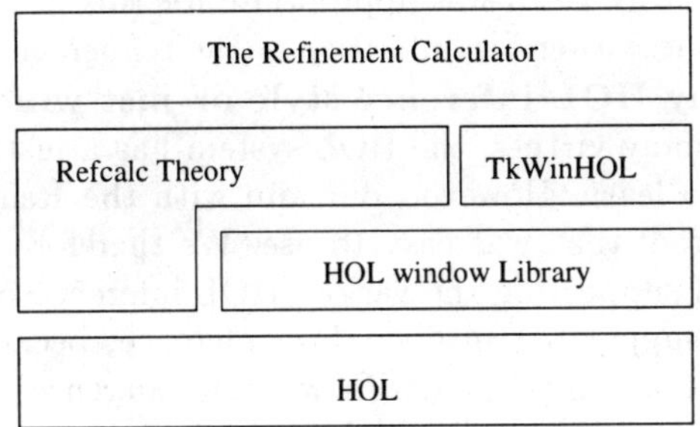

Fig. 1. The layered design of the Refinement Calculator

Library simplified the development of refinement proofs in HOL, it was soon clear that a more sophisticated tool was needed. Certain aspects of working with window inference in HOL are tedious. Fortunately these aspects can be simplified significantly by using a GUI such as TkWinHOL.

Since TkWinHOL was designed with the Refinement Calculator in mind, it was designed to be extendible (more information of how the tool can be extended is given in Sect. 5.3). For example, one can add new menu choices and bind these to HOL commands. This is largely what the Refinement Calculator does, i.e. it adds a number of menu choices that are bound to HOL commands that have been defined in the Refcalc theory. This is indicated in Fig. 1 by placing the Refinement Calculator layer on top of both the Refcalc and TkWinHOL layers. Actually the layered design stretches beyond this as well. One can add further layers above the refinement calculator by defining new menus and binding them to new HOL commands etc. The *data refinement* support discussed in Sect. 6.4 is an example of such an additional layer.

3.2　Design Choices

While developing the Refinement Calculator we had to make a number of choices concerning how to proceed. Below we summarise the major design choices we were faced with:

- **To use an existing theorem prover or build our own:** The choice here was whether or not to use an existing theorem prover, and if so, which one. We chose to use an existing system to avoid the obvious difficulty of building a sufficiently powerful theorem prover of our own, particularly one with the high degree of reliability deemed necessary for the project to have been worth while.

  Among existing theorem provers, HOL was selected for its reliability and flexibility, but primarily because earlier work by members of the team had placed us in the possession of a HOL formalisation of the refinement calculus. Sect. 4.1 gives a more detailed description of the HOL system itself, further illustrating both why HOL was appropriate for this project, and why HOL was chosen for the earlier formalisation of the refinement calculus.

- **To support any HOL inference style or just window inference:** In contrast to its many virtues, the HOL system has a justified reputation for being difficult to learn. However, our aim with the Refinement Calculator was to build a tool that was easy to use. We therefore chose to avoid the full generality of supporting the various HOL inference mechanisms, and to concentrate on supporting just window inference, because a great deal of reasoning can be accomplished with window inference simply by selecting subterms of an expression and applying one of a relatively small set of transformations. These actions are handled naturally in a GUI by selection with the mouse and choosing options from a menu.

  Furthermore, the structure imposed on proofs by this restricted discipline of transformation results in proofs that are easy to browse. The recording and browsing of proofs is described in Sect. 5.2.

- **Choosing an interface building tool:** We began by examining an earlier window inference based GUI for HOL built using the Centaur tool [26], and then by experimenting with other interface tools including the Cornell Synthesiser Generator [24] and Tcl/Tk [23]. In the end, we adopted Tcl/Tk as the implementation vehicle for our own graphical interface.

  Tcl/Tk consists of a general purpose scripting language (Tcl), together with a powerful set of widgets (Tk). The main strengths of Tcl/Tk are its light weight, generality, and ease of use. Unlike the other tools examined, Tcl/Tk offers no particular enhancements for manipulating syntax trees. This lack of specialisation, however, proved in some ways to be a blessing. For example, it was relatively easy for us to implement a freely re-associating selection mechanism (described in Sect. 5.1), which necessarily ignores the syntax tree of the underlying HOL term.

  The Expect [18] library for Tcl is used to manage the communication between HOL and the user interface. At present we use Expect only for coupling the interface and the theorem prover, but this feature may prove useful in future versions of the refinement calculator for interfacing the system with external oracles. The use of oracles to guide HOL proofs has already been demonstrated by a system integrating HOL and a computer algebra system [16].

4   HOL and Window Inference

The previous section described the overall design of the Refinement Calculator. The use of the HOL theorem proving system [9] and its accompanying window Library [12] form a major part of that design. Together they make up the underlying engine for the formal manipulation of expressions. In this section we describe HOL and the window Library in more detail, giving a more complete rational for their choice as components of the Refinement Calculator.

4.1　The HOL System

The HOL theorem prover has a number of features that distinguish it from competing systems and make it particularly well suited for use as a component in a tool like the Refinement Calculator. One of these is the architecture of the HOL system. For reasons explained below, this architecture allows us to place a particularly high degree of trust in tools built using HOL. We felt that the effort required to build the Refinement Calculator would only be justified if it was possible to confidently trust the results proved with it.

HOL is built using the *LCF architecture* for theorem provers, named for LCF [10], the first system of this kind. Tools with the LCF architecture are built around a small, trusted core implementing the abstract data-type of theorems in the logic of the tool. The axioms of the logic are constants of the data-type, while the inference rules are represented by functions that return elements of the data-type. Modus ponens, for example, would be implemented as a function taking two theorems of the form 'P' and '$P \Rightarrow Q$', and returning a theorem of the form 'Q'. The data-type may also include functions to make new theorems that extend the logic, for example by defining new constants. No other ways of creating a theorem are included in the signature of the data-type. The remainder of the system is built around this core. The architecture ensures our freedom to further extend the system with features needed for the Refinement Calculator without jeopardising its soundness. Any errors we might make can result only in a failure to produce a theorem, or in the production of a theorem other than the one intended, they can not result in the production of a 'nontheorem'.

Another relevant strength of HOL was the existing experience of embedding languages in the system. This gave us sources of both inspiration and comparison. The two common approaches to embedding languages in HOL are known as 'deep embedding' and 'shallow embedding'. In a *deep embedding*, the syntax of the language to be embedded is used to define a type of terms in that language. A function is then defined to map terms to their meaning. Deep embeddings offer opportunities to prove abstract and theoretical properties of a language that shallow embeddings do not. In a *shallow embedding* no new type of terms of the language to be embedded is defined. Instead, terms in the embedded language are identified with terms in the logic by extra-logical parsing and pretty-printing functions. We chose a shallow embedding for the practical reason that it allows us to identify types in the embedded language with types in the HOL logic. This means we can make the programming language of our refinement system strongly typed by inheriting the HOL type system rather than constructing our own. It also means that we can reuse existing HOL theories describing numbers, arrays, lists, and other data-types for the types of variables in our language.

The logic of the HOL system (Church's simple theory of types [7]) makes an excellent choice for formalising program refinement for several reasons. Firstly, the predicate transformer semantics commonly associated with program refinement is naturally modelled in higher-order logic. Predicate transformers are functions from predicates to predicates. It is possible to define such things in higher-order logics, but not in first-order logics. As a result, we are able to safely define the predicate transformer semantics of our target programming language as a definitional extension of higher-order logic; while similar tools based on first order logic, like PRT [6], need extend their logic with new axioms to achieve the same effect. Secondly, to avoid limiting our tool to trivial applications, we needed a logic with sufficient abstraction mechanisms to specify both complex refinement problems and the complex data-types needed to solve them. Part of our motivation for selecting the HOL system was that the power of the abstraction mechanisms of its logic had already been well demonstrated in the field of hardware verification [19]. Temporarily setting aside our already noted preference for a higher-order (and hence typed) logic; we observe that when choosing an expressive logic, a decision must be made between using a set theory or a type theory. The programming language used with the Refinement Calculator is typed, as is the HOL logic, and – as noted before – by opting for a shallow embedding we were able to avoid modelling the types of the programming language by inheriting the types of the logic. If we had chosen a theorem prover based on a set theoretic logic, we would also have had to model the type system of the target programming language within the logic ourselves.

4.2 Window Inference

Window inference is a transformational style of reasoning proposed by Robinson and Staples [25], and later generalised by Grundy [14]. A transformational proof begins with an term E, and proceeds by applying transformations that preserve some desired relationship, R. The proof ends when E has been transformed into another term E' that has some required property. The result is a theorem of the form $\vdash E\ R\ E'$. It is often the case with such proofs that the form of the solution E' is not known at the outset, but is discovered as part of the proof. Program refinement, for example, is an instance of transformational reasoning where E is a specification, R is refinement, and the property that E' must have is executability.

In contrast to this, most mechanised proof assistants support a style of reasoning that is best described as *goal directed*. In such systems a solution is proposed at the beginning of the proof, and the proof is a check that the proposed solution is valid. Although logical variables can be used with some systems to avoid the need to state the solution at the outset, the transformational approach of window inference would seem to be a closer fit to the program refinement process we want to support. As with most theorem provers, the usual interface to HOL is goal directed. However, the LCF architecture of HOL has allowed the implementation of an additional window inference based interface in the form of the HOL window Library [12]. Furthermore, the window Library is designed to be extended with databases of rules to support reasoning in various problem domains. This facility has been exploited to provide direct support for reasoning in the refinement calculus.

The distinguishing feature of window inference is that it allows users to transform a term by restricting attention to a subterm and transforming it. The remainder of the term is left unchanged. While transforming a subterm, it is possible to make assumptions based on its context. For example, if we want to transform the term $A \wedge B$; this may be done by transforming the subterm A under the assumption of B. We may assume B, because if it were false the entire term would be false regardless of A. It is also possible to select and transform assumptions to derive new ones which can then be used in subsequent transformations. The HOL implementation of window inference also allows the temporary conjecture of additional assumptions. If used, these *conjectures* must be proved – by transforming them to true – before the proof as a whole is considered complete.

Reasoning in window inference is conducted with a stack of windows. Each window has a *focus*, F, which is the term to be transformed; a set of formulae, Γ, called the *assumptions*, which can be assumed in the context of the window; and a *relation*, R, which is the relation to be preserved between the focus and any term to which it is transformed. (In the HOL implementation of window inference R must be a preorder.) Such a window will be written as follows:

$$! \; \Gamma$$
$$R * F$$

Four kinds of operations are permitted as part of a window inference proof:

1. A proof is begun by creating a window stack of depth one.
2. A transformation can be applied to the focus of the top window on the stack, providing it preserves the relationship associated with the window.
3. A new window can be pushed onto the stack. The focus of the new window must be a subterm of the focus or an assumption of the previous window. The assumptions and relation of the new window depend on the position of the new focus within the previous window. This operation is called *opening* a window.
4. The top window of a stack can be removed. The relationship established between the initial and final focus of that window is used to transform the window underneath. This operation is called *closing* a window.

Each level in the stack stores a theorem, called the *window theorem*, which records the reasoning done with that window.

5 TkWinHOL – A Tool for Transformational Reasoning

One of the design objectives of the Refinement Calculator was that it be easy to use. While the window Library made the stepwise derivation of programs with HOL a possibility, the tedious nature of the textual interface it provided ensured that the refinement of practical programming examples remained as challenging as ever. This section describes TkWinHOL, a graphical user interface for the window Library designed to help solve this problem. In TkWinHOL, work is carried out by selecting some information on the screen and then choosing an operation to perform through a menu selection or button click.

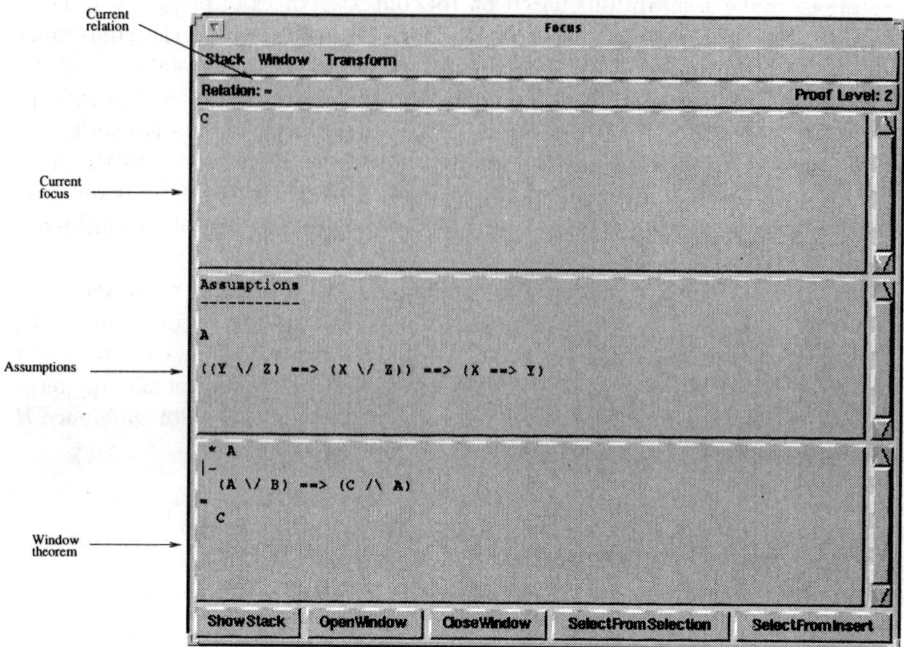

Fig. 2. The appearance of the focus (primary) window of TkWinHOL

5.1   Working with TkWinHOL

Once TkWinHOL is started the user is presented with three windows. Two of these are of less importance, offering a simple editor and a session window for entering commands directly to HOL. The third window (the focus window – see Fig. 2) presents a view of the current state of the active window stack. Thus different types of information (i.e. focus, context, window theorem and the relation preserved) about the current stack are displayed in different regions of the window.

Opening windows With TkWinHOL, a new window is opened on a subterm of the current focus by selecting the required subterm with the mouse, and then pressing the *OpenWindow* button. For further convenience, TkWinHOL can also deal with associativity when selecting subterms. For example, consider the focus

$$(A \wedge B) \wedge C$$

where the gray area denotes a text segment selected with the mouse. Pressing the *OpenWindow* button in this case causes the focus to be automatically re-associated and a new window to be opened with focus $B \wedge C$. These selection features greatly facilitate the use of window inference.

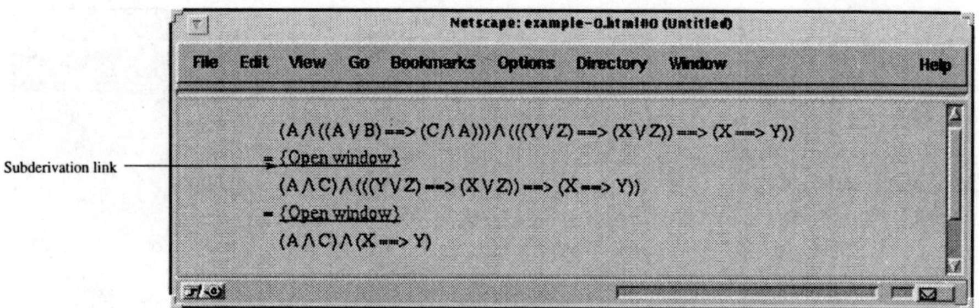

Fig. 3. An example of a browsable proof script

The selection mechanism is implemented by generating extra hidden information about the syntactic structure of the focus during pretty-printing. This information is then used by TkWinHOL to generate paths that access the HOL representation of selected subterms; such paths are required by the underlying HOL window Library.

Executing transformation commands The commands of the HOL window Library are bound to buttons and menus. For many parameters, the information required can be supplied by using the mouse to select relevant information already present on the screen. If that is not enough, information is entered through dialogs. The dialogs are not modal, thus it is possible to use information already present on the screen to fill some of the fields in any given dialog. The kind of information one typically has to fill in is the names of HOL theorems used for rewriting etc.

The use and role of some of the important transformation commands provided by the tool are described in Sects. 6 and 7.

5.2   Proof Visualisation

HOL proofs are not easy to read or understand. Therefore we have created a way of generating and presenting structured scripts that document proofs performed using TkWinHOL. Fig. 3 shows an example of such a proof script. The basic idea is to present proofs in a hierarchical and browsable format [13] (using HTML) so that the reader can focus on the parts of the proof that are of particular interest. For example, consider the logical term

$$(A \wedge ((A \vee B) \Rightarrow (C \wedge A))) \wedge (((Y \vee Z) \Rightarrow (X \vee Z)) \Rightarrow (X \Rightarrow Y))$$

$$(1)$$

which can easily be shown to be equivalent to the term

$$(A \wedge C) \wedge (X \Rightarrow Y)$$

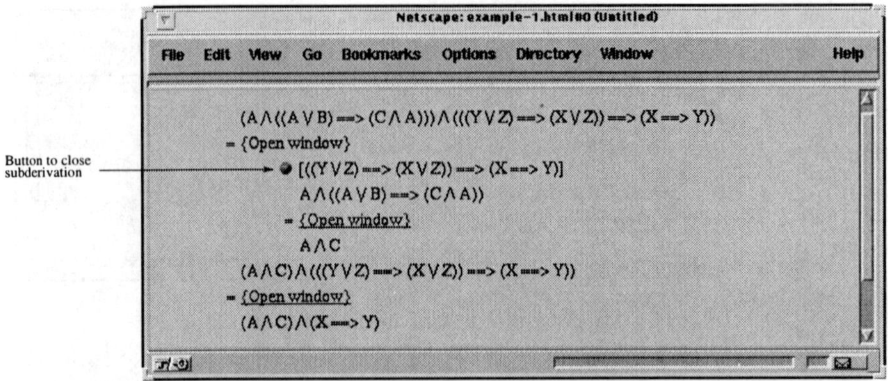

Fig. 4. A browsable proof script with a subderivation link expanded

To prove the equivalence one could start by focussing on the left conjunct of (1) and transform that subterm into $(A \wedge C)$. After that, one could similarly transform the right conjunct of (1) to $(X \Rightarrow Y)$. This is exactly the proof strategy shown in Fig. 3. At the level of abstraction shown in Fig. 3 both the subderivations mentioned above are hidden under subderivation links. Hidden subderivations are created when the user opens a new window, i.e. every time the scope of interest is reduced by the user.

By clicking one of the links, the proof script is expanded to contain the subderivation chosen. Fig. 4 shows the appearance of the proof script once the upper subderivation of Fig. 3 has been expanded. Clicking the button indicated in Fig. 4 recreates the state of Fig. 3. Alternatively, clicking one of the subderivation links in Fig. 4 would expand the script even further.

The main use of proof scripts such as these is to record proofs in a readable and easily distributable (through the WWW) format for pedagogic purposes. A detailed description of the proof scripting mechanism is available as a TUCS technical report [15].

5.3   Extending TkWinHOL

Our original goal was to build a tool supporting mechanically verified program derivation. Thus, emphasis has been placed on making TkWinHOL easy to customise for the needs of particular HOL theories (like those for program refinement [29] or lattice theory [17]). In such specific theories one usually wants to use a higher-level notation (abstract syntax) for the interaction with HOL. This can be achieved in TkWinHOL by adding a theory specific parser and pretty printer. For example, when developing refinement proofs it is more convenient to use a programming language syntax (described in the next section) rather than the relatively clumsy syntax defined by the underlying HOL theory.

Users of TkWinHOL can further customise the tool by adding theory specific menu choices bound to specialised HOL commands. Theory specific menu choices, and parsers and pretty printers, can be packaged together with the theories they support in a system of loadable libraries called 'extensions'. The Refinement Calculator is an example of such a TkWinHOL extension.

6 The Refinement Calculator

6.1 Programming Notation

The Refinement Calculator is an extension of TkWinHOL and supports a customised program notation with the following syntax:

$$
\begin{aligned}
Prog ::= &\ \textbf{program } Name \ \textbf{var } v : Type \cdot Com \\
Com ::= &\ Com;\ Com \\
&\ |\quad \{BTerm\} \\
&\ |\quad v := Term \\
&\ |\quad v := v' \bullet BTerm \\
&\ |\quad \textbf{if } BTerm \ \textbf{then } Com \ \textbf{else } Com \ \textbf{fi} \\
&\ |\quad \textbf{do } BTerm \rightarrow Com \ \textbf{od} \\
&\ |\quad |[\ \textbf{var } v : Type \cdot Com\ ]|
\end{aligned}
$$

Here, v represents a (list of) variable name(s), $Type$ represents a corresponding (list of) HOL type(s), $Term$ represents any HOL term, and $BTerm$ represents any boolean-valued HOL term. Because the tool uses a shallow embedding of the refinement calculus in HOL, the types and terms can be any such objects known to the HOL system when the derivation is being performed. In this way, the standard HOL notation for types and terms is embedded in the program-specific notation.

The programming notation is basically Dijkstra's guarded commands language extended with *assertions* and *nondeterministic assignment*. Occurrence of an assertion, $\{p\}$, in a program allows us to assume that p holds at that point in the program. Thus assertions can be used to carry context information in programs. A nondeterministic assignment, $x := x' \bullet p$, specifies that x is assigned some value x' satisfying predicate p. The final construct, $|[\ \textbf{var } v : Type \cdot Com\ ]|$, represents a block with local variable v.

6.2 Window Opening Rules for Refinement

A program may be specified with a statement $\{pre\}; v := v' \bullet post$. Derivation of a program satisfying this involves transforming the specification while preserving the refinement relation. Monotonicity of the program constructs allows us to focus on subprograms as, for example, the following window opening rules for sequential composition show:

$$
\frac{\vdash S \sqsubseteq S'}{\vdash S;T \sqsubseteq S';T}\ R1
\qquad\qquad
\frac{\vdash T \sqsubseteq T'}{\vdash S;T \sqsubseteq S;T'}\ R2
$$

Cond Introduction:

$\vdash\ S\ \sqsubseteq\ \textbf{if}\ G\ \textbf{then}\ S\ \textbf{else}\ S$

Block Introduction:

$\vdash\ v := v' \bullet post\ \sqsubseteq\ \|[\ \textbf{var}\ x : T \cdot x := E;\ v, x := v', x' \bullet post\ ]\|$

Loop Introduction 1:

$$\frac{\begin{array}{c} \vdash\ pre \Rightarrow inv \\ \vdash\ inv \wedge \neg G \Rightarrow post[v' := v] \end{array}}{\begin{array}{l} \vdash\ \{pre\}; v := v' \bullet post\ \sqsubseteq \\ \quad \textbf{do}\ G \to \{inv \wedge G\}; v := v'.inv[v := v'] \wedge E[v := v'] < E\ \textbf{od} \end{array}}\ [v\ \text{not free in}\ post]$$

Loop Introduction 2:

$$\frac{\begin{array}{l} \vdash\ pre \Rightarrow inv \\ \vdash\ (inv \wedge G \wedge E = e) \ll Body \gg (inv \wedge E < e) \\ \vdash\ inv \wedge \neg G \Rightarrow post[v' := v] \end{array}}{\vdash\ \{pre\}; v := v' \bullet post\ \sqsubseteq\ \textbf{do}\ G \to \{inv \wedge G\}; Body\ \textbf{od}}\ [v\ \text{not free in}\ post]$$

Trailing Assignment:

$$\begin{array}{l} \vdash\ v, x := v', x' \bullet post\ \wedge\ x' = E \\ \quad \sqsubseteq\ v := v' \bullet post;\ x := E[v' := v] \end{array} \quad \left[\begin{array}{l} x'\ \text{not free in}\ post, \\ v\ \text{not free in}\ E \end{array}\right]$$

Fig. 5. Refinement Transformations

The context information provided by assertions can be used directly when we focus on the right hand side of an assignment or the postcondition of a nondeterministic assignment as shown by the following rules:

$$\frac{p \vdash e1 = e2}{\vdash \{p\};\ x := e1\ \sqsubseteq\ \{p\};\ x := e2}\ R3$$

$$\frac{p \vdash q1 \Leftarrow q2}{\vdash \{p\}; x := x' \bullet q1\ \sqsubseteq\ \{p\}; x := x' \bullet q2}\ R4$$

Notice that, in the first case, the relation to be preserved on the expression is equality while, in the second case, the relation to be preserved on the postcondition is reverse implication.

The refinement calculator includes a complete set of window opening rules for selecting the subexpressions of the programming language defined in Sect. 6.1. The combination of these rules and the general window opening rules for navigating though terms in the HOL logic means that users of the refinement calculator can choose to focus their attention on any subterm of a program they wish to examine.

6.3 Refinement Transformations

The Refinement Calculator provides a menu of transformations for refining programs; some of these are represented as rules in Fig. 5. The refinement transformations are all derived from the HOL semantics of programs and refinement [1, 29, 31]. Some of the rules in Fig. 5 require the user to provide arguments (that possibly may not appear in the current focus) before the transformation is applied; for example, the *Cond Introduction* rule requires a guard G to be supplied while the *Loop Introduction 1* rule requires a guard G, an invariant *inv* and a variant E. When such a transformation is selected by the user of the tool, a dialogue box appears with input fields for each of the arguments; these must be filled out appropriately by the user before the tool applies the transformation to the focus.

Some of the rules have assumptions and side-conditions. When a rule with assumptions is applied, the assumptions will be instantiated appropriately and added to the current window context as conjectures to be established later. A transformation will fail if its side-conditions are not satisfied (failure of a transformation leaves the current window unchanged.)

Note from Fig. 5 that two *Loop Introduction* rules are used, the difference between them being that *Loop Introduction 1* determines the loop body from the invariant and the variant, while *Loop Introduction 2* requires the loop body to be supplied explicitly by the user. In the tool, both these rules are provided as a single command; if the *Body* field is left empty by the user in the dialogue box that appears, then *Loop Introduction 1* gets applied, otherwise *Loop Introduction 2* gets applied.

A term of the form $pre \ll prog \gg post$, as used in the *Loop Introduction 2* rule, describes a total-correctness assertion stating that *prog* is guaranteed to establish *post* when executed in initial state satisfying *pre*. The Refinement Calculator provides transformations for converting assertions of this form into boolean terms. The calculator provides a single command for propagating context information as assertion statements. This command uses theorems such as the following:

$$\vdash \quad \textbf{if } G \textbf{ then } S \textbf{ else } T \textbf{ fi} \quad \sqsubseteq \quad \textbf{if } G \textbf{ then } \{G\}; S \textbf{ else } \{\neg G\}; T \textbf{ fi}$$
$$\vdash \quad x := E \quad \sqsubseteq \quad x := E; \{x = E\} \quad [x \text{ not free in } E]$$

The calculator also provides commands for rewriting the focus with equations selected from the assumptions, for transforming the focus with user-supplied HOL theorems, and for replacing a focus F (while preserving relation R) by a focus F', where F' is supplied by the user, generating the conjecture $F \, R \, F'$.

The role of the rules of Fig. 5 will be seen more clearly in Sect. 7 where they are applied to an example. The full program syntax and list of refinement rules supported by the Refinement Calculator may be found in its manual and tutorial [4].

6.4   Data refinement

Data refinement is a special case of refinement where abstract program variables are replaced with more concrete ones. Typically, 'more concrete' means more easily or efficiently implementable. Since all occurrences of the abstract variables are to be replaced, the transformation affects the entire scope of their declaration. In the Refinement Calculator, the scope of declaration is the block statement. Thus, for the abstract variables a and concrete variables c, data refinement is the following transformation:

$$\lvert[\mathbf{var}\ a \cdot S]\rvert \sqsubseteq \lvert[\mathbf{var}\ c \cdot T]\rvert$$

where the abstract statement S refers to the variables v, a, and the concrete statement T to v, c (v represents the state variables that are not data-refined and are thus common to both S and T). An *abstraction relation* R, relating the abstract and concrete variables, is required in order to perform a data refinement transformation. It has been shown [21, 29] that the concrete statement T can be calculated from the abstract statement S and the abstraction relation R. This allows us to define a function $\mathcal{D}_R$ such that

$$\lvert[\mathbf{var}\ a \cdot S]\rvert \sqsubseteq \lvert[\mathbf{var}\ c \cdot \mathcal{D}_R(S)]\rvert$$

The function is defined recursively over the structure of program notation as, for example, in the case of sequential composition:

$$\mathcal{D}_R(S_1; S_2) \;\widehat{=}\; \mathcal{D}_R(S_1); \mathcal{D}_R(S_2)$$

For the basic statements like assignment, $\mathcal{D}_R$ gives the corresponding concrete statement as follows (assuming that R is of the form $(\lambda a, c, v.abs)$):

$$\mathcal{D}_R(v, a := v', a' \bullet post) \;\widehat{=}\;$$
$$v, c := v', c' \bullet (\forall a.abs \Rightarrow (\exists a'.abs[a, c, v := a', c', v'] \land post))$$

In the Refinement Calculator, the *Data-Refinement* command implements the function $\mathcal{D}_R$. Given variables to be replaced, a, new variables, c, and the abstraction relation R, it calculates the new focus, $\lvert[\mathbf{var}\ c \cdot T]\rvert$, from the old focus, $\lvert[\mathbf{var}\ a \cdot S]\rvert$, and automatically proves the correctness of the refinement, possibly adding conjectures. Note that the transformation need not necessarily replace all variables declared in the block. Furthermore, if the list of variables to be replaced is empty, the rule implements *superposition refinement*, where only new variables are added under the relation R.

7   Example Derivation: Sorting an Array

We present the outline of an example derivation that may be carried out using the Refinement Calculator. The example involves the derivation of a program that sorts an array of numbers.

Arrays may be modelled by defining a new HOL type; assume $(\alpha)array$ is the type of arrays that are polymorphic on α, and that the following functions on arrays have been defined:

$size : (\alpha)array \to num$
$lookup : (\alpha)array \to num \to \alpha$
$swap : (\alpha)array \to num \to num \to (\alpha)array$

An array a is indexed from 0 to $(size\ a) - 1$. The ith element of a is given by $(lookup\ a\ i)$, while $(swap\ a\ i\ j)$ represents the array a with the values at positions i and j swapped around.

Let *sorted* and *perm* be defined as follows:

$$sorted\ (a : (num)array)\ (r : num \to bool) =$$
$$(\forall i \cdot (r\ i) \Rightarrow i < (size\ a)) \land$$
$$(\forall ij \cdot (r\ i) \land (r\ j) \land i < j \Rightarrow (lookup\ a\ i) \leq (lookup\ a\ j))$$
$$perm\ (a1 : (num)array)\ (a2 : (num)array) =$$
$$(size\ a1) = (size\ a2) \land$$
$$(\exists f \cdot (injective\ f) \land$$
$$(\forall i \cdot i < (size\ a1) \Rightarrow (f\ i) < (size\ a1) \land$$
$$(lookup\ a1\ i) = (lookup\ a2\ (f\ i))))$$

The term $(sorted\ a\ r)$ states that that projection of array a whose indices satisfy r is sorted, e.g., $(sorted\ a\ (\lambda i \cdot 5 \leq i \leq 10))$ says that the array slice 5..10 of a is sorted. The term $(perm\ a1\ a2)$ specifies that array $a2$ is a permutation of array $a1$. The sorting program is then specified as:

program *Sort* **var** $a : (num)array \cdot$
$$\{a = a0\};\ a := a' \bullet (sorted\ a'\ (\lambda i \cdot i < (size\ a0))) \land (perm\ a0\ a')$$

Here, $a0$ is a specification constant, not a program variable. This specification (in ASCII form) is provided as input to the tool in order to begin a program derivation; the specification will then appear in the focus region of the tool (see Fig. 2).

Using the mouse to focus on the highlighted body of the specification yields the following window:

$$\sqsubseteq \ * \ \{a = a0\};$$
$$a := a' \bullet (sorted\ a'\ (\lambda i \cdot i < (size\ a0))) \land (perm\ a0\ a')$$

We shall implement this using an insertion sort algorithm which requires a pair of nested loops. The body of the outer loop will ascend the array ensuring that the array slice $0..k-1$, where k is the current position, is sorted. Before introducing the loop, we must introduce a local variable k that will be used to ascend the array. This is introduced and initialised by applying the *Block Introduction* transformation to the highlighted statement above, with $x : T$ instantiated by $k : num$ and E instantiated by 0. This results in the focus:

$$\sqsubseteq \ * \ \{a = a0\};$$
$$|[\ \textbf{var}\ k : num \cdot k := 0;$$
$$a, k := a', k' \bullet (sorted\ a'\ (\lambda i \cdot i < (size\ a0))) \land (perm\ a0\ a')$$
$$]|$$

Now, we execute the command that propagates assertions yielding:

$$\sqsubseteq \; * \; |[\; \mathbf{var} \; k : num \cdot k := 0;$$

$$\{(k = 0) \wedge (a = a0)\};$$
$$a, k := a', k' \bullet (sorted \; a' \; (\lambda i \cdot i < (size \; a0))) \wedge (perm \; a0 \; a')$$

$$]|$$

Next, we select the highlighted part of this block using the mouse, open a window on it, and apply the *Loop Introduction* command. This causes a dialogue box to appear with fields for loop guard, body, invariant and variant. We supply the following values for the guard, invariant and variant and leave the *Body* field blank:

$Guard :$ $k < (size \; a)$
$Invariant :$ $k \leq (size \; a) \wedge (sorted \; a \; (\lambda i \cdot i < k)) \wedge (perm \; a0 \; a)$
$Variant :$ $(size \; a) - k$

This transformation results in the following window:

$$\sqsubseteq \; * \; \mathbf{do} \; k < (size \; a) \rightarrow$$

$$\{k < (size \; a) \wedge (sorted \; a \; (\lambda i \cdot i < k)) \wedge (perm \; a0 \; a)\};$$
$$a, k := a', k' \bullet (sorted \; a' \; (\lambda i \cdot i < k')) \wedge (perm \; a0 \; a') \wedge$$
$$((size \; a') - k') < ((size \; a) - k)$$

$$\mathbf{od}$$

This step also generates two conjectures which are displayed in the context region of the tool. These state respectively that the invariant should hold initially and the invariant and the negated guard should establish the original postcondition:

$$(\forall k, a \cdot (k = 0) \wedge (a = a0) \Rightarrow$$
$$k \leq (size \; a) \wedge (sorted \; a \; (\lambda i \cdot i < k)) \wedge (perm \; a0 \; a))$$
$$(\forall k, a \cdot \neg(k < (size \; a)) \wedge k \leq (size \; a)$$
$$\wedge (sorted \; a \; (\lambda i \cdot i < k)) \wedge (perm \; a0 \; a)$$
$$\Rightarrow (sorted \; a \; (\lambda i \cdot i < (size \; a0))) \wedge (perm \; a0 \; a))$$

At this stage in the derivation, we can choose either to attempt to discharge the conjectures[6] or to continue refining the body of the loop. Conjectures such as this may be established by opening windows on the conjectures, with backward implication as the relation to be preserved, then transforming the focus to true.

We proceed with the refinement of the loop body:

$$\sqsubseteq \; * \; \{k < (size \; a) \wedge (sorted \; a \; (\lambda i \cdot i < k)) \wedge (perm \; a0 \; a)\};$$
$$a, k := a', k' \bullet (sorted \; a' \; (\lambda i \cdot i < k')) \wedge (perm \; a0 \; a') \wedge$$
$$((size \; a') - k') < ((size \; a) - k)$$

Focussing on the highlighted term yields:

$$\Leftarrow \; * \; (sorted \; a' \; (\lambda i \cdot i < k')) \wedge (perm \; a0 \; a') \wedge$$
$$((size \; a') - k') < ((size \; a) - k)$$

---

[6] If we postpone establishing the conjectures, then what we have is a refinement theorem with the conjectures as assumptions.

Notice that the relation to be preserved is reverse implication. We strengthen this term by replacing the highlighted subterm with $k' = k + 1$, yielding the following focus:

$$\Leftarrow \ * \ \boxed{(sorted\ a'\ (\lambda i \cdot i < k')) \land (perm\ a0\ a')} \ \land$$
$$k' = k + 1$$

along with the conjecture

$$k' = k + 1 \ \Rightarrow \ ((size\ a') - k') < ((size\ a) - k)$$

This conjecture may discharged by opening a window on it and transforming it to true using appropriate arithmetic theorems (this also requires $(perm\ a0\ a)$ and $(perm\ a0\ a')$ which will appear as contextual assumptions). We skip these steps and continue with the main derivation. Focussing on the highlighted subterm of the previous focus yields

$$\Leftarrow \ * \ (sorted\ a'\ (\lambda i \cdot i < \boxed{k'})) \land (perm\ a0\ a')$$

with $k' = k + 1$ as a context assumption. Applying the *Rewrite* command to the focus using this assumption yields:

$$\Leftarrow \ * \ (sorted\ a'\ (\lambda i \cdot i < (k + 1))) \land (perm\ a0\ a')$$

Closing (twice) gives us:

$$\sqsubseteq \ * \ \{k < (size\ a) \land (sorted\ a\ (\lambda i \cdot i < k)) \land (perm\ a0\ a)\};$$
$$\boxed{a, k := a', k' \bullet (sorted\ a'\ (\lambda i \cdot i < (k + 1))) \land (perm\ a0\ a') \land}$$
$$\boxed{k' = k + 1}$$

Applying the *Trailing Assignment* transformation yields:

$$\sqsubseteq \ * \ \boxed{\{k < (size\ a) \land (sorted\ a\ (\lambda i \cdot i < k)) \land (perm\ a0\ a)\};}$$
$$\boxed{a := a' \bullet (sorted\ a'\ (\lambda i \cdot i < (k + 1))) \land (perm\ a0\ a')};$$
$$k := k + 1$$

So, under the assumption that the array is sorted between 0 and $k - 1$, the highlighted specification must establish that the array is sorted between 0 and k. It will do this by shuffling the element of a at position k down the array to the appropriate position. The following arguments will be used to perform the introduction of the inner loop:

$$Guard: \quad l > 0 \land (lookup\ a\ l) < (lookup\ a\ (l - 1))$$
$$Body: \quad a := (swap\ l\ (l - 1)\ a);\ l := l - 1$$
$$Invariant: l \le k \land (sorted\ a\ (\lambda i \cdot i < l \lor (i > l \land i \le k))) \land$$
$$l < k \Rightarrow (lookup\ a\ l) < (lookup\ a\ (l + 1)) \land$$
$$(perm\ a0\ a)$$
$$Variant: \quad l$$

This invariant states that the array slice $0..k$, with the lth position excluded, is sorted, and that the value at the lth position is less than the value at the $(l+1)^{th}$ position. We focus on the highlighted part of the previous focus and introduce l as a local variable in the same way that k was introduced. We then apply the *Loop Introduction* command with the above arguments yielding the focus

$$\sqsubseteq \ * \ \textbf{do } l > 0 \wedge (lookup\ a\ l) < (lookup\ a\ (l-1)) \rightarrow$$
$$a := (swap\ l\ (l-1)\ a);$$
$$l := l - 1$$
$$\textbf{od}$$

along with some appropriate conjectures.

Closing several windows gives us an implementation of sorting:

$$\sqsubseteq \ * \ |[\ \textbf{var } k : num \cdot k := 0;$$
$$\textbf{do } k < (size\ a) \rightarrow$$
$$|[\ \textbf{var } l : num \cdot l := k;$$
$$\textbf{do } l > 0 \wedge (lookup\ a\ l) < (lookup\ a\ (l-1)) \rightarrow$$
$$a := (swap\ l\ (l-1)\ a)$$
$$l := l - 1$$
$$\textbf{od }]|;$$
$$k := k + 1$$
$$\textbf{od }]|$$

8   Conclusions

We have described the Refinement Calculator, a tool for the derivation of provably correct programs. The calculator is an extension of TkWinHOL, a tool that provides a graphical user interface to window inference proof in HOL. The architecture of TkWinHOL has allowed us to make use of several existing packages to reduce the effort involved in implementing the Refinement Calculator, i.e. HOL, the window Library and the Refcalc theory.

Our experience has shown that the window inference style of proof is well suited to program derivation using the refinement calculus. Furthermore, window inference does not preclude the use of the more traditional verification style of program derivation, rather both styles can be easily mixed. For example, the conjectures generated by the application of *Loop Introduction 2* rule of Fig. 5 are exactly those proof obligations used in loop verification. We claim several other advantages for our approach:

- Without the GUI described here, a detailed knowledge of HOL and its meta-language (ML) is required in order to use the HOL window Library and the Refcalc theory. Even with such detailed knowledge, focussing on subterms is particularly difficult when working directly with the HOL window Library. The GUI improves the usability of the HOL window Library and the Refcalc theory considerably by providing visually-based access to subterms and context assumptions as well as an abstract programming language syntax, and menus and dialogue boxes for the application of transformations.

- The use of HOL as an underlying proof engine gives us a high degree of confidence in the reliability of the tool.
- Existing and future HOL libraries supporting various data structures or more automated proof can easily be used with the tool.
- The programmability of the tool allows us to abstract many similar transformation rules into single user commands. A simple example of this is the way the two loop-introduction rules of Fig. 5 are accessed by a single user command. This reduces the number of different commands that the user needs to be aware of.

One disadvantage of our approach is that defining new transformations requires knowledge HOL and ML. This is because transformations are not provided to the tool in the form of inference rules as used in Fig. 5 but rather are implemented as ML functions. It should be possible to overcome this to an extent by allowing a sequence of existing transformations to be combined into a single transformation. Another disadvantage is that the target programming language syntax is currently hardwired into the tool, making it impossible for users to extend the syntax. This shortcoming, however, is not integral to the tool. It should be possible for us to provide the same facilities via general purpose parser and pretty printer tools.

Independently of the tool described here, a refinement tool called PRT has been developed by a group at the University of Queensland [6]. PRT is built on top of the Ergo theorem prover [27], which also supports the window inference style of reasoning. This tool is similar to the Refinement Calculator though an important difference is that PRT uses a purpose-built logic in which commands, predicates, program variables, and logic variables are treated as separate syntactic classes avoiding the need for a parsing and pretty-printing layer between the user and the proof engine. The Refinement Calculator, on the other hand, uses only a conservative extension of the HOL logic, giving us a higher degree of confidence in its soundness.

Currently the Refinement Calculator supports the refinement of sequential programs, including the application of data refinement, as well as the browsable presentation of proofs. Future plans include extending the tool to support procedures and modules as well as the *action system* formalism [3], an extension of the refinement calculus dealing with parallel and distributed systems. We are also investigating ways of further increasing the usability and extendibility of the tool.

References

1. R. Back and J. von Wright. Refinement concepts formalized in higher order logic. *Formal Aspects of Computing*, 2:247–272, 1990.
2. R.-J. R. Back. On correct refinement of programs. *Journal of Computer and System Sciences*, 23(1):49–68, Feb. 1981.
3. R.-J. R. Back and R. Kurki-Suonio. Decentralisation of process nets with centralised control. In *2nd ACM SIGACT-SIGOPS Symp. on Principles of Distributed Computing*, pages 131–142, 1983.

4. M. Butler, T. Långbacka, R. Rukšėnas, and J. von Wright. Refinement Calculator tutorial and manual. Draft — available upon request.

5. M. J. Butler and T. Långbacka. Program derivation using the refinement calculator. In von Wright et al. [30], pages 93–108.

6. D. Carrington, I. Hayes, R. Nickson, G. Watson, and J. Welsh. A tool for developing correct programs by refinement. In H. Jifeng, editor, *BCS FACS 7th Refinement Workshop*. Springer-Verlag, July 1996.

7. A. Church. A formulation of the simple theory of types. *Journal of Symbolic Logic*, 5:56–68, 1940.

8. E. W. Dijkstra. *A Discipline of Programming*. Prentice Hall Series in Automatic Computation. Prentice Hall, 1976.

9. M. J. C. Gordon and T. F. Melham, editors. *Introduction to HOL: A theorem proving environment for higher order logic*. Cambridge University Press, 1993.

10. M. J. C. Gordon, R. Milner, and C. P. Wadsworth. *Edinburgh LCF: A Mechanised Logic of Computation*, volume 78 of *Lecture Notes in Computer Science*. Springer-Verlag, 1979.

11. L. J. Groves, R. G. Nickson, and M. Utting. A tactic driven refinement tool. In C. B. Jones, B. T. Denvir, and R. C. F. Shaw, editors, *5th Refinement Workshop*, Workshops in Computing, pages 272–297, London, Jan. 1992. BCS-FACS, Springer-Verlag.

12. J. Grundy. The HOL window library. In *The HOL System*, volume Libraries. SRI International, Cambridge Research Center, Millers Yard, Mill Lane, Cambridge CB2 1RQ, England, 2.01 edition, July 1991.

13. J. Grundy. A browsable format for proof presentation. *Mathesis Universalis*, 1(2), Spring 1996. Available from `http://www.pip.com.pl/MathUniversalis/2/grundy/mu.html`.

14. J. Grundy. Transformational hierarchical reasoning. *The Computer Journal*, 39(4):291–302, 1996.

15. J. Grundy and T. Långbacka. Towards a browsable record of HOL proofs. Technical Report 7, Turku Centre for Computer Science, Lemminkäisenkatu 14A, 20520 Turku, Finland, May 1996. Available from `http://www.tucs.abo.fi/publications/techreports/TR7.ps.gz`.

16. J. Harrison and L. Théry. Extending the HOL theorem prover with a computer algebra system to reason about the reals. In J. J. Joyce and C.-J. H. Seger, editors, *Higher Order Logic Theorem Proving and Its Applications: 6th International Workshop*, volume 780 of *Lecture Notes in Computer Science*, pages 174–184, Vancouver, Canada, Aug. 1993. Springer-Verlag.

17. L. Laibinis. Using lattice theory in higher order logic. In von Wright et al. [30], pages 315–330.

18. D. Libes. *Exploring Expect: A Tcl-based Toolkit for Automating Interactive Programs*. O'Reilly & Associates, 1995.

19. T. F. Melham. *Higher Order Logic and Hardware Verification*, volume 31 of *Cambridge Tracts in Theoretical Computer Science*. Cambridge University Press, 1993.

20. C. C. Morgan. *Programming from Specifications*. Prentice Hall, 2 edition, 1994.

21. C. C. Morgan and P. H. B. Gardiner. Data refinement by calculation. *Acta Informatica*, 27(6):481–503, 1990.

22. J. M. Morris. A theoretical basis for stepwise refinement and the programming calculus. *Science of Computer Programming*, 9(3):287–306, Dec. 1987.

23. J. K. Ousterhout. *Tcl and the Tk Toolkit*. Addison Wesley, 1994.

24. T. W. Reps and T. Teitelbaum, editors. *The Synthesizer Generator. A System for Constructing Language-Based Editors*. Springer-Verlag, 1988.

25. P. J. Robinson and J. Staples. Formalizing a hierarchical structure of practical mathematical reasoning. *Journal of Logic and Computation*, 3(1):47–61, Feb. 1993.

26. L. Théry. An X-windows interface for the window inference system. In M. Gordon, ed., The Final Report on DSTO Grant: Improved Interface for HOL, Internal Report. University of Cambridge, Computer Laboratory, New Museums Site, Pembroke Street, Cambridge CB2 3QG, England, 1993.

27. M. Utting and K. Whitwell. Ergo user manual. Technical Report 93-19, Software Verification Research Institute, University of Queensland, QLD 4072, Australia, Oct. 1993.

28. T. Vickers. An overview of a refinement editor. In *5th Australian Software Engineering Conference*, pages 39–44, Sydney, May 1990. IREE.

29. J. von Wright. Program refinement by theorem prover. In D. Till and R. C. F. Shaw, editors, *6th Refinement Workshop*, Workshops in Computing, pages 121–150, London, Jan. 1994. BCS-FACS, Springer-Verlag.

30. J. von Wright, J. Grundy, and J. Harrison, editors. *Theorem Proving in Higher Order Logics: 9th International Conference*, volume 1125 of *Lecture Notes in Computer Science*, Turku, Finland, Aug. 1996.

31. J. von Wright, J. Hekanaho, P. Luostarinen, and T. Långbacka. Mechanising some advanced refinement concepts. *Formal Methods in Systems Design*, 3:49–81, 1993.

A Modular Approach to Defining Composition Operators

Steven Butler

Software Verification Research Centre,
School of Information Technology,
The University of Queensland, Australia 4072
email: seb@it.uq.edu.au

Abstract. In object-oriented systems, collections of interlinked objects work together towards achieving some goal. The aim of this paper is to introduce and define a formal semantics of a suite of composition operators for specifying object interactions. Composition operators are specified using a modular parametric approach which highlights the generality of such operators. Composition frameworks that capture the common features of operators use specification fragments to form different composition operators. The fragments may also be used in the specification of multiple operators, highlighting such operators' similarities.

1 Introduction

In object-oriented systems, collections of interlinked objects transform and interact with each other, moving the system towards achieving some goal. These interactions present an interesting challenge: how to determine a system's meaning, based on the meaning of its component objects and the way they interact. The aim of this paper is to introduce and define a formal semantics of a suite of composition operators used for specifying object interactions.

A variety of protocols have been proposed for specifying and implementing object interaction.

- Programming languages such as Eiffel [9] and Smalltalk [5] capture object interaction via information passing and operation invocation. Objects performing operations may send relevant environmental information to another object and invoke one of that object's own operations.
- Process algebras such as CSP [8] present operators for capturing concurrency, choice, etc. These operators work at the process (as distinct from the event) level, constructing new processes out of existing processes. Other examples of global composition operators are given in Abadi and Lamport [1] and Duke *et al.* [3].
- The Object-Z specification language [4, 7, 10] provides several operators for composing operations. The operators work at the operation (as distinct from the object) level and effectively construct new system operations out of the existing operations defined on the underlying objects.

In this paper, a compositional model of operations is specified that follows the approach to object interaction used in Object-Z. While Object-Z forms a basis of the work presented here, the compositional model defines a general model of composition in an object-oriented system and unlike the work in Griffiths [6] it is not intended to be specifically a semantics of composition in Object-Z.

This paper refines the compositional model presented in Butler and Duke [2]. A modular parametric approach for specifying the composition operators improves upon the earlier work by highlighting the generality of such operators. Common features of each composition operator are extracted and distilled into composition frameworks. Then, component specification fragments are used in a framework to form a composition operator. This modular parametric approach allows the reuse, where appropriate, of the composition fragments.

1.1 The Object Model

Informally, an object-oriented system is a collection of objects. Objects encapsulate two things. First, they have state, which is a collection of variable names bound to values. Second, they have operations that can change the object's state.

An object in a system is known by its unique identity. A variable inside an object can be uniquely identified by a combination of the identity of the object that contains it, and the variable's name.

Operations defined in objects have a signature (i.e. the interface between the operation and its environment), a pre-condition and a post-condition. An attempt to use an operation to change the system from any particular state is called an invocation of that operation. Operation invocations are either successful thereby transforming the system, or unsuccessful, in which case the system remains unchanged.

The external effect of a successful operation invocation is an assignment of values to the variable names in the operation's signature (i.e. a signature binding). An operation interacts with other operations, and with its environment through its external effect. Unlike the variable names in objects, those in an operation's signature must have a role, which signifies the direction of information flow. Input variable names are interpreted as taking their values from the environment of the operation while output variable names are interpreted as supplying their values to the environment of the operation.

Along with its external effect, an operation invocation has some effect on a subset of one or more objects' state variables (a post-frame). The values bound to the variables in the post-frame after a successful invocation of the operation are determined by the operation's post-condition. An operation invocation is only successful when the objects' state satisfies the pre-condition of the operation. If the operation's pre-condition is not satisfied, the operation invocation is said to be blocked. The pre-condition of an operation is defined over some subset of one or more objects' state variables, known as the operation's pre-frame. A subset of variables (i.e. a frame) given some value binding is known as a sub-state.

A pre- or post-condition is usually satisfied by more than one state binding. An operation invocation causes a transformation of the values of variables in the operation's post-frame to one of the post-states that satisfies the post-condition. The operation invocation only succeeds if the variables in the operation invocation's pre-frame are in a state that satisfies the pre-condition. The modification of the post-frame variables combined with the necessary pre-state is the internal effect of a successful operation invocation.

Primitive operations are defined without the use of composition operators. Composition operators define rules for constructing new operations from existing operations. Taking the primitive operations as a basis, new and more complex operations can repeatedly be constructed using the composition operators.

Two frameworks for specifying composition operators are modelled in this paper: The first, operation choice, creates a new operation such that a successful invocation of that operation has the same effect as a successful invocation of one of its constituent operations. Only one operation's effect is felt for any particular successful operation invocation.

The second, operation collaboration, creates a new operation such that a successful invocation of that operation is a collaboration of successful invocations from both constituent operations. The manner in which the operation invocations collaborate is determined by the composition operator.

Some collaborations model communication of information between the external effects of the constituent operations. Communication can take two forms: external and inter-object.

External communication occurs when the signatures of the two operations have common variable names. The common variables must take (for inputs) or give (for outputs) the same value from (to) the environment. This form of communication is external because it involves the environment of the constructed operation. The common signature variables in the constituent operations are effectively merged into a single signature variable in the constructed operation.

Inter-object communication between operations happens when a signature variable from each of the constituent operations are bound to the same value, but are hidden from the external effect of the constructed operation and thus its environment.

Object-Z has a suite of different composition operators, some of which provide a model of inter-object communication. The informal view of inter-object communication in Object-Z is a flow of information from output signature variables in one operation to the input signature variables in another operation. Output variables are denoted by a variable name ending in a shriek (e.g. $x!$); similarly, input variables are denoted by a variable name ending in a query (e.g. $x?$). Inter-object communication models the flow of information from output variables to input variables with the same base name (e.g. from $x!$ to $x?$). The query and shriek forms of a variable name are *complementary*.

Four composition operators that are commonly used in Object-Z are modelled in this paper:

1. Angelic choice is an example of operation choice. A successful invocation of an operation constructed using angelic choice is the same as a successful invocation of one of its constituent operations. If only one of the con-

stituent operations satisfies its pre-condition, then it is invoked. If both operations satisfy their pre-conditions, then the operation invoked is chosen non-deterministically. If neither operation satisfies its pre-condition, then the invocation of the constructed operation is unsuccessful.

2. Conjunction is a form of operation collaboration. Invoking the constructed operation results in a simultaneous invocation of both constituent operations. If both constituent operations try to modify an overlapping piece of the system, then they must agree on how it changes or the operation invocation is unsuccessful. Conjunction does not model inter-object communication, but identically named variables in the external effects of constituent operations communicate externally, through the use of the environment.

3. Parallel is also a form of operation collaboration. As in conjunction, an invocation of a parallel constructed operation results in a simultaneous invocation of both constituent operations. Parallel differs from conjunction because, as well as having external communication, it also models inter-object communication between the signature bindings of constituent operations. Parallel models a bi-directional communication protocol; information flows from output variables in either operation to complementary input variables in the other operation. The signature variables that perform inter-object communication are hidden and the remaining variables with common names in the signature of each constituent operation communicate externally as in conjunction.

4. Sequential composition is also a form of operation collaboration, but the internal effects of the constituent operations are not invoked simultaneously. The first operation is invoked, changing the state of the system; then from that new state the second operation is invoked, further changing the state of the system. If the invocation of either of the two constituent operations is unsuccessful, then the invocation of the constructed operation is also unsuccessful. The double state transition caused by the internal effect of the constructed operation is atomic. Sequential composition also models inter-object communication between signature bindings, however, information flows only from the output variables of the first operation to the input variables of the second operation; it is uni-directional. As in parallel, the signature variables that perform inter-object communication are hidden, leaving the remaining variables in the signature to communicate according to the model of external communication.

In Section 2, an abstract model of an operation is introduced. An operation is sufficiently abstract to model both primitive operations, and operations that are defined as a composition of other operations.

Section 3 introduces the concept of operation choice, develops a framework for specifying operation choice operators and includes a specification of the angelic choice operator, inspired by the operator of the same name in Object-Z.

Section 4 examines operation collaboration and develops a framework for specifying operators that cause operations to collaborate. Inside this framework, three composition operators, conjunction, parallel and sequential composition, also taken from Object-Z are specified. Finally, two operators, one analogous to the Z piping operator and another which models piping in the opposite direction are specified using the collaboration framework. These operators demonstrate the advantages of a modular, parametric specification framework because they are defined through the reuse of composition fragments specified for other operators.

The Z specification language is used as the semantic meta-language for constructing semantic domains.

2 Operations

In this section, a formal model for operations is developed. Section 2.1 introduces the basic components of objects and operations: variable names and values. Section 2.2 examines the effect an operation invocation has on its environment and Section 2.3 discusses the internal transitions that are caused by operation invocations. Finally, Section 2.4 combines the external and internal effects, defining operations.

2.1 Variables and Values

Operations are defined in terms of how they can change the state of objects in a system, and thus the entire system's state. The state of an object is represented as a binding of variable names to their values; both variable names and value are modelled as given types.

$$[Name, Value]$$

Variable names in operations often have a role. In Object-Z a variable, say x, can have either an input role ($x?$), output role ($x!$), or have no role (just x). This is modelled by partitioning the given type $Name$ into input, output and no role sets. A mapping that shows complementary names will allow the development of a communication model between objects, through the use of input and output variables in operations.

Variable names may be bound to either a reference to another object, or a primitive, unstructured value (e.g. $\mathbb{N}$ or a given type). The given type, $Name$, is thus partitioned into reference name and primitive name sets.

In a consistent object-oriented system, a reference variable refers to an object, and a primitive variable has some value (that is not an object), so $Value$ is also partitioned into two sets: object identity values ($\mathbb{O}$), and primitive values ($\mathbb{V}$). Note that the identity of an object is persistent throughout the lifetime of a system; only an object's internal state may change as the system transforms.

$$\begin{array}{|l}
Reference, Primitive, Input, Output, Norole : \mathbb{P}\, Name \\
\mathbb{O}, \mathbb{V} : \mathbb{P}\, Value \\
\hline
\langle Reference, Primitive \rangle \text{ partitions } Name \\
\langle Input, Output, Norole \rangle \text{ partitions } Name \\
\langle \mathbb{O}, \mathbb{V} \rangle \text{ partitions } Value
\end{array}$$

Complementary input and output variable names are modelled by the function *comp_name*. Because variables with no role are not used for communication, it is sufficient that such variables are the complement of themselves. Note that the definition says only that for every name, there is exactly one corresponding name, but nothing else; in particular, the notion of concrete names like $x?$ and $x!$ are not directly modelled at this level of abstraction.

$$\begin{array}{|l}
comp_name : Name \rightarrowtail Name \\
\hline
comp_name^{\sim} = comp_name \\
comp_name(\!|\ Input\ |\!) = Output \\
\forall\, n : Norole \bullet comp_name(n) = n
\end{array}$$

2.2 External View of an Operation

A binding gives values to a set of names. The set of names can consist of both reference names and primitive names, so the set of values that they are bound to might consist of both object identities and primitive values.

A binding is any partial function from names to values with the proviso that a reference name may only be bound to an object identity and a primitive name may only be bound to a primitive value. The term "binding" refers exclusively to functions from this constrained set.

$$Binding == \{b : Name \nrightarrow Value \mid b(\!|\ Reference\ |\!) \subseteq \mathbb{O} \\
b(\!|\ Primitive\ |\!) \subseteq \mathbb{V}\}$$

The signature of an operation (i.e. the interface between the operation and its environment) is a set of names. Note that an operation signature can consist of a mixture of input and output names.

$$Sig == \mathbb{P}(Input \cup Output)$$

An auxiliary function, *comp* gives the complement of an entire signature.

$$\begin{array}{|l}
comp : Sig \rightarrow Sig \\
\hline
comp = \lambda\, s : Sig \bullet comp_name(\!|\ s\ |\!)
\end{array}$$

During operation invocation, the names in the signature of an operation are given a specific value; this is modelled by *SigBinding*, which is a binding of names in a signature to values.

$$SigBinding == \{sb : Binding \mid \operatorname{dom} sb \in Sig\}$$

2.3 Internal View of an Operation

A variable in a system is identified by a combination of the identity of the object that contains it, and its name. The object identity is needed because several objects in a system may contain variables with the same name.

When an operation is invoked, its effects are localised to changing the values of some subset of system variables (possibly spanning more than one object). Subsets of variables in a system are modelled by *Frame*. Note that *Frame* captures the property that object variables have no role – variables with a role are reserved for operation signatures.

$$Frame == \mathbb{O} \leftrightarrow Norole$$

A *SubState* binds identified variables (i.e. a frame) to values. The variables in objects must be bound to appropriate values (i.e. reference names are bound to object identities and primitive names are bound to primitive values).

$$SubState == \{st : (\mathbb{O} \times Norole) \nrightarrow Value \mid$$
$$\forall idvar : \operatorname{dom} st \bullet \{second(idvar) \mapsto st(idvar)\} \in Binding\}$$

A system has to satisfy some pre-condition before an operation invocation is successful. After a successful invocation of the operation, the system conforms to the requirements of the operation's post-condition. In object-oriented systems an operation's pre- and post-conditions are generally only concerned with some subset of the state of the entire system (i.e. the operation's pre- and post-frames). Outside of an operation's pre- and post-frames, the system may be in any pre-state, and change (or stay the same) as it sees fit. In this respect, operations model a *partial* effect on system state; two operations can be invoked simultaneously without contradiction if each has a pre-condition depending on disjoint subsets of the system state, and a post-condition modifying disjoint subsets of the system state.

A single *Internal* effect (i.e. some state transition) that an operation invocation causes in a system is defined by two sub-state bindings. The sub-state *pre* represents the binding that some subset of the system must have for the internal effect to be applicable. The sub-state *post* represents the binding that some subset of the system must have after the successful invocation of the operation.

The frame *cond* (read conditional) provides a convenient way to access the variables which are required to have some pre-state for the operation to be enabled. Similarly the frame *mod* (read modified) is the set of variables that become bound to the values defined by the post-state of the internal effect.

```
┌─ Internal ──────────────────────────────────
│ pre, post : SubState
│ cond, mod : Frame
├─────────────────────────────────────────────
│ dom pre = cond
│ dom post = mod
└─────────────────────────────────────────────
```

It is important to note that the sub-state changed by the operation and the sub-state needed for that change to occur need not be related. An example is an operation that has pre-condition $x < 10$, but whose post-condition is simply $y' = 2$ (following the Z convention of using primed ($'$) variables to represent post-condition variables and unprimed variables to represent pre-condition variables). In this example, the post-value of y is not related in any way to the pre-value of x, yet, unless the value of x before the operation satisfies the pre-condition, the operation cannot proceed. Suppose the post-condition of the operation is $y' = y + 1$; in this case, the pre-state of the operation is related to the post-state (i.e. the value of y in each state is related) so the pre-value of y will also be a part of the pre-state (along with the condition that $x < 10$).

2.4 Operations and their Effects

An operation invocation involves some external effect (signature binding), seen by the environment, together with an internal effect, which modifies a sub-state. A particular external effect can be related to many different internal transitions and vice versa.

Each pair in an *Effect* relation corresponds to a single operation invocation. The first element in the pair is the signature binding given to the invocation and the second element determines both the success of the invocation, and the effect it has on the system's state, if successful.

$$Effect == SigBinding \leftrightarrow Internal$$

The *Operation* schema encapsulates the signature and all possible invocations of an operation. The specification of an *Operation* is sufficiently abstract that it models both primitive operations and operations that are constructed from other operations with composition operators.

```
┌─ Operation ─────────────────────────────────────
│ sig : Sig
│ effect : Effect
├──────────────────────────────────────────────────
│ ∀ e : effect • dom first(e) = sig
└──────────────────────────────────────────────────
```

So they can be statically typed, operations have a static signature. Every signature binding in an operation's *effect* relation is a binding of values to the same signature. The *Operation* schema models the semantics of all object-oriented operations, including operations created through the use of composition operators and basic operations that are defined in isolation from other operations. The implication is that operations defined through composition of component operations must have the property that for each invocation of the operation, its signature is the same; therefore it is important that composition operators are specified so that any new operations that they create do not violate this constraint.

While the signature of an operation is constant across invocations, the pre- and post-frames of an operation can be different for each invocation. There are no constraints on the pre- and post-frames of individual operation invocations. Consider the case where an operation is defined as a choice between two operations defined in two separate objects; if an invocation of the constructed operation results in an invocation of the first operation, then the sub-state of the system that is modified is in the object that defines the first operation. Similarly, if the invocation of the constructed operation results in an invocation of the second operation, then the sub-state of the system that is modified is in the object that defines the second operation. Since the two objects are distinct, the state variables inside each are disjoint subsets of the system state; hence different invocations of the constructed operation can modify different portions of the system state.

Given a signature binding, an operation is blocked if there is no related internal effect whose pre-state exactly matches a sub-state of the system. As an example, two operations are defined below. The operation *block* is always blocked, and the operation *skip* is always enabled but explicitly changes nothing. To simplify their definition, both operations have an empty signature[1].

$$block == (\mu\ Operation \mid sig = \varnothing \wedge effect = \varnothing)$$
$$skip == (\mu\ Operation \mid sig = \varnothing$$
$$effect = \{(\varnothing, (\mu\ Internal \mid pre = \varnothing \wedge post = \varnothing))\})$$

Now consider the *effects* of the *block* and *skip* operations. The *block* operation has no signature bindings related to internal effects (i.e. the *effect* relation is empty) so it can never do anything. The *skip* operation has only one signature binding (an empty one), related to an internal transition. The internal transition's pre-condition applies to an empty subset of the system state, so it is always enabled, and its post-condition also applies to an empty subset of the system state, so no state ever needs to be modified to satisfy the post-condition of the operation.

In the next two sections, the problem of operation composition is discussed, and formal definitions for a number of composition operators are developed. In Section 3, a framework for composing operations using choice is developed. Section 4 introduces another framework, this time for operation composition through collaboration.

Choice operators choose between operations. A successful invocation of a choice-composed operation has the same effect as a successful invocation of one of its constituent operations.

With operation collaboration, the effects of two operation invocations collaborate so that for any successful invocation of the constructed operation, a successful invocation from both component operations will work together to form the overall effect of the new operation. The operation collaboration frame-

---

[1] It is possible to define a variant of the given *block* and *skip* operations for any operation signature.

work uses functions for separately composing external and internal effects of constituent operations, and a policy for relating the composed partial effects.

3   Operation Choice

In this section, the framework for operation choice is developed, and an example operator, angelic choice, defined. A choice composition function takes two operation effects and produces a new operation effect. Choice composing functions must satisfy the condition that, in the domain of the two argument effects and the resulting effect, every signature binding has the same signature.

$$ChoiceComposer == \{f : Effect \times Effect \nrightarrow Effect \mid$$
$$\forall\, e_1, e_2, e_3 : Effect \bullet ((e_1, e_2), e_3) \in f \Leftrightarrow$$
$$\forall\, sb_1, sb_2 \in \mathrm{dom}(e_1 \cup e_2 \cup e_3) \bullet \mathrm{dom}\, sb_1 = \mathrm{dom}\, sb_2\}$$

The choice operator framework constructs new operators of the form

$$ChoiceOperator == Operation \times Operation \nrightarrow Operation$$

The choice operators are partial because they cannot combine two operations unless they have the same signature.

The effects of the two operations are combined with the composing function and the result is the effect of the new operation. The signature of the new operation is the common signature of the two operations. The restriction on the type of the *ChoiceComposer* function guarantees that operations composed with choice framework generated operators satisfy the predicate of the *Operation* schema.

$$\begin{array}{l}
choice : ChoiceComposer \rightarrow ChoiceOperator \\
\hline
choice = \lambda\, choicec : ChoiceComposer \bullet \\
\quad \lambda\, op_1, op_2 : Operation \mid op_1.sig = op_2.sig \bullet \\
\quad\quad (\mu\, Operation \mid sig = op_1.sig \\
\quad\quad\quad\quad effect = choicec(op_1.effect, op_2.effect))
\end{array}$$

Angelic choice allows two operations to be combined so that, from some pre-state, if at least one of the operations is enabled, so too is the constructed operation. If only one of the component operations is enabled, then an invocation of the constructed operation results in a successful invocation of the enabled component operation. In the case that both operations are enabled, angelic choice non-deterministically invokes one of the component operations. If both component operations are blocked, then the constructed operation is also blocked.

The composition function for the angelic choice operator is now defined as the union of the effects of the component operations. The angelic composition function only composes effects that satisfy the condition on the domain of *ChoiceComposer* functions.

$$
\begin{array}{|l}
\hline
angelic\_composer : ChoiceComposer \\
\hline
angelic\_composer = \lambda\, e_1, e_2 : Effect \mid \\
\quad (e_1, e_2) \in \mathrm{dom}\ ChoiceComposer \bullet e_1 \cup e_2 \\
\end{array}
$$

The composition operator for angelic choice ($[\,]$) is defined as an application of the *choice* composition function to the angelic composer function. The restriction that choice compositions only work for operations with identical signatures is captured by the predicate in the choice framework.

$$
\begin{array}{|l}
\hline
\_\,[\,]\,\_ : ChoiceOperator \\
\hline
[\,] = choice(angelic\_composer) \\
\end{array}
$$

4   Operation Collaboration

In this section, a framework for operation composition through collaboration is developed. In the specification of an operation, a distinction is made between an operation's internal and external effect, with the two related. For operation collaboration, this distinction proves extremely useful because the process of composition can also be separated into the internal effect of the composition, and its external effect. With the internal and external effects of operations composed separately, a compositional framework is needed to combine the fragments into a complete operation.

First, some abstractions are introduced:

In operation collaboration, the signature of a composite operation is determined by the signature of its constituents.

$$SigCompose == Sig \times Sig \rightarrow Sig$$

defines the type of the functions used by the composition framework for composing operation signatures.

Further,

$$SigBindingCompose == SigBinding \times SigBinding \nrightarrow SigBinding$$

defines the type of the functions used by the collaboration framework for composing signature bindings; these functions are partial because many signature binding pairs cannot be combined. In conjunction, for example, a pair can only be combined if identical names in the signature of each are bound to identical values.

Finally,

$$InternalCompose == Internal \times Internal \nrightarrow Internal$$

defines the type of the functions used by the composition framework for composing the internal effects of operations; it too is partial, because in most composition operators, many pairs of internal effects cannot be composed. Again, taking conjunction as an example, two internal effects are only compatible if they agree on the values given to any common variables in their pre- and post-states.

The compositional framework takes as arguments the functions that describe how to compose the different parts of operations and produces a collaboration operator. So, define

$$Collaboration == SigCompose \times SigBindingCompose \times InternalCompose$$

and

$$Collaborator == Operation \times Operation \rightarrow Operation$$

for this purpose.

The collaboration framework does not work for all combinations of compositional fragments; only fragments that are consistent with the predicate imposed upon an operation (i.e. signature bindings in the effect of an operation are consistent with the operation's signature) are useful within the collaboration framework. In fact, this consistency problem suggests that using separate composition functions for signatures and signature bindings is redundant. Any one of the signature bindings constructed for the new operation encapsulates the constructed operation's signature. Unfortunately, composition sometimes results in an operation that is never enabled; the resultant operation is analogous to the earlier defined operation *block*, perhaps with a different signature. Since the *block* operation has no possible invocations, and thus no signature bindings, it is impossible to derive its signature from its effect relation.

Rather than exclude blocked operations from the compositional model, a signature composition function gives the signature of all composite operations. When a constructed operation isn't always blocked, the signature composition function is redundant because the domain of any of the new operation's signature bindings give its signature. Two functions are consistent only if every signature binding that is constructed by applying the signature binding composition function to two signature bindings has the same signature as that gained by applying the signature composition function to the signatures of the same two signature bindings.

The effect of a new operation is constructed by relating the results of the signature binding and internal effect composition functions; the compositional framework provides the "glue" that determines which internal and external effects are related.

Signature binding and internal effect fragments are constructed from the fragments in each component operation through their respective composition functions. Inclusion in the new operation's effect relation requires that a signature binding, say sb and internal effect, say int meet certain conditions. First, a signature binding sb_1 in the first component operation's effect relation and a signature binding sb_2 in the second component operation's effect relation must, when composed, yield the signature binding, sb. Second, for the internal effects, an int_1 and int_2 belonging to the effect relation of the first and second component operations respectively must, when composed, yield the internal effect, int.

Finally, sb_1 must relate to int_1 in the effect relation of the first component operation and sb_2 must relate to int_2 in the effect relation of the second component operation.

$$
\begin{array}{l}
collaborate : Collaboration \nrightarrow Collaborator \\
\hline
collaborate = \lambda\, sc : SigCompose;\ sbc : SigBindingCompose; \\
\quad intc : InternalCompose \mid \\
\qquad (\forall\, sb_1, sb_2 : SigBinding \mid (sb_1, sb_2) \in \operatorname{dom} sbc \bullet \\
\qquad \operatorname{dom} sbc(sb_1, sb_2) = sc(\operatorname{dom} sb_1, \operatorname{dom} sb_2)) \bullet \\
\quad \lambda\, op_1, op_2 : Operation \bullet \\
\qquad (\mu\, Operation \mid \\
\qquad\quad sig = sc(op_1.sig, op_2.sig) \\
\qquad\quad effect = \{ sb_1, sb_2 : SigBinding;\ i_1, i_2 : Internal \mid \\
\qquad\qquad (sb_1, i_1) \in op_1.effect \wedge (sb_2, i_2) \in op_2.effect \\
\qquad\qquad (sb_1, sb_2) \in \operatorname{dom} sbc \wedge (i_1, i_2) \in \operatorname{dom} intc \bullet \\
\qquad\qquad (sbc(sb_1, sb_2), intc(i_1, i_2)) \})
\end{array}
$$

The remainder of this section demonstrates the creation of composition operators by specifying compositional fragments and then plugging them into the compositional framework.

4.1   The Conjunction Operator

In Object-Z, the conjunction operator models the effect of a simultaneous occurrence of two operations. The constructed operation is only enabled if both constituent operations are enabled given the pre-state of the system, and their post-states agree where their frames overlap. In addition, if both operations communicate with their environment using identical names, then all such names are bound to the same value in the signature bindings of each operation.

The formal definition of conjunction uses composition functions for signatures, signature bindings and internal effects. First,

$$
\begin{array}{l}
conj\_sig : SigCompose \\
\hline
conj\_sig = \lambda\, s_1, s_2 : Sig \bullet s_1 \cup s_2
\end{array}
$$

defines the signature of any two operations combined using conjunction. The signature of the new operation is the union of the signatures of its constituent operations.

Next, the signature binding composition function is defined. Unlike signatures, not all signature binding pairs can be combined. In the case of conjunction, the only condition on combining signature bindings is that common names in each are bound to the same values. The new signature binding for compatible signature bindings is the union of the constituent signature bindings. It is trivial to show that the signature binding composition function is consistent with the signature composition function, a property required by the compositional framework.

$$
\begin{array}{|l}
conj\_sb\_compose : SigBindingCompose \\
\hline
conj\_sb\_compose = \lambda\, sb_1, sb_2 : SigBinding \mid \\
\quad sb_1 \cup sb_2 \in SigBinding \bullet sb_1 \cup sb_2
\end{array}
$$

With the composition functions for the external effects defined, all that remains is to define the composition function for the internal effects, and then put them all together using the compositional framework. For conjunction, the internal effects of the constituent operations can be composed if they place conditions on disjoint subsets of state in their pre- and post-frames, or, if the frames overlap, the parts that overlap are bound to the same value. To conjoin two internal effects is to have both occur simultaneously. The pre-state of the new internal effect is the union of the pre-states of the constituent effects and similarly for the post-states. The pre- and post-frames (*cond* and *mod* respectively) of the new internal effect follow from the predicate placed on *Internal*.

$$
\begin{array}{|l}
simul : InternalCompose \\
\hline
simul = \lambda\, i_1, i_2 : Internal \mid \\
\quad i_1.pre \cup i_2.pre \in SubState \wedge i_1.post \cup i_2.post \in SubState \bullet \\
\qquad (\mu\, Internal \mid pre = i_1.pre \cup i_2.pre \wedge post = i_1.post \cup i_2.post)
\end{array}
$$

Finally, the conjunction operator is specified as the application of the compositional fragments inside the compositional framework. The framework relates signature binding compositions to the appropriate internal effect compositions, producing a new operation from two existing operations.

$$
\begin{array}{|l}
\_ \wedge \_ : Collaborator \\
\hline
\wedge = collaborate(conj\_sig, conj\_sb\_compose, simul)
\end{array}
$$

4.2 The Parallel Operator

On the whole, the parallel operator is like conjunction. However, the parallel operator also models bi-directional inter-object communication between the external effects of operations.

As for conjunction, the signature of operations constructed using the parallel operator are defined. The parallel operator models inter-object communication between complementary input and output signature variable names. Such names are hidden from the external view of the constructed operation. The signature of an operation composed with the parallel operator is the union of its constituents' signatures, with the communicating variable names hidden. In the definition of *par_sig*, complementary names are determined via the *comp* function.

$$
\begin{array}{|l}
par\_sig : SigCompose \\
\hline
par\_sig = \lambda\, s_1, s_2 : Sig \bullet (s_1 \setminus comp(s_2)) \cup (s_2 \setminus comp(s_1))
\end{array}
$$

Next, the signature binding composition function is defined for signature bindings that are compatible under the parallel operator. Assuming communicating variables are hidden as described above, any remaining variables with the same name in the signature of each operation are bound to the same values. In addition, for those variables that are hidden, the complementary names in each operation are also bound to the same values. Again, these requirements are specified using the *comp* and *comp_name* functions to determine the complementary names. The parallel composition of those compatible signature bindings is then the union of the signature bindings from the constituent operations, after first hiding the names used for communication between the operations. As with conjunction, it is trivial to show that the signature binding and signature composition functions for parallel are consistent, as required by the compositional framework.

$$par\_sb\_compose : SigBindingCompose$$

$$
\begin{aligned}
&par\_sb\_compose = \lambda\, sb_1, sb_2 : SigBinding \mid \\
&\quad (comp(\mathrm{dom}\, sb_2) \mathbin{\lhd} sb_1 \cup comp(\mathrm{dom}\, sb_1) \mathbin{\lhd} sb_2 \in SigBinding \\
&\quad \forall\, n_1 : \mathrm{dom}\, sb_1;\ n_2 : \mathrm{dom}\, sb_2 \mid \\
&\qquad comp\_name(n_1) = n_2 \bullet sb_1(n_1) = sb_2(n_2)) \bullet \\
&\qquad\quad comp(\mathrm{dom}\, sb_2) \mathbin{\lhd} sb_1 \cup comp(\mathrm{dom}\, sb_1) \mathbin{\lhd} sb_2
\end{aligned}
$$

The internal effect of operations defined using the parallel operator is constructed as for operation conjunction. The similarity between the two operators is highlighted by the use of the function *simul*, developed for conjunction, to specify the construction of new internal effects for parallel. The ability to reuse fragments is another advantage of using a compositional framework approach to specifying the composition operators.

With all the fragments defined, the parallel operator is defined as an application of the compositional framework, similarly to conjunction. The only difference is the signature and signature binding composition functions supplied to the framework specify that certain variable names communicate and are hidden from the external view of the new operation.

$$\_ \parallel \_ : Collaborator$$

$$\parallel = collaborate(par\_sig, par\_sb\_compose, simul)$$

4.3   The Sequential Composition Operator

The sequential composition operator has the effect of two operations being invoked in a sequential order (i.e. one operation is invoked, and from the post-state of that operation the second operation is invoked). The sequence of two state transitions becomes a single atomic transition. If either invocation is unsuccessful, then the entire invocation of the constructed operation is also unsuccessful. Sequential composition also models inter-object communication between signature bindings, only it is uni-directional – from the first operation's output variables to variables with complementary input names in the second operation.

These communicating signature variables are hidden from the environment, and the remaining signature variables in the constituent operations that have matching names are bound to the same value and merged in the constructed operation, as with the conjunction operator.

First, the signature composition function for the sequential composition operator is defined. The communicating output names from the first operation and communicating input names from the second operation are hidden from the signatures of their respective operations. The signature of the new operation is the union of the resulting two signatures. Again, the *comp* function is used to determine which signature variables from each constituent operation's signature are hidden from the constructed operation's signature.

$$
\begin{array}{|l}
seq\_sig : SigCompose \\
\hline
seq\_sig = \lambda\, s_1, s_2 : Sig \bullet \\
\quad (s_1 \setminus comp(s_2 \cap Input)) \cup (s_2 \setminus comp(s_1 \cap Output))
\end{array}
$$

Next, a function that gives the signature bindings that result from sequentially composing compatible signature bindings is specified. Compatible signature bindings must satisfy two conditions: first, with the communicating names hidden from both signature bindings, the union of the remaining signature bindings is a signature binding (i.e. identical names from each are bound to the same value, as in conjunction); second, the output variables from the first signature binding that have a complementary variable in the inputs to the second signature binding are bound to the same value. Again, it is trivial to show that *seq_sb_compose* is consistent with *seq_sig* in the context of the compositional framework.

$$
\begin{array}{|l}
seq\_sb\_compose : SigBindingCompose \\
\hline
seq\_sb\_compose = \lambda\, sb_1, sb_2 : SigBinding \mid \\
\quad (comp(\mathrm{dom}\, sb_2 \cap Input) \lhd sb_1 \cup \\
\quad comp(\mathrm{dom}\, sb_1 \cap Output) \lhd sb_2 \in SigBinding \\
\quad \forall\, n_1 : (\mathrm{dom}\, sb_1 \cap Output);\ n_2 : \mathrm{dom}\, sb_2 \mid \\
\qquad comp\_name(n_1) = n_2 \bullet sb_1(n_1) = sb_2(n_2)) \bullet \\
\qquad\quad comp(\mathrm{dom}\, sb_2 \cap Input) \lhd sb_1 \cup \\
\qquad\quad comp(\mathrm{dom}\, sb_1 \cap Output) \lhd sb_2
\end{array}
$$

Sequential composition combines internal transitions differently to the conjunction and parallel operators. In conjunction and parallel, both operations transform parts of the system simultaneously, requiring agreement on pre- and post-states if there is any overlap. Sequential works differently: although both operations have an effect on the state, their affects are applied in sequence (although the resulting effect is atomic). Instead of the entire overlap of pre- and post-states agreeing, as in conjunction and parallel, the post-state of the first operation must agree with the pre-state of the second operation, where their frames overlap. Outside of this overlapping, the frames of the pre-states and the frames of the post-states from each operation must also agree. The pre-state of a sequentially constructed transition is the pre-state of the second internal transition functionally overridden by the pre-state of the first internal transition. The post-state is defined similarly, except it is the post-state of the first that is overridden by the post-state of the second.

$$
\begin{array}{|l}
\quad sequence : InternalCompose \\
\hline
\quad sequence = \lambda\, i_1, i_2 : Internal \mid (\textbf{let } ovlp == i_1.mod \cap i_2.cond \bullet \\
\qquad (ovlp \lhd i_1.post) = (ovlp \lhd i_2.pre) \\
\qquad (ovlp \ntriangleleft i_1.pre) \cup (ovlp \ntriangleleft i_2.pre) \in SubState \\
\qquad (ovlp \ntriangleleft i_1.post) \cup (ovlp \ntriangleleft i_2.post) \in SubState) \bullet \\
\qquad\qquad (\mu\, Internal \mid pre = i_2.pre \oplus i_1.pre \wedge post = i_1.post \oplus i_2.post)
\end{array}
$$

This definition shows that where the post-state of the first internal transition and the pre-state of the second internal transition give bindings to disjoint names, *sequence* is equivalent to *simul*.

Given these sequential compositional fragments, the sequential composition operator is defined by plugging them into the composition framework.

$$
\begin{array}{|l}
\quad \_ \mathbin{\raisebox{0.2ex}{\scriptsize$\stackrel{\circ}{9}$}} \_ : Collaborator \\
\hline
\quad \mathbin{\raisebox{0.2ex}{\scriptsize$\stackrel{\circ}{9}$}} = collaborate(seq\_sig, seq\_compat, sequence)
\end{array}
$$

In the next subsection, compositional fragments developed for the sequential composition operator are combined with those used by the parallel and conjunction operators, producing more composition operators through reuse.

4.4   Exploiting Modular Composition

The modular compositional framework presented in this section gives a well-defined structure to the specification of composition operators; the internal and external effects of composition operators are specified in isolation, allowing reasoning about each without the complication of their effect on each other. Then, the composition framework provides a uniform way of combining the internal and external effects into a composition operator. All collaboration operators relate their internal and external effects in the same way, so this reasoning is not repeated for every collaboration operator definition.

As seen with the parallel and conjunction operators, modular reasoning about internal and external effects isn't the only advantage of using a compositional framework. The parallel operator's internal effects are composed in exactly the same way as the conjunction operator's; the only difference between the two operators is that the parallel operator provides a communication model for external effects. Because the internal effects of each operator are the same, the same internal composition function is used for each.

Here, the advantages of reuse are exploited further with the production of two new composition operators, using minimal new specifications.

The Z specification language provides a mechanism for schemas to communicate through a uni-directional piping operator ($\gg$). The external effect of Z piping is analogous to the sequential composition operator's external effect, but its internal effect is that each operation causes a simultaneous change in state. A similar piping operator is defined using the external composition functions of sequential and the internal composition function of conjunction.

$$\_ \gg \_ \ : Collaborator$$

$$\gg = collaborate(seq\_sig,\ seq\_sb\_compose,\ simul)$$

A piping operator that works in reverse is trivial to specify as a forward piping of operations, with the arguments reversed.

$$\_ \ll \_ \ : Collaborator$$

$$\ll = \lambda\, op_1,\, op_2 : Operation \bullet op_2 \gg op_1$$

5 Conclusion

This paper demonstrates the use of modular parametric frameworks for specifying object-oriented composition operators. The frameworks serve to highlight similarities and differences between composition operators. The essence of a class of composition operators is captured by a framework, while specialized functions define the individuality of specific operators. Various operators can use the same specialized function as part of their definition.

The definition of operations presented in this paper models both primitive and composed operations uniformly. The advantage is that the definitions for the composition operators need not distinguish between the two types of operations.

The extension of this modular parametric framework to a full object model has been anticipated; hence the inclusion of object identities. Such a complete model is beyond the scope of this paper.

The formal approach to composition, when related to informal development methods, could provide an effective formal basis for reasoning about systems developed using such methods.

Acknowledgements This paper was written under the supervision of Roger Duke. Thanks to Steve Atkinson for proofreading and suggestions for improvement.

References

1. M. Abadi and L. Lamport. Composing specifications. In J.W. de Bakker, W.-P. de Roever, and G. Rozenberg, editors, *Proc. REX Workshop on Stepwise Refinement of Distributed Systems*, volume 430 of *Lect. Notes in Computer Science*, pages 1–41. Springer-Verlag, 1990.

2. S. Butler and R. Duke. Defining Composition Operators for Object Interaction. Technical Report 96-12, Software Verification Research Centre, Dept. of Computer Science, Univ. of Queensland, 1996. Accepted in the journal *Object Oriented Systems*.

3. R. Duke, C. Bailes, and G. Smith. A Blocking Model for Reactive Objects. *Formal Aspects of Computing*, 8:347–368, 1996.

4. R. Duke, G. Rose, and G. Smith. Object-Z: A specification language advocated for the description of standards. *Computer Standards & Interfaces*, 17:511–533, 1995.

5. A. Goldberg and D. Robson. *Smalltalk-80: the language*. Addison-Wesley, 1989.

6. A. Griffiths. An Extended Semantic Foundation for Object-Z. Technical Report 95-39, Software Verification Research Centre, Dept. of Computer Science, Univ. of Queensland, 1995.

7. A Griffiths and G Rose. A Semantic Foundation for Object Identity in Formal Specification. Technical Report 95-38, Software Verification Research Centre, Dept. of Computer Science, Univ. of Queensland, 1995. To appear in the journal *Object Oriented Systems*.

8. C.A.R. Hoare. *Communicating Sequential Processes*. International Series in Computer Science. Prentice-Hall, 1985.

9. B. Meyer. *Eiffel: The language*. Prentice-Hall Object-Oriented Series. Prentice-Hall, 1992.

10. G. Rose. Object-Z. In S. Stepney, R. Barden, and D. Cooper, editors, *Object Orientation in Z*, Workshops in Computing, pages 59–77. Springer-Verlag, 1992.

A Semantics for Recursive Operations in Object-Z

Alena Griffiths

Software Verification Research Centre,
School of Information Technology,
University of Queensland. Australia. 4072.

Abstract. Although Object-Z began as an extension to Z, the introduction of a reference semantics for object identity, and support for the notion of strict modularity, has resulted in a semantics that is quite different to Z. One consequence of this change is that it is now possible to give meaning to recursively defined operations, something not supported in Z. This paper briefly surveys the treatment of recursion in a variety of model-oriented specification languages. An outline of the semantics of Object-Z operations is presented, and we focus particularly on the issue of recursively defined operations. The denotation of a system operation is defined as the effect of the operation on the entire collection of objects in the system. We also describe an alternative representation of a system operation, in which a system operation is described by the operation call-graph that it induces. It is conjectured that the two models are equivalent, and a proof sketch of the conjecture is provided. We expect the result to be useful in allowing the more intuitive framework of call graphs to be used to reason about object interaction.

1   Introduction

Recursion and iteration are used widely in programming to define procedures which alter the program state. Support for recursion in model-oriented specification languages is, however, fairly limited. When there is no attempt made to divide the system state into modules or objects, then it is usually straightforward to find predicates that describe the required state change in a non-recursive manner. For this reason, the omission of a recursive or iterative construct from traditional specification languages is justifiable. However, where it is desired to partition the state into modules (as in object-oriented systems), and where we seek to describe system transitions in terms of existing operations on the individual module states, then some iterative or recursive construct is necessary. Yet, as this paper shows, a brief survey of model-oriented, object-oriented, specification languages reveals that their support for recursively defined operations is also weak.

The aim of this paper is to describe the semantics of operations in Object-Z, and to show how this semantics takes account of recursive operations. An alternative representation of system operations as call graphs is presented, and it is conjectured that the representations are equivalent. These contributions are useful in terms of the twin goals of the Object-Z specification language: to be an expressive modelling language, and to have the formal foundations on which a theory of reasoning and refinement can be developed.

Object-Z is an object-oriented formal specification language, developed at the University of Queensland [6, 23]. Recent work [9, 11] on the formal semantics for this language has been characterised by two significant departures from the Z semantic model. The first major change was the introduction of a reference semantics for object identity [5, 11]. In this model, the denotation of an object variable is an object identity, and so an *object map* associating object identities with the object values they refer to becomes necessary. The meaning of a specification is given by the set of object maps, representing snapshots in the system's evolution, together with a set of labelled transitions on the object maps, which represent possible system evolutions. This model, although useful for the representation of object sharing and recursive object references, presents real challenges in the area of operation composition. The difficulty arises because operations defined in a class affect not only the local object, but if they call operations in other objects, they potentially affect the entire object map. It is difficult to interpret local definitions in terms of the system-wide transitions in which they participate.

This problem motivated the second departure from traditional Z semantics, which is the notion of strict modularity [9]. Strict modularity is the idea that classes should be given a meaning that is independent of the overall specification in which they appear. It requires a composition function that, assuming well-formedness, will take these independent meanings and compose them to yield a set of object maps and transitions.

This paper is structured as follows. In Section 2 the issue of recursion in specification is considered, and the inclusion of a recursive construct in structured specification languages is motivated. Section 3 reviews the main ideas underlying Object-Z semantics, providing an overview of the necessary background for the remainder of the paper.

Section 4 provides an informal presentation of operation composition in Object-Z. First, a naive, algorithmic description of composition is presented. This description does not account for recursive operations. Moreover, because the semantic function is itself described recursively, it is possibly undefined for infinite object maps. An amended composition algorithm is then described. The new description handles recursive operations, but the semantic description is mathematically unsatisfactory because it is unclear whether the recursion in the metalanguage is well-founded. It is intuitively clear however, that when the operation call-graph induced by the system transition is finite, the algorithmic description is adequate.

To address the problems of definedness inherent in the algorithmic description, Section 5 presents the formal semantics of composition as a set of *Effect*s. An *Effect* is a tuple that records the effect of a message on an object map, including the inputs required and the outputs produced. For a given specification, the effect set is developed incrementally, and defined as the limit of that construction. As such, the effect set does not suffer from the mathematical inadequacies of the previous informal algorithmic description.

Section 6 examines the relationship between the two different representations of a system transition: the call-graph representation described informally in Section 4, and the set of *Effect* described in Section 5. We conjecture that the system transitions described by the effect set are exactly those for which there exists a finite call-graph representation. A proof sketch is provided, and we comment on the utility of this result.

2 Recursion in Specification

2.1 Terminology

In this paper, an operation is a mechanism for altering the system state, possibly parametrised with inputs and outputs. Care should be taken to distinguish between operations and side-effect free functions, which do not change the system state. Recursion in the latter kind of construct has already received much attention in the literature [19, 20] and will not be considered in this paper.

The focus of this paper is on recursion, but we acknowledge that a language that supports iteration is as expressive as one that supports recursion. Hence, in surveying languages to determine their support for recursion, we will also comment on whether they support iteration.

2.2 Non-object-oriented Languages

The best known languages for the specification of sequential systems are Z and VDM.

In Z, operations are defined using the schema calculus. The scope rules of Z are such that the definition of a schema is interpreted in an environment not yet enriched with the name of the schema being defined. As such, the schema name cannot be referred to in the definition of the schema itself, and so self-recursive definitions are not permitted. Moreover, the principle of definition before use applies, and so mutual recursion is also not permitted. These rules apply in the semantics of Z as articulated by Spivey [26], and in the developing standard [1].

There is no support for imperative code fragments in Z, and so there is no construct to support iteration.

VDM-SL (Vienna Development Method — Specification Language) is a broad-spectrum language, supporting the expression of system requirements via abstract specification or imperative code.

In VDM, operations may be defined implicitly (i.e. abstractly) or explicitly (i.e. imperatively). Implicit operation are defined by a pre- and post-condition. These conditions must be predicates, and so implicit operation definitions may not include recursive operation calls.

Associated with the definition of a VDM implicit operation Op are two auxiliary functions $pre-Op$ and $post-Op$ [13, p 136]. These functions return boolean values, and so they are valid formulae. There appears to be nothing in the syntax or semantics to prevent the formula *post−Op* appearing in the definition of the postcondition for *Op*, and so in this way, it seems that recursive implicit operation definitions can be constructed. In contrast with the attention paid to issues of soundness in the semantics of recursive VDM functions [20, 19], we are unaware of similar work in relation to recursive VDM operations. It is therefore unclear whether the semantics of VDM was developed with the recursive use of post-conditions in mind.

Imperative code is not permitted in implicit operation definitions, and so implicit operations may not be defined using iteration.

With regard to explicit operation definitions, imperative code can be used to define the body of an operation. Recursive operation calls are valid statements, and iteration is also supported.

In summary:

- recursive operations are:
  - not permitted in Z;
  - permitted in abstract-style VDM specifications (although some soundness issues may be unresolved);
  - permitted in imperative-style VDM specifications; and,
- iteration is:
  - not supported in Z;
  - not supported in abstract-style VDM specifications;
  - supported in imperative-style VDM specifications.

2.3   Motivating Example

The previous section showed that support for recursion and iteration is either poor or non-existent in both Z and abstract-style VDM. Some might defend this, arguing that while recursion and iteration are necessary in imperative programs, they are unnecessary in the abstract specifications for which Z and abstract-style VDM are intended.

This subsection challenges that argument. The following example demonstrates that where it is desired to partition the system state into modules (or objects), and where access to the substate is available only via a published interface, then recursion or iteration is actually necessary to express certain system transitions.

Example. Suppose *buf*1 and *buf*2 are two devices with the functionality of a buffer. A channel is required to transfer the contents of *buf*1 to *buf*2. Our task is to specify the channel. We are given the following Z specification of the buffers.

Buffers contain elements of the given type *Item*.

[*Item*]

The state of *buf*1 is captured by a sequence of items:

$$\begin{array}{|l}
\hline
State_{buf1} \\
\hline
stack_1 : \text{seq } Item \\
\hline
\end{array}$$

There are operations to decide whether the buffer is empty, to push an item onto the buffer, and to pop an item from the buffer.

$$\begin{array}{|l}
\hline
Empty_{buf1} \\
\hline
\Xi State_{buf1} \\
\hline
\#stack_1 = 0 \\
\hline
\end{array}$$

$$\begin{array}{|l}
\hline
Push_{buf1} \\
\hline
\Delta State_{buf1} \\
item? : Item \\
\hline
stack_1' = \langle item? \rangle \frown stack_1 \\
\hline
\end{array}
\qquad
\begin{array}{|l}
\hline
Pop_{buf1} \\
\hline
\Delta State_{buf1} \\
item! : Item \\
\hline
stack_1 = \langle item! \rangle \frown stack_1' \\
\hline
\end{array}$$

A specification of the second device $buf2$ is achieved by renaming the variable $stack_1$ to $stack_2$, and by renaming the schemas $State_{buf1}$, $Empty_{buf1}$, $Push_{buf1}$ and Pop_{buf1} to be $State_{buf2}$, $Empty_{buf2}$, $Push_{buf2}$ and Pop_{buf2} respectively.

Now suppose that the device $buf1$ is accessible only via the published operations $Empty_{buf1}$, $Push_{buf1}$ and Pop_{buf1}, although the state schema $State_{buf1}$ will be used to denote the buffer object itself. Similar provisions apply to $buf2$.

Our task then is to specify a channel that provides an operation $Transfer$ allowing the contents of $buf1$ to be emptied into $buf2$. The state of the channel may be specified as follows.

$$\begin{array}{|l}
\hline
Channel \\
\hline
State_{buf1} \\
State_{buf2} \\
\hline
\end{array}$$

One approach to specifying the operation $Transfer$ might be the following.

$$\begin{array}{|l}
\hline
Transfer_1 \\
\hline
\Delta State_{buf1} \\
\Delta State_{buf2} \\
\hline
stack_2' = reverse(stack_1) \frown stack_2 \\
stack_1' = \langle \rangle \\
\hline
\end{array}$$

The schema $Transfer_1$, although a succint way of expressing the requirements of the operation $Transfer$, is unsatisfactory for two reasons. Firstly, it fails to take advantage of the fact that the actions of removing and adding items from and to the buffers $buf1$ and $buf2$ have already been captured by the abstractions Pop_{buf1} and $Push_{buf2}$ respectively. Secondly, by referring directly to the states of $buf1$ and $buf2$ it violates an important principle of modular design — the principle of information hiding[1].

A better specification, it is submitted, would use the existing specifications for $buf1$ and $buf2$ and build a specification of the *Transfer* operation from the specifications of the published interface of the buffers.

For example, if it was permitted to specify operations recursively in Z (which it is not), then one way of specifying the *Transfer* operation might be as follows.

$$Transfer_2 \mathrel{\widehat{=}} (Empty_{buf1} \wedge [\Xi State_{buf2}])$$
$$\vee$$
$$((Pop_{buf1} \wedge Push_{buf2}[item!/item?]) \mathbin{\raise1pt{;}} Transfer_2)$$

Alternatively, if an iterative construct was available, *Transfer* could be defined using that construct.

The example shown is intended to illustrate the point that if the buffers are permitted to be arbitrarily large, and if the principle of access only via published interface is enforced, then it is impossible to find a finite specification of the *Transfer* operation without a recursive or iterative construct.

2.4   Object-oriented Languages

By the argument advanced in the previous section, object-oriented formal specification languages seeking to enforce information hiding should certainly support either iteration or recursion. Three of the best known[2] object-oriented formal specification languages are Z^{++} [16, 18, 15], VDM^{++} [14, 16, 22] and Object-Z. The semantics of specification languages such as VDM and Z are well-defined and readily accessible. There is a wealth of literature available. In contrast, the semantics of object-oriented specification languages is still a relatively new area. In fact, it is not clear that any of these languages can claim to have a complete, accessible, formal semantics. As such, it is difficult to say exactly what features these languages support. In the following survey, the most recent published work was used to determine support for recursive operations.

In Z^{++}, recursive operations are explicitly forbidden by the semantic description of the language contained in [16, p 339]. Fragments of imperative code appear to be allowed in method definitions [16, p 338], although it is not clear that the semantics [16, p 339] accommodates these code fragments. The extent of support for iteration is therefore unclear.

---

[1] In [12], Hayes suggests that the rule against direct reference to the abstract state of a module should be relaxed during the initial specification phase. In the method he suggests, direct references are replaced with calls to module operations during refinement. We support this philosophy, and in [10] we show how direct reference to the internal state of objects is supported in Object-Z. However, the point of the example in this section is that, were it desired to express the requirements of the *Transfer* operations solely in terms of calls to operations in the module interface, this would not be possible without some recursive or iterative construct.

[2] The decision as to which languages are "best known" is subjective. A survey of most object-oriented formal specification languages can be found in [17].

In VDM^{++}, explicit language restrictions in [16, p 374] preclude the definition of methods in a mutual or self-recursive manner. However, fragments of imperative code are included in the language. As with Z^{++}, the extent of semantic support for iteration is unclear.

Object-Z does not include imperative code, and so iteration is forbidden. Object-Z specifications that feature recursively-defined operations have appeared in the literature (e.g. [3]), but previous work on the formal semantics of Object-Z [2, 4, 9, 11, 24, 25] has not accommodated this feature.

The remainder of this paper presents a semantics for operations in Object-Z that accommodates recursion.

3 Object-Z Semantics — An Overview

This section provides a brief overview of the semantics of Object-Z. A more detailed account is available in [8].

3.1 Object Identity

Underlying the semantics of Object-Z (and likewise of most object-oriented programming languages [7, 21, 27]) is the idea that if a variable a is declared to be of type class A, then the value of a is an object identity [5, 11]. This in turn requires the existence of an implicit object map, which is a function mapping identities to the underlying values (i.e. the attributes and operations) of objects they identify. The values may change over time, but the identities persist.

3.2 The Object Map Model

Within the object map model, a system specified by an Object-Z specification consists of a collection of objects. If one were to take a snapshot of the system at some point in its evolution, the objects collectively denote some system value. The possible values that the system denoted by a specification can adopt are defined by the following schema:

$$\begin{array}{|l}
\hline
_Sys_val \underline{\hspace{8cm}} \\
sys_ob : \mathbb{O} \\
object_map : \mathbb{O} \nrightarrow Object_val \\
\hline
sys_ob \in \mathrm{dom}\, object_map \\
\hline
\end{array}$$

In the above definition $\mathbb{O}$ denotes the set of all object identities in a system. The understanding is that the value of a system at some point in time is given by an *object_map*. This is a function mapping object identities to object values, where the value of an object consists of the value of each attribute of the object, together with a list of operations enabled at that point. One of the objects in the *object_map* is identified as the system object. The schema variable *sys_ob* captures this idea. In the kind of systems that this model supports, all inter-actions of a system with its environment are via the interface provided by the system object. Moreover, although the constitution of the variable *object_map* may change with time, there is an implicit requirement that the system object should persist (i.e. the system object should always belong to the object map, although the internal value of the system object may change).

Within the object map model, the behaviour of the system is modelled by a set of labelled transitions from one system value to another. The transition is labelled with the operation that takes the system from one point to the next, and the parameters for that operation. Formally:

$$Sys\_trans == Sys\_val \times Sys\_label \times Sys\_val$$
$$Sys\_label == Valuation \times Name \times Valuation$$

In the above definition, the set *Valuation* defines a binding from variable names to values. Thus, the understanding is that *Sys_label* consists of a binding for input parameters, the name of the system operation that is called, and a binding for output parameters.

An Object-Z specification denotes: (i) the set of all possible system snapshots (elements of the set *Sys_val*); (ii) a subset of which represent possible initial configurations; and (iii) a set of labelled transitions that collectively model the behaviour of systems. Formally:

$$
\begin{array}{|l}
\hline
\;\underline{\;Spec\_meaning\;} \\
\quad values : \mathbb{P}\ Sys\_val \\
\quad initial\_values : \mathbb{P}\ Sys\_val \\
\quad behaviour : \mathbb{P}\ Sys\_trans \\
\hline
\quad initial\_values \subseteq values \\
\quad \forall\,(s_1, l, s_2) : behaviour \bullet \{s_1, s_2\} \subseteq values \\
\hline
\end{array}
$$

3.3   Strict Modularity

Strict modularity is the idea that a class should be given a meaning that is inde-pendent of the overall specification in which it is used. A goal of this philosophy is to develop a basis for modular reasoning, and ultimately, reuse of formally verified components.

Traditionally, the meaning of a class A could only be given a modular meaning if objects of class A did not contain other objects. If an object $a : A$ included an attribute $b : B$ (where B is a class and b is an object identity), then the meaning of A would include (or rely on) the meaning of B. This means that if one were to revise the specification by supplying a new class B with a different meaning, then one would need to reprove any theorems about class A, because the meaning of class A would also have changed.

Strict modularity allows us to distinguish between local properties of objects, that follow from the strictly modular meaning of a class, and system properties, that describe how the objects behave as a system. When a class is reused, then assuming the specification is well-formed, properties about constituent objects still hold. Only system properties need be reproved.

Object identity facilitates strict modularity, because object attributes can be represented as object references. Changes in the value of a referenced object are not a local concern.

The main challenge in finding a strictly modular meaning for classes lies in finding a denotation for operations. This is because the operations defined in a class may prescribe the invocation of an operation on one of the referenced objects. As such, the effect of such operations is not confined to the referencing object — it is not inherently modular.

3.4   Operations Have Two Parts

In order to give a strictly modular meaning to classes, a two-part notion of operations has been adopted. The first part is internal, and describes a relation on the state of the object. The second part is external and describes an interaction with the environment. Either part may be null, so that every operation can be considered uniformly.

The external interaction takes the form of messages sent to other objects. The variables that denote the identity of objects to which messages are sent are recorded, and so is the structure of the interaction. By structure, we mean two things. Firstly, there is the name of the operation invoked on an object. Secondly, messages may be sent concurrently, sequentially, or in a variety of other ways. For example, the simplest kind of message is a simple message such as $a.Op$. In this message, a is a variable whose value is an object identity, and Op is the name of an operation defined in the class to which a belongs. In the semantics, variables belong to the set *Word* and operation names belong to the set *Name*.

The formal definition of a message is:

$$
\begin{aligned}
Message \ ::= \ &skip \\
| \ &simple \langle\!\langle Word \times Name \rangle\!\rangle \\
| \ ¶llel \langle\!\langle Message \times Message \rangle\!\rangle \\
| \ &choice \langle\!\langle Message \times Message \rangle\!\rangle \\
| \ &sequence \langle\!\langle Message \times Message \rangle\!\rangle \\
| \ &hide \langle\!\langle Message \times (\mathbb{F}\ Word) \rangle\!\rangle \\
| \ &rename \langle\!\langle Message \times (Word \nrightarrow Word) \rangle\!\rangle
\end{aligned}
$$

The value of a message includes a binding that associates object identities with the variables used in the message. The value of a message is defined as follows:

$$
\begin{array}{|l}
\hline
\ Message\_val \\
\hline
\ msg : Message \\
\ var\_vals : Valuation \\
\hline
\ vars(msg) = \mathrm{dom}\ var\_vals \\
\hline
\end{array}
$$

In the above definition the function *vars* (defined at [8, p 27]) extracts the variables used in a message. For example, the variables in the message $a.Op_1 \; \mathring{,} \; b.Op_2$ are $\{a, b\}$.

3.5   The Meaning of a Class

A class defines the value and behaviour of constituent objects.

An object value is the underlying value of an object at some point in its evolution. This comprises the value of the object's state, together with a list of operations that are enabled for this value of the object's state.

$$
\begin{array}{|l|}
\hline
\_\,Object\_val \\
\hline
state : Valuation \\
ops : Name \twoheadrightarrow En\_op\_val \\
\hline
\forall\, eov : \mathrm{ran}\; ops \bullet eov.pre\_val = state \\
\hline
\end{array}
$$

The formal definition of *En_op_val* (enabled operation value) can be found at [8, p 28]. Intuitively, it defines possible internal transitions and the message associated with each internal transition.

Object behaviour is modelled as a set of transitions that denote the evolution of an object from one value to another. The transitions are labelled, and the label is a 4-tuple consisting of a binding for the input parameters, an operation name, the external interaction which is associated with that operation (which is a message value), and a binding for the output parameters. Formally:

$$Object\_trans == Object\_val \times Object\_label \times Object\_val$$
$$Object\_label == Valuation \times Name \times Message\_val \times Valuation$$

The meaning of a class is a set of possible object values, a distinguished subset of which comprise the objects satisfying the initialisation predicate, and a set of transitions. In short, a class describes the structure and behaviour of constituent objects. We have:

$$
\begin{array}{|l|}
\hline
\_\,Class\_meaning \\
\hline
object\_values : \mathbb{P}\; Object\_val \\
initial\_object\_values : \mathbb{P}\; Object\_val \\
behaviour : \mathbb{P}\; Object\_trans \\
\hline
initial\_object\_values \subseteq object\_values \\
\forall(o_1, l, o_2) : behaviour \bullet \{o_1, o_2\} \in object\_values \\
\hline
\end{array}
$$

Algebraically, the similarities between the meaning of a class and the meaning of a specification are evident. The main differences are that: (i) classes describe objects and specifications describe systems (sets of objects); and, (ii) the label on an object transition includes a message to model interaction with other objects, but a system transition interacts only with its environment and so the label on a system transition does not include a message.

3.6   A Compositional Semantics

The semantics of Object-Z is compositional. In the context of strict modularity, this means that there exists some function that takes the strictly modular meaning of classes (a set of bindings for *Class_meaning*), together with the knowledge that one of those classes is the class of the system object, and produces the meaning of a specification (a binding for *Spec_meaning*).

In practice, the composition occurs in two steps. Firstly, the set of possible system values is calculated. This set represents all snapshots of the system defined by the specification, at each point in its evolution. The function *sys_vals*, defined at [8, p 39], describes this set. The second step involves calculating the set of possible transitions. As one might expect, it is this calculation that defines the manner in which the external messages associated with individual object transitions are interpreted in the broader system context.

It is this second step of the composition that is the subject of the remainder of the paper.

4   Operation Composition — An Algorithmic Description

This section describes operation composition in Object-Z in a structured but informal fashion.

4.1   A Naive Composition Algorithm

We require a decision procedure that uses the modular meaning of classes to decide whether the transition $(OM, (in, n, out), OM')$ is a possible system transition. In the transition, OM and OM' represent the system object maps before and after the transition. The variable n is the name of the operation invoked, and the variables *in* and *out* represent the bindings for the input and output parameters. The handling of input and output will not be considered further in this paper (see [8, Sect. 8] for details). The decision process described by the naive algorithm is pictured in Figure 1.

Naive Algorithm. Given two object maps OM and OM', $(OM, (\_, n, \_), OM')$ is a system transition if:

1. the system object so_{id} of OM is the system object of OM';
2. the system object value so_{val} has an operation n in its public interface;
3. the operation n produces an internal transition i_trans_n that changes so_{val} to so'_{val};
4. associated with the internal transition i_trans_n is an external message msg_n;
5. the effect of invoking msg_n on the remaining object map REM_OM (this is OM with the pair (so_{id}, so_{val}) removed) is to transform REM_OM to REM_OM' (this transition is labelled $effect\_of(msg_n)$ in Figure 1):

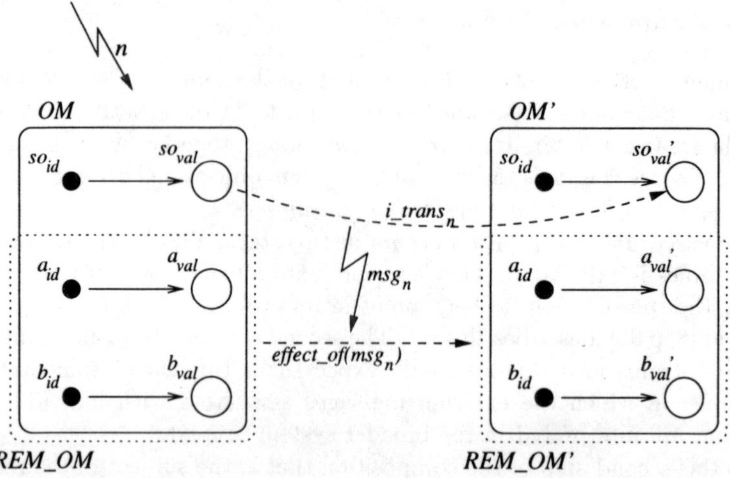

Fig. 1. A Naive Algorithm for Operation Composition.

(a) to see how $effect\_of(msg_n)$ can be calculated, consider the simple case where msg_n is a simple call of the form $object\_identity.operation\_name$. Substitute $object\_identity$ for so_{id}, $operation\_name$ for n, and REM_OM for OM, and apply the naive algorithm again.

6. OM' is equal to the union of REM_OM' and the set containing the pair (so_{id}, so'_{val}).

End.

The naive algorithm is described recursively (Step 5(a) involves a recursive call). It should be clear that if the naive algorithm is applied to finite object maps, then the recursive description is well-defined, because successive recursive calls are applied to arguments of smaller cardinality, and cardinality is a well-founded relation.

The restriction to finite object maps is not necessarily a welcome one. But, if cardinality is not a suitable measure, then either some other well-founded measure needs to be constructed, or the semantic function must be defined as the least fixed point of a functional corresponding to the above definition. In either case, removing the limitation of finiteness would require more work to ensure well-definedness.

Aside from the issue of finite object maps, the above definition has a more fundamental limitation. That limitation is that recursion is not permitted. To see this, note that on each application of the naive algorithm, the object undergoing the internal transition is removed from the object map before the effect of the message is considered. As such, a message whose "effect" includes a direct (i.e. self-recursive) call to the original object, or even an indirect (i.e. via a chain of intermediate methods) call, cannot be described by the above definition. Furthermore, an operation defined in terms of another operation on the same object is also forbidden. In this sense, the definition is clearly deficient.

4.2 Handling Recursion

To handle recursion, a simple generalisation of the above composition algorithm can be applied. The generalised composition algorithm is described in Figure 2.

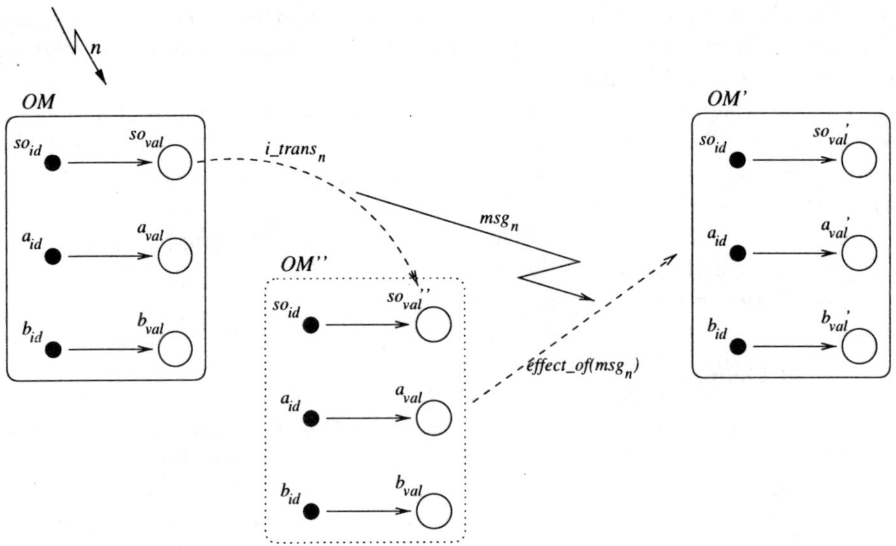

Fig. 2. A Generalised Algorithm for Operation Composition.

General Algorithm. Given two object maps OM and OM', $(OM, (\_, n, \_), OM')$ is a system transition if:

1. the system object so_{id} of OM is the system object of OM';
2. the system object value so_{val} has an operation n in its public interface;
3. the operation n produces an internal transition i_trans_n that changes so_{val} to so''_{val};
4. associated with the internal transition i_trans_n is an external message msg_n;
5. the effect of invoking msg_n on the object map OM'' (this is OM with the pair (so_{id}, so_{val}) replaced by (so_{id}, so''_{val})) is to transform OM'' to OM' (this transition is labelled $effect\_of(msg_n)$ in Figure 2):
   (a) to see how $effect\_of(msg_n)$ can be calculated, consider the simple case where msg_n is a simple call of the form *object_identity.operation_name*. Substitute *object_identity* for so_{id}, *operation_name* for n, and OM'' for OM, and apply the general algorithm again.

End.

The differences between the naive and general algorithm occur in Steps 3, 5 and 6 (the sixth step is removed in the general algorithm). The simple algorithm applied the internal transition to the system object value, and then removed the system object before considering the effect of the message. The general algorithm applies the internal transition to the system object value, but then considers the effect of the message on the entire, updated, object map.

In the general algorithm, meaning can be given to messages that directly or indirectly affect the object from which they originated, because the entire object map is available for consideration. As such, the generalised algorithm gives meaning to self or mutually recursive operations.

The naive algorithm had problems with definedness if infinite object maps were allowed. The generalised algorithm is potentially ill-defined even if the object maps are finite, as successive recursive calls are applied to arguments which are not necessarily "smaller". Intuitively though, it seems clear that if a system operation induces a graph of method calls, and the graph is finite, then the general algorithm should describe the effect of that operation. This suggests a representation of system operations as call graphs.

4.3   Call Graphs

A call graph is a pair (N, E), where N is the set of nodes and E is the set of edges. A node $n \in N$ is a pair (id, op) where id is an object identity and op is the name of an operation invoked on id. An edge $e \in E$ is an ordered, labelled triple (n_1, k, n_2) where n_1 and n_2 are nodes and $k \in \mathbb{N}$. Edges model operation calls. A graph includes an edge $((id_{n_1}, op_{n_1}), k, (id_{n_2}, op_{n_2}))$ if, as a result of operation op_{n_1} being invoked on object id_{n_1}, a message is sent to object id_{n_2} invoking operation op_{n_2}. The element k is a positive number recording the call level. For example, an edge corresponding to a call resulting directly from the system operation would have call level '1'. Calls resulting from that call would have call level '2'. By labelling edges in this manner, different call numbers are used to label edges corresponding to successive recursive calls. As such, even though a graph may contain cycles, it corresponds to an operation describable by the general algorithm if it is finite. The relationship between operations and call graphs is one-to-many. This allows for differing inputs and non-determinism.

Some examples of operation call graphs are given in Figure 3.

The scenarios that Graphs 1, 2 and 3 of Figure 3 model can be described as follows:

Graph 1. System operation n invoked on the system object so_{id} produces a message. The effect of the message is to invoke operation OpA on object a_{id}. The message associated with opA is the null message $skip$ (modelling no further interaction) and so (a_{id}, OpA) is a leaf node[3].

---

[3] We use the term "leaf" node to describe a node associated with a $skip$ message. Due to the possibility of recursion, this is slightly different to the definition of a leaf node used in traditional graph theory, although the intuition is similar.

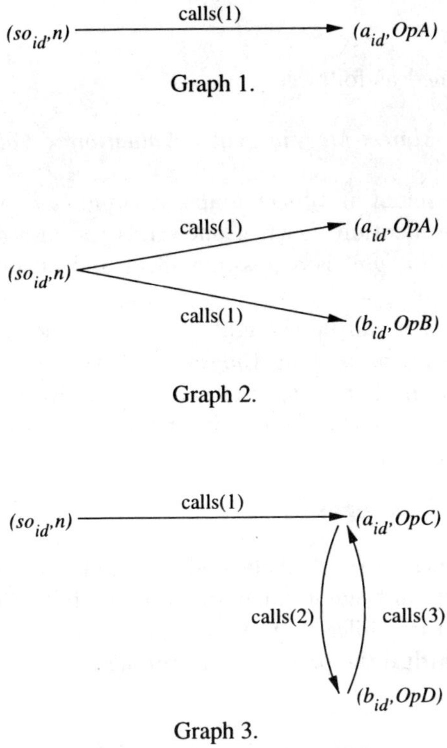

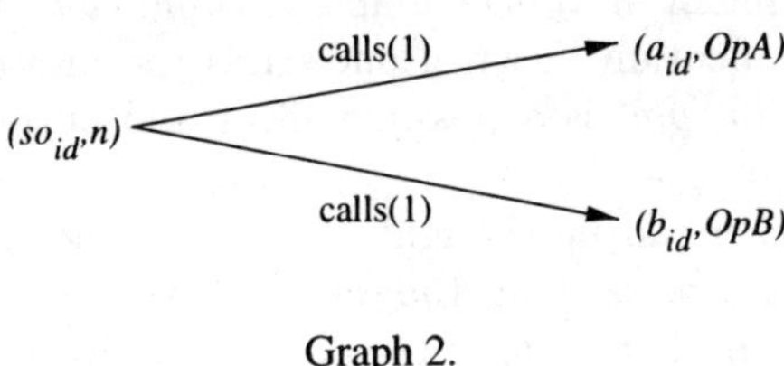

Graph 1.

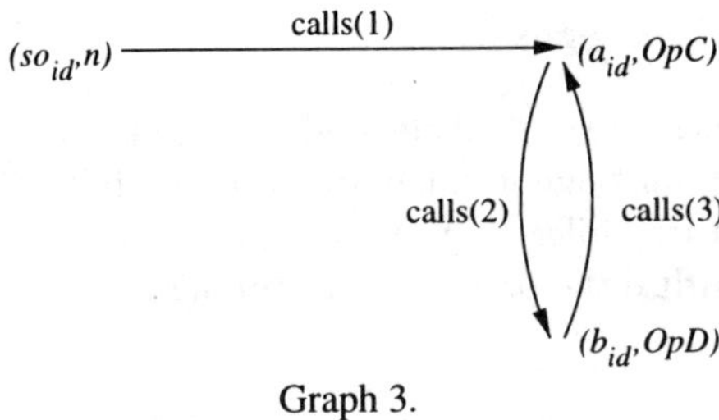

Graph 2.

Graph 3.

Fig. 3. Some Operation Call Graphs.

Graph 2. System operation n invoked on the system object so_{id} produces a message. The effect of the message is to send messages to a_{id} invoking OpA, and to b_{id} invoking OpB. Whether the messages are sent concurrently or sequentially is not recorded in the graph.

Graph 3. System operation n invoked on the system object so_{id} produces a message. The effect of the message is to invoke operation OpC on object a_{id}. The effect of OpC is to invoke OpD on b_{id}. This in turn produces a call invoking OpC on a_{id} (again). On this final call, however, a *skip* message is produced.

5   Calculating the Effect Set

In order to overcome the uncertainties of definedness inherent in the recursive definition of the previous section, this section develops an alternative model of operation composition as a set of *Effects*.

5.1   An Effect

The set *Effect* is defined as follows:

$$Effect == Map \times Map \times Message\_val \times Valuation \times Valuation$$

where *Map* is the set of all object maps. A tuple (om, om', m, in, out) is in the effect set of a specification if, when one sends the message m to the object map om, with inputs in, om' is a possible effect and outputs out are possible outputs.

A system operation n can be thought of as a message to the system object telling it to invoke operation n. Thus the set of all system transitions is given by those effects that are simple messages to the system object. As such, the effect set of a specification is sufficient to describe all system transitions.

5.2   The Effect of Messages

The calculation of a message's effect depends on the kind of message. For example, the effect of a *skip* message is calculated differently to the effect of a *simple* message, which is in turn different from the calculation for a *parallel* message, etc. In order to standardise the presentation, the signature for all effect functions will be:

$$X\_msg\_effect : Spec\_struct \rightarrow ((\mathbb{P}\ Effect) \nrightarrow (\mathbb{P}\ Effect))$$

In the above template, a *Spec_struct* is a semantic object that includes the modular meaning of classes. Further details about *Spec_struct* are irrelevant here (see [8, p 35]). For some specification ss, $X\_msg\_effect(ss)$ is a partial function that takes a set of known effects, and produces another set of effects. $X\_msg\_effect(ss)$ is a function because, except in the case of a *skip* message, the effect of all messages depends directly or indirectly on the effect of other messages.

The *skip* Message. The function *skip_msg_effect* interprets the effect of the *skip* message on an object map. The message *skip* has no effect on persistent objects in the object map. This may appear straightforward, but it is important to note that it semantically enforces the notion that the value of an object will not change unless an operation is expressly invoked on that object. We have:

$$skip\_msg\_effect : Spec\_struct \rightarrow ((\mathbb{P}\ Effect) \nrightarrow (\mathbb{P}\ Effect))$$

$$
\begin{aligned}
skip\_msg\_effect = {}&\lambda\, ss : Spec\_struct \bullet \lambda\, known : (\mathbb{P}\ Effect) \bullet \\
&\{(om, om', mv, in, out) : Effect \mid \\
&\quad mv.msg = skip \\
&\quad \forall\, oid : (\mathrm{dom}\ om) \cap (\mathrm{dom}\ om') \bullet \\
&\qquad om(oid) = om'(oid)\}
\end{aligned}
$$

This definition insists that objects whose identities appear in both object maps have equivalent values. The reason the object maps are not simply equated is that identities may have been passed in or out of the system in the internal state transition associated with the sending of this *skip* message. This would require an expansion or contraction of the object map, but must not change the value of objects that persist. For any specification, the requirement that the expansion or contraction happens according to specification is ensured by other parts of the semantics (see [8, p 44]). The matter is not pursued here.

The *simple* Message. A simple message identifies an object and invokes an operation on it. The function *simple_msg_effect* defines a set of *Effect* that captures the semantics of the operation call. This function formalises the intuition that the possible evolutions of a system are a composition of the individual behaviours of constituent objects, and the interactions between those objects.

The key idea is that the message identifies an object, and that the object must undergo a transition defined in the class to which the object belongs. This idea is captured by the predicate labelled (i) in the definition below. The transition will have an associated external message (denoted by m_l in the definition), which must in turn produce a known effect (captured by the predicate (ii)). The known effect involves the updated object map *new_om* and the final object map *om'*. Both of these object maps include the object *oid* that was the subject of the internal transition. As such, the known effect can potentially involve *oid*. In other words, recursion is permitted.

The definition is:

$$simple\_msg\_effect : Spec\_struct \to ((\mathbb{P}\, Effect) \nrightarrow (\mathbb{P}\, Effect))$$

$$
\begin{aligned}
&simple\_msg\_effect = \lambda\, ss : Spec\_struct \bullet \lambda\, known : (\mathbb{P}\, Effect) \bullet \\
&\quad \{(om, om', mv, in, out) : Effect \mid \\
&\qquad mv.msg \in \operatorname{ran} simple \\
&\qquad \exists\, w : Word;\ n_l : Name;\ oid : \mathbb{O} \bullet \\
&\qquad\quad mv.msg = simple(w, n_l) \\
&\qquad\quad mv.var\_vals(w) = oid \\
&\qquad\quad oid \in \operatorname{dom} om \wedge oid \in \operatorname{dom} om' \\
&\qquad\quad \exists\, new\_ob\_val : Object\_val; \\
&\qquad\qquad in_l, out_l : Valuation \mid (in \subseteq in_l \wedge out \subseteq out_l); \\
&\qquad\qquad m_l : Message\_val \bullet \\
&\qquad\qquad\quad (om(oid), (in_l, n_l, m_l, out_l), new\_ob\_val) \in \\
&\qquad\qquad\quad (class\_interpret(ss.cdict(name\_of\_class(oid)))).behaviour \\
&\hfill\text{(i)} \\
&\qquad\qquad \operatorname{\textbf{let}} new\_om == om \oplus \{oid \mapsto new\_ob\_val\} \bullet \\
&\qquad\qquad\quad (new\_om, om', m_l, in, out) \in known\} \\
&\hfill\text{(ii)}
\end{aligned}
$$

The *sequence* Message. The sequential composition operator involves an interesting idea. That is, at one level we view $Op_1 \, \mathbin{\raise0.3ex\hbox{$\scriptstyle\circ$}} \, Op_2$ as an atomic transition.

To decide that this transition is valid however, it is necessary to show the existence of an intermediate system value om'' that results from the application of Op_1, and in which Op_2 can be applied. Each of these transitions must be known effects.

In Object-Z, complementary variables in the output of Op_1 and input of Op_2 are equated. We have:

$$
\begin{array}{l}
sequence\_msg\_effect : Spec\_struct \to ((\mathbb{P}\,Effect) \nrightarrow (\mathbb{P}\,Effect)) \\
\hline
sequence\_msg\_effect = \lambda\,ss : Spec\_struct \bullet \lambda\,known : (\mathbb{P}\,Effect) \bullet \\
\quad \{(om, om', mv, in, out) : Effect \mid \\
\qquad mv.msg \in \operatorname{ran} sequence \\
\qquad \mathbf{let}\ m_1 == (\mu\,Message\_val \mid msg = \pi_1(sequence^\sim(mv.msg)) \\
\qquad\qquad\qquad\qquad\qquad var\_vals = vars(msg) \lhd mv.var\_vals); \\
\qquad\qquad m_2 == (\mu\,Message\_val \mid msg = \pi_2(sequence^\sim(mv.msg)) \\
\qquad\qquad\qquad\qquad\qquad var\_vals = vars(msg) \lhd mv.var\_vals) \bullet \\
\qquad \exists\,om'' : Map; \\
\qquad\quad in_1, out_1, in_2, out_2 : Valuation \mid in_2\ \underline{compatible\_with}\ out_1 \bullet \\
\qquad\quad (om, om'', m_1, in, out_1) \in known \\
\qquad\quad (om'', om', m_2, in_2, out) \in known\}
\end{array}
$$

Other Messages The semantics of Object-Z, as articulated in [8], supports four other message primitives: the *parallel*, *choice*, *hiding* and *rename* message. The last two constructs tailor the interface of a message. The messages described so far in this paper should provide the reader with a feel for the treatment of operation composition. The semantics of the remaining operators are not presented in this paper, but full definitions are available in [8, pp 45–48].

5.3   Building the Effect Set

We are now in a position to define the effect set for a given specification. The definition is structured as follows:

- First, the function $\mathcal{E}_0 \in Spec\_struct \to (\mathbb{P}\,Effect)$ is defined. This is the set of effects resulting from *skip* messages, where the set of known effects is $\varnothing$;
- Then, define the function $\mathcal{E}_{i+1}$. This is the set of effects already known, together with effects resulting from the composition of known effects.
- Finally, the union of all functions $\mathcal{E}_i$ gives the function $\mathcal{E}$, as required.

Beginning with the first stage of the definition, we have:

$$
\begin{array}{l}
\mathcal{E}_0 : Spec\_struct \to (\mathbb{P}\,Effect) \\
\hline
\mathcal{E}_0 = \lambda\,ss : Spec\_struct \bullet \\
\quad skip\_msg\_effect(ss)(\varnothing)
\end{array}
$$

The next stage of the definition is:

$$\mathcal{E}_{i+1} : Spec\_struct \to (\mathbb{P}\, Effect)$$

$$
\begin{aligned}
\mathcal{E}_{i+1} = \lambda\, ss : Spec\_struct \bullet \\
\mathcal{E}_i(ss) \cup \\
simple\_msg\_effect(ss)(\mathcal{E}_i(ss)) \cup \\
parallel\_msg\_effect(ss)(\mathcal{E}_i(ss)) \cup \\
choice\_msg\_effect(ss)(\mathcal{E}_i(ss)) \cup \\
sequence\_msg\_effect(ss)(\mathcal{E}_i(ss)) \cup \\
hide\_msg\_effect(ss)(\mathcal{E}_i(ss)) \cup \\
rename\_msg\_effect(ss)(\mathcal{E}_i(ss))
\end{aligned}
$$

Finally, the overall effect set of a specification ss is defined as:

$$\mathcal{E}(ss) = \bigcup\{i : \mathbb{N} \bullet \mathcal{E}_i(ss)\}$$

Allowable system transitions are defined by those effects that are simple messages to the system object. As such, the effect set describes operation composition.

5.4 Discussion

The definitions presented in this section describe the composition of overall system transitions from individual object transitions and associated messages. Co-operation amongst objects is ensured by only incorporating the effect of an internal transition when the environment is receptive to the external message, and by insisting upon agreement between input/output parameters from different object transitions. Although we have not dwelt upon the issue in this paper, a few remarks about the expressive power of this semantics are in order.

The model of an operation used in this paper is quite simple and does not, on the face of it, accommodate distributed operators, more complex transitions, or direct reference to the internal state of other objects. With a few extra definitions, these phenomena can all be supported by the two-part model of operations. A more detailed discussion of this issue can be found in [10].

A fundamental limitation, however, is that the characterisation of object behaviour as a set of transitions implies that the level of granularity of a single process is the object level. In short, it is not possible to have two active processes within the same object. With regard to recursion, this means that it must be possible to "unfold" the recursive operation to reveal a sequence of atomic state transitions — it must be possible to draw a call-graph. An operation on object a in which the input parameters are derived via an indirect call to an attribute of a, and where the internal transition changes the value of that attribute of a, are beyond the scope of this model.

6 Relating Effects and Call-graphs

Section 4 presented an algorithmic, informal view of operation composition, which led to a representation of operations as the set of call graphs they induce.

Section 5 presented a mathematical characterisation of operation composition, where operations were represented by their effect on the system object map. This section explores the relationship between the two models.

The set of effects for a given specification ss is the set $\mathcal{E}(ss)$. This set is the limit of a chain of sets $\mathcal{E}_i(ss)$. It is conjectured that this limit is the least fixed point of a functional corresponding to the informal algorithmic description of the composition presented in Section 4. In other words, we conjecture that a system transition is described by an effect in $\mathcal{E}(ss)$ exactly when the system transition has a representation as a finite call graph.

The following proof sketch outlines the reasoning behind this conjecture.

Proof Sketch. Suppose a system transition st is described by an effect e in $\mathcal{E}(ss)$. Then there exists some $i \in \mathbb{N}$ such that $e \in \mathcal{E}_i(ss)$. Then any subsidiary operation calls that st induces must, according to the X_msg_effect functions, produce an effect that belongs to some $\mathcal{E}_j(ss)$ where $j < i$. Similarly, any subsidiary calls induced by those effects must produce an effect in some $\mathcal{E}_k(ss)$, where $k < j < i$, and so on. As i is finite, there is a finite call graph associated with st.

Suppose now that there is a system transition st that produces the call graph (N, E). Assign each of the nodes a depth index $d \in \mathbb{N}$. For example, leaf nodes have a depth of 0. Nodes that have a single path to a leaf node have a depth equal to the length of the path. Nodes having multiple paths to leaf nodes have a depth that is the maximum of the lengths of each path. Because the graph is finite, there will be a node with a maximum depth n (and it will be the node corresponding to the original system operation call). Consider all nodes with depth 0. These nodes correspond to operations that produce *skip* messages, the effect of which is described by an element of the set $\mathcal{E}_0(ss)$. Nodes at depth 1 correspond to operations that produce a message that results in another operation that produces a *skip* message. In other words, the effect associated with nodes at depth 1 is an element of the set $\mathcal{E}_1(ss)$. We expect to able to show by induction that a node with depth i corresponds to an effect in $\mathcal{E}_i(ss)$, and so the original system operation call corresponds to an effect in $\mathcal{E}_n(ss)$, which means it is an effect in $\mathcal{E}(ss)$. $\square$

A task for future work is to formalise the algorithmic description, so that the inductive lemma, and hence the conjecture, can be proved more formally. The utility of this equivalence is that we expect call graphs to provide an intuitive basis against which the soundness of inference rules about object interaction could be demonstrated. For example, the fact that an operation Op called in parallel with a *skip* message reduces to the operation Op is intuitively clear from a call-graph representation of operations, since the call graphs are the same.

7   Conclusion

This paper has considered the issue of recursively defined operations in Object-Z. First, the use of recursion in Z and VDM was considered. There appears to be little support for recursively defined operations both in Z, and in the abstract part of the VDM language spectrum. It could possibly be argued that recursion is too operational a concept to be used in abstract model-oriented specifications, but an example was given to motivate the need for recursion in specifications where it is sought to partition the state into modules or objects. For these reasons, it was suggested that object-oriented formal specification languages should certainly support recursion. Surprisingly, support there also appears to be weak.

To address this problem, a semantics for operation composition in Object-Z was presented. Composition was first described in an informal, algorithmic fashion. This description suggested a representation of system operations as the call graphs that the system call induces. The description, although reasonably intuitive, is mathematically unappealing because it is not clear that the recursion in the semantic meta-language is well-founded.

An alternative representation of operation composition as a set of effects was also presented. This definition does not suffer from the mathematical uncertainties of the algorithmic definition, but is slightly less intuitive.

Finally, it was conjectured that the two representations are equivalent. That is, an operation call has a representation as a finite call graph according to the algorithmic description exactly when it has a representation as an element of the effect set. We expect this result to be useful in allowing us to use the more intuitive framework of call graphs as a basis for demonstrating the soundness of behavioural inference rules.

8 Acknowledgements

This work was conducted under the supervision of Roger Duke. Discussions with Peter Lindsay, Peter Kearney and Paul Strooper were also helpful. Thanks to the anonymous referees whose comments improved the quality of the final version.

References

1. S. Brien, T. King, J. Nicholls, J. Woodcock, and J. Wordsworth. Z Base Standard — Version 1.0. Technical report, Oxford University Computing Laboratory - Programming Research Group, November 1992.
2. S. Butler and R. Duke. Defining Composition Operators for Object Interaction. *Object Oriented Systems*, 1996. Accepted for publication, to appear.
3. J. S. Dong, R. Duke, and G. Rose. An Object-Oriented Approach to the Semantics of Programming Languges. In G Gupta, editor, *Proceedings of the 17th Australasian Computer Science Conference (ACSC'17)*, pages 767–775, January 1994.
4. D. Duke. *Object-Oriented Formal Specification*. PhD thesis, University of Queensland, 1991.
5. R. Duke and G. Rose. Modelling Object Identity. In *Proceedings 16th Australian Computer Science Conference (ACSC-16)*, pages 93–100, February 1993.
6. R. Duke, G. Rose, and G. Smith. Object-Z: a Specification Language Advocated for the Description of Standards. *Computer Standards and Interfaces*, 17(5):511–533, 1995.
7. A. Goldberg. *Smalltalk-80 : the language*. Addison-Wesley, 1989.

8. A. Griffiths. A Semantics for a Simple Sub-language of Object-Z. Technical Report 96-33, Software Verification Research Centre, The University of Queensland. Australia. 4072, December 1996.

9. A. Griffiths. An Extended Semantic Foundation for Object-Z. In *1996 Asia-Pacific Software Engineering Conference*, pages 194–205. IEEE Computer Society Press, 1996. The proceedings for the conference are being reprinted. This may result in a page number change for this citation.

10. A. Griffiths. Object-oriented Operations Have Two Parts. Technical Report 97-20, Software Verification Research Centre, School of Information Technology, The University of Queensland. Australia. 4072, March 1997.

11. A. Griffiths and G. Rose. A Semantic Foundation for Object Identity in Formal Specification. *Object Oriented Systems*, 2:195–215, December 1995.

12. I. Hayes. Supporting Module Reuse in Refinement. *Science of Computer Programming*, 27(2):175–184, September 1996.

13. J. Dawes. *The VDM-SL Reference Guide*. Pitman Publishing, 1991.

14. S. Kent and R. Moore. An Axiomatic Semantics for VDM^{++}: OO Aspects. This is a technical report of the ESPRIT Project 6500 Afrodite., July 1993.

15. K. Lano. Z^{++}, an object-oriented extension to Z. In J Nicholls, editor, *Z User Meeting, Oxford, UK*, Workshops in Computing. Springer-Verlag, 1991.

16. K. Lano. *Formal Object-Oriented Development*. Formal Approaches to Computing and Information Technology. Springer-Verlag, 1995.

17. K. Lano and H. Haughton, editors. *Object-Oriented Specification Case Studies*. Prentice-Hall, 1994.

18. K. Lano and H. Haughton. The Z^{++} Manual. Technical report, Imperial College, 1994.

19. P. G. Larsen and B. S. Hansen. Semantics of Under-determined Expressions. *Formal Aspects of Computing*, 8(1):47–66, 1996.

20. P. G. Larsen and W. Pawlowski. The Formal Semantics of ISO VDM-SL. *Computer Standards and Interfaces*, 17(5):585–602, September 1995.

21. B. Meyer. *Eiffel : the language*. Prentice Hall Object-Oriented Series. Prentice-Hall, 1992.

22. S. Mitra. Object-oriented Specification in VDM^{++}. In Kevin Lano and Howard Haughton, editors, *Object-Oriented Specification Case Studies*, The Object-Oriented Series, chapter 6, pages 130 – 136. Prentice Hall, 1993.

23. G. Rose. Object-Z. In R Barden S Stepney and D Cooper, editors, *Object Orientation in Z*, Workshops in Computing, pages 59–77. Springer-Verlag, 1992.

24. G. Smith. *An Object-Oriented Approach to Formal Specification*. PhD thesis, Department of Computer Science, University of Queensland, 1993.

25. G. Smith. Extending $\mathcal{W}$ for Object-Z. In J P Bowen and M G Hinchey, editors, *Ninth International Conference of Z Users*, volume 967 of *Lecture Notes in Computer Science*, pages 276–295. Springer-Verlag, 1995.

26. J. M. Spivey. *Understanding Z: A specification language and its formal semantics*, volume 3 of *Cambridge Tracts in Theoretical Comput. Sci.* Cambridge University Press, UK, 1988.

27. B. Stroustrup. *The C++ Programming Language*. Addison-Wesley, 1986.

An Algorithm for Pattern-Matching Mathematical Expressions

David Hemer

Software Verification Research Centre
School of Information Technology,
The University of Queensland, Brisbane 4072, Australia

Abstract. Development of formally verified software can be made more viable by making use of reusable components, which include not only code fragments but also lemmas, proofs and proof tactics. By making use of formal parameters in the specification language, components can be generalised, and therefore modifiable to meet the software developers needs. This also allows us to build a pattern-matching based search mechanism, for finding suitable components, which doesn't require any formal reasoning support.

1 Introduction

1.1 Motivation

The idea of using reusable components as building blocks in software development was introduced by McIlroy [7]. The idea is analogous to the development of complex electronic systems where the electrical engineer selects suitable components from a catalog of standard components, and combining them to build the complete system. When a component meeting the engineer's need cannot be found, the engineer might typically modify one that closely models their requirements.

When applied in a traditional software development process, the reusable components are typically code fragments, but may include other kinds of software such as designs. These components normally have an interface (or specification) which presents an abstract view of the component, hiding unnecessary implementation details. The specification of the components can either be formal or informal.

The benefits of using reusable components of this kind are obvious. For example rather than having to reinvent commonly used algorithms (such as sorting and searching algorithms) and data structures, the software engineer can simply select them from a library of components. However the use of such components is only practical if the amount of effort in using a component (or components) to develop a program (or part of a program) is less than the effort to develop the program from scratch. Two important issues that effect whether or not the above condition is true are the ability to: modify components to meet the user's needs and find suitable components within a large collection.

Within the domain of formally verified software, the scope of reusable components can be extended to include mathematical lemmas, proofs and proof tactics. The interface for reusable code fragments is a formal specification which uses the same mathematical language as the other components listed above. Within this kind of framework we can not only reuse pieces of code, but we can reuse formal proofs, which typically are very time consuming, tedious and often require advanced theorem proving skills from the software engineer (indeed formal proof seems to be the major hurdle in development of formally verified software). For these reasons, the benefits of reusing software components in formal software development are potentially far greater than within a more traditional development process.

By including formal parameters in the mathematical language that is used to specify component interfaces, we can generalise components. This allows the software engineer to modify components by instantiating the parameters with suitable values. Also by using this approach, we are able to implement an effective searching mechanism which uses pattern-matching to find suitable components. Unlike other approaches which involve matching of formally specified software components [9, 1], our approach doesn't require formal reasoning support.

1.2   This paper

In the following paper we look at an algorithm for pattern-matching parameterised Z-like expressions. The syntax for these expressions is introduced in Section 2. In Section 3 we describe how parameterised expressions are "instantiated": that is where the formal parameters of an expression are "filled in" with actual values. In Section 4 a top-level specification for matching expressions is given in terms of the instantiate function. In Section 5 this top-level specification is refined. Finally in Section 6 we briefly discuss how this approach can be extended to component-wise matching and module level matching. Appendix A contains a justification that the algorithm given in Section 5 is indeed a correct refinement of the top-level specification.

2   Modelling Mathematical Constructs

In the following section the Z language [8] is used to model a mathematical language that includes sets, terms and formulae. Each of these three groups can contain parameters, which can be instantiated to certain values. The model used here is based on the mathematical language used in the CARE language[2], but it can also be seen to model (a subset of) the Z language - in particular the non-schema parts of the language.

Fig. 1 shows a number of examples of mathematical expressions represented in a tree notation based on the abstract syntax given below. These examples are described as we proceed through the following section.

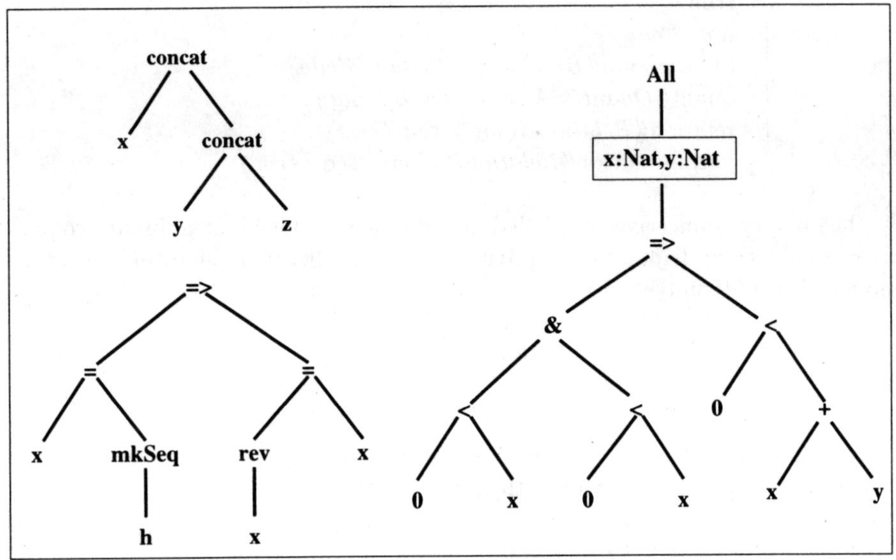

Fig. 1. Examples of mathematical expressions

2.1   Terms

A term can be a placeholder, a variable, a non-parametric function of a parametric function.

$$[\mathit{Var}, \mathit{FunctionName}, \mathit{FunctionParam}]$$

$$
\begin{aligned}
\mathit{Term} \;::=\; &\mathrm{ph}\langle\!\langle \mathbb{N}_1 \rangle\!\rangle \\
\mid\; &\mathrm{var}\langle\!\langle \mathit{Var} \rangle\!\rangle \\
\mid\; &\mathrm{fnapplic}\langle\!\langle \mathit{FunctionName} \times \mathrm{seq}\; \mathit{Term} \rangle\!\rangle \\
\mid\; &\mathrm{termparam}\langle\!\langle \mathit{FunctionParam} \times \mathrm{seq}\; \mathit{Term} \rangle\!\rangle
\end{aligned}
$$

Note that symbolic placeholders (e.g. **x**, **y**) are used in examples, but for convenience of specification they are modelled here as non-zero natural numbers.

Example: Fig. 1 shows the term $concat(x, concat(y, z))$ represented in tree notation.

2.2   Formulae

The formulae modelled here are binary connectives, 'true', negation, quantified formulae, non-parametric relations and parametric relations.

$$[\mathit{RelationName}, \mathit{RelationParam}]$$

$$Fmla ::= \text{true}$$
$$| \quad \text{not}\langle\!\langle Fmla \rangle\!\rangle$$
$$| \quad \text{binaryConn}\langle\!\langle BinConn \times Fmla \times Fmla \rangle\!\rangle$$
$$| \quad \text{quant}\langle\!\langle Quant \times VarDeclars \times Fmla \rangle\!\rangle$$
$$| \quad \text{relation}\langle\!\langle RelationName \times \text{seq } Term \rangle\!\rangle$$
$$| \quad \text{paramrelation}\langle\!\langle RelationParam \times \text{seq } Term \rangle\!\rangle$$

The binary connectives modelled include the standard logical binary connectives conjunction, disjunction, equivalence and implication. Quantifiers that are modelled include universal and existential quantifiers.

$$BinConn == \{`\wedge', `\vee', `\Leftrightarrow', `\Rightarrow', \ldots\}$$
$$Quant == \{`\forall', `\exists', \ldots\}$$

Variables are declared within a *variable declaration list*, where each variable is associated with a set of values that it may take.

$$VarDeclars == \text{seq}_1(Var \times Set)$$

The functions *sig* and *varSet* which return the signature and set of variables respectively of a variable declaration list will be used later.

$$sig : VarDeclars \to \text{seq}_1 \, Set$$
$$varSet : VarDeclars \to \mathbb{F} \, Vars$$

Example: Fig. 1 shows the formulae $x = mkSeq(h) \Rightarrow rev(x) = x$ and $\forall x : Nat, y : Nat \bullet 0 < x \wedge 0 < y \Rightarrow 0 < x + y$ represented in tree notation.

2.3   Sets

Sets that are modelled are given sets (such as the set of natural numbers or the set of characters etc), constructed sets and parametric sets.

$$[MathTypeName, TypeParam]$$

$$Set ::= \text{given}\langle\!\langle MathTypeName \rangle\!\rangle$$
$$| \quad \text{constructor}\langle\!\langle TypeConstructor \times \text{seq}_1 \, Set \rangle\!\rangle$$
$$| \quad \text{paramset}\langle\!\langle TypeParam \rangle\!\rangle$$

Constructed sets are built from one or more other sets using some type constructor; examples of type constructors include sequences, finite power sets and power sets.

$$TypeConstructor == \{`\text{seq}', `\mathbb{P}', `\mathbb{F}', \ldots\}$$

2.4 Auxiliary functions

The functions *freeVars*, *renameVars* and *replacePlaceHolders* will be overloaded for the different kinds of mathematical expressions.

freeVars returns the set of free variables in a mathematical expression, e.g. for terms:

$$\mid \ \mathit{freeVars} : \mathit{Term} \to \mathbb{F}\ \mathit{Var}$$

renameVars renames the free variables in a given expression in accordance with a given mapping, e.g. for terms:

$$\mid \ \mathit{renameVars} : \mathit{Term} \times (\mathit{Var} \nrightarrow \mathit{Var}) \to \mathit{Term}$$

substPlaceHolders replaces placeholders by the corresponding expression from a list, e.g. for terms:

$$\mid \ \mathit{substPlaceHolders} : \mathit{Term} \times (\mathrm{seq}\ \mathit{Term}) \nrightarrow \mathit{Term}$$

3   Instantiation of Expressions

A *formal parameter instantiation* is essentially a mapping from formal parameters to mathematical expressions. An instantiation of a parameter p to an expression e is represented by $p \rightsquigarrow e$. Where the parameter is a relation or function, the placeholders are also given. For example the instantiation $f(a, b) \rightsquigarrow a + b$ maps the binary function parameter f with argument placeholders a and b to the expression $a + b$. An instantiation is defined here as three separate mappings, one each for the functional parameters, relational parameters and set parameters.

```
┌─ Inst ──────────────────────────────────────────────
│ finsts : FunctionParam ⇸ Term
│ rinsts : RelationParam ⇸ Fmla
│ sinsts : TypeParam ⇸ Set
└─────────────────────────────────────────────────────
```

Instantiating an expression with the *trivial instantiation* defined below leaves the expression unchanged.

```
┌─ TrivInst : Inst ──────────────────────────────────
├─────────────────────────────────────────────────────
│ TrivInst.finsts = ∅ ∧ TrivInst.rinsts = ∅ ∧ TrivInsts.sinsts = ∅
```

The function *freeVars* can be defined for an instantiation. It returns the union of the free variables in the expressions appearing in the range of the three instantiation maps.

$$\mid \ \mathit{freeVars} : \mathit{Inst} \to \mathbb{F}\ \mathit{Var}$$

Two instantiations are mergeable if they agree for common parameters.

$$areMergeableInsts : \mathbb{P}(Inst \times Inst \times Inst)$$

$$\forall\, u \in \mathrm{dom}\; i_1.finsts \cap \mathrm{dom}\; i_2.finsts \bullet i_1.finsts(u) =_\alpha i_2.finsts(u) \wedge$$
$$\forall\, v \in \mathrm{dom}\; i_1.rinsts \cap \mathrm{dom}\; i_2.rinsts \bullet i_1.rinsts(v) =_\alpha i_2.rinsts(v) \wedge$$
$$\forall\, w \in \mathrm{dom}\; i_1.sinsts \cap \mathrm{dom}\; i_2.sinsts \bullet i_1.sinsts(w) =_\alpha i_2.sinsts(s)$$
$$\Rightarrow (i_1, i_2, i) \in areMergeableInst$$
$$\mathbf{where}\; i.finsts = i_1.finsts \oplus i_2.finsts \wedge i.rinsts = i_1.rinsts \oplus i_2.rinsts$$
$$i.sinsts = i_1.sinsts \oplus i_2.sinsts$$

Here we use the relation $=_\alpha$ which stands for α-equivalence. In general two expressions are α-equivalent if they are equivalent up to renaming of bound variables.

When applying instantiations to expressions we shall only be interested in those which instantiate all of the parameters within the given expression (that is we are not interested in so-called *partial* instantiations). The relation *CompleteInst* shall be used to define the domain of the instantiate functions; it is overloaded for the various kinds of expressions. For example, for terms it is defined as:

$$\_CompleteInst\_ : Inst \leftrightarrow Term$$

$$i\; CompleteInst\; t \Leftrightarrow fparams(t) \subseteq \mathrm{dom}\; i.finsts$$
$$\wedge\; rparams(t) \subseteq \mathrm{dom}\; i.rinsts \wedge tparams(t) \subseteq \mathrm{dom}\; i.sinsts$$

The functions *fparams*, *rparams* and *tparams* return the set of functional, relational and type parameters respectively of an expression.

3.1 Instantiating terms

Terms are instantiated by considering the different cases as follows:

- placeholders[1] and unbound variables are left unchanged
- non-parametric functions are instantiated by instantiating the argument list
- parametric functions of the form $f(a_1, .., a_m)$ are instantiated by instantiating f, and then replacing any placeholders j, where $j \in 1..m$ in the resulting term by the result of instantiating a_j.

$$instantiate : Term \times Inst \nrightarrow Term$$

$$\mathrm{dom}\; instantiate = \{t : Term;\; i : Inst \mid i\; CompleteInst\; t\}$$
$$instantiate(\mathrm{ph}\; n, i) = \mathrm{ph}\; n$$
$$instantiate(\mathrm{var}\; v, i) = \mathrm{var}\; v$$
$$instantiate(\mathrm{fnapplic}(f, as), i) = \mathrm{fnapplic}(f, instantiate(as, i))$$
$$instantiate(\mathrm{termparam}(f, as), i) = substPlaceHolders(e, m)$$
$$\mathbf{where}$$
$$i.finsts(f) = e$$
$$\wedge\; m = \{j : 1..\#as \bullet j \mapsto instantiate(as(j), i)\}$$

---

[1] Normally terms which are to be instantiated would not contain placeholders; however this case is included for completeness

Example: Fig. 2 gives an example of how the term $f(g(1), 0)$ is mapped to $(1 + 1) + 0$ under the instantiation map $f(x, y) \rightsquigarrow x + y$, $g(z) \rightsquigarrow z + 1$. In the first step, $f(g(1), 0)$ is instantiated to $x + y$ with the placeholders x and y being replaced by $g(1)$ and 0 respectively. In the second step $g(1)$ is instantiated to $z + 1$ and the placeholder z is replaced by 1.

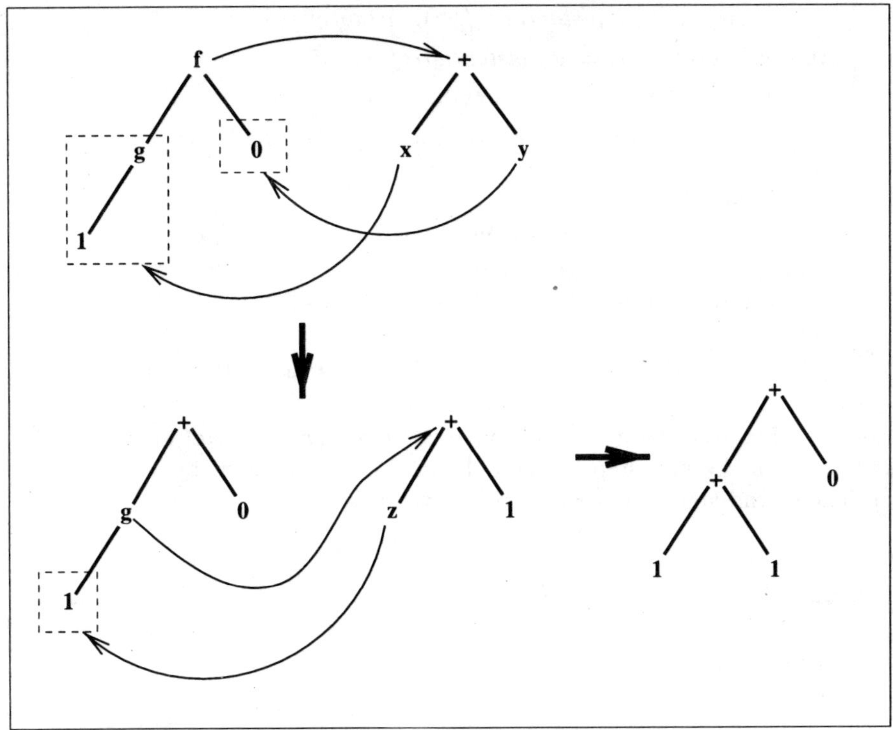

Fig. 2. Instantiating the term $f(g(1), 0)$ with $f(x, y) \rightsquigarrow x + y$, $g(z) \rightsquigarrow z + 1$

A sequence of terms can be instantiated by instantiating the individual members of the sequence.

$$instantiate : \text{seq } Term \times Inst \nrightarrow \text{seq } Term$$

$$\text{dom } instantiate = \{es : \text{seq } Term; \; i : Inst \mid i \; CompleteInst \; es\}$$

$$instantiate(es, i) = \{j : 1 \mathrel{..} \#es \bullet j \mapsto instantiate(es(j), i)\}$$

3.2 Instantiating formulae

The instantiation process for binary connectives, negations and non-parametric relations is straightforward. The process for instantiation of parametric relations is similar to process for parametric functions. To instantiate quantified formulae, the bound variables must first be renamed to avoid any potential name clashes with free variables occuring in the instantiation. The body is then instantiated.

$$instantiate : Fmla \times Inst \rightarrow Fmla$$

$$\mathrm{dom}\, instantiate = \{f : Fmla;\ i : Inst \mid i\ CompleteInst\ f\}$$

$$instantiate(\mathrm{binaryConn}(b, f_1, f_2), i) =$$
$$\qquad \mathrm{binaryConn}(b,\, instantiate(f_1, i),\, instantiate(f_2, i))$$

$$instantiate(\mathrm{not}\, f, i) = \mathrm{not}\, instantiate(f, i)$$

$$instantiate(\mathrm{relation}(r, es), i) = \mathrm{relation}(r,\, instantiate(es, i))$$

$$instantiate(\mathrm{paramrelation}(r, es), i) = substPlaceHolders(f, m)$$
where
$$\qquad i.rinsts(r) = f\ \wedge$$
$$\qquad m = \{j : 1\,..\,\#es \bullet j \mapsto instantiate(as(j), i)\}$$

$$(varSet(vs_1) \cap free\,Vars(i)) = \varnothing \Rightarrow$$
$$\qquad instantiate(\mathrm{quant}(q, vs, f), i) = \mathrm{quant}(q, vs_1, f_1)$$
where $vs_1 =_\alpha instantiate(vs, i)\ \wedge$
$$f_1 = instantiate(rename\,Vars(f, \{j : 1\,..\,\#vs \bullet \mathrm{first}\,vs(j) \mapsto \mathrm{first}\,vs_1(j)\}), i)$$

Example: Fig. 3 shows how the formula $\forall x : E \bullet P(x)$ is instantiated by $P(a) \rightsquigarrow a < x * x$. In the first step the bound variable x is replaced by a new variable $x1$. The second step shows $P(x1)$ being instantiated to $x1 < x * x$.

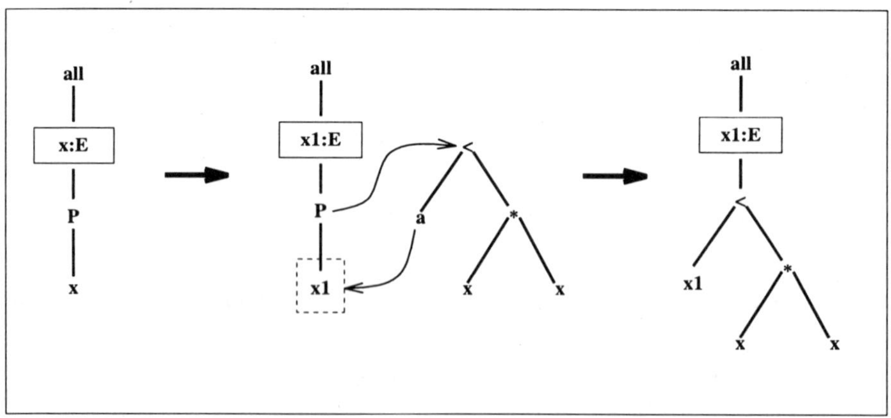

Fig. 3. Instantiating a formula

A variable declaration list is instantiated by instantiating each of the sets in the list.

$$| \quad instantiate : VarDeclars \times Inst \rightarrow VarDeclars$$

3.3   Instantiating sets

Instantiation of sets is straightforward.

$$| \quad instantiate : Set \times Inst \rightarrow Set$$

4   Top-level Specification of Matching

Two mathematical expressions p and t are said to match if there is an instantiation i such that instantiating p with i yields t. We refer to the expression p which possibly contains parameters as the *pattern*, while t is referred to as the *target*. The top level specification of *match* for terms is given by:

$$\begin{array}{|l}
match : \mathbb{P}(\,Term \times \,Term \times Inst) \\
\hline
\forall (p, t, i) \in match \bullet instantiate(p, i) =_\alpha t
\end{array}$$

match is overloaded for lists of terms, formulae and sets as well. The definition for these cases is similar to above.

5   An Algorithm for Matching

In the following section we show that the top-level specification for matching expressions can be refined by providing an algorithm which can then easily be implemented. A similar algorithm, used to match mathematical terms in **mural** is described in [5].

5.1   Informal description

In the following section we describe an algorithm for matching mathematical expressions which works on structural induction on the pattern. The algorithm for matching terms is described by considering the different kinds of patterns (i.e. variables, non-parameteric functions and parameteric functions):

case 1: $p = x$, where x is a variable.

   p matches only the term x with a trivial instantiation.

case 2: $p = f(a_1, \ldots, a_m)$, where f is a non-parametric function.

   p matches targets of the form $f(b_1, \ldots, b_m)$, where a_j matches b_j for each j. The set of matches can be formed by merging the sets formed by matching a_j against b_j for each j.

case 3: $p = f(a_1, \ldots, a_m)$, where f is a parametric function.

The set of matches of p with an arbitrary expression e is formed by considering all disjoint sets of subtrees of e for which each subtree e_{sub} in the set matches an a_j for some $j \in 1 \ldots m$. The sets of instantiations formed by matching each e_{sub} in the disjoint set with the corresponding a_j are merged to form a set of instantiations. These are then merged with the instantiation $\{f(x_1, \ldots, x_m) \rightsquigarrow e'\}$, where e' is formed by replacing each subtree by the corresponding placeholder.

The complete set of instantiations is formed by taking the union of instantiation sets for all disjoint sets of subtrees, and finally filtering out any instantiations which contain free variables.

Example: In Fig. 4 an example of term matching is given. The pattern $hd(base, x)$, where hd and $base$ are both parameters, is matched with $cons(nil, x)$. A match is found by firstly replacing the subterms nil and x of $cons(nil, x)$ by placeholders, with resulting instantiation $base \rightsquigarrow x$ and then instantiating hd to the resulting term, i.e. $hd(a, b) \rightsquigarrow cons(a, b)$

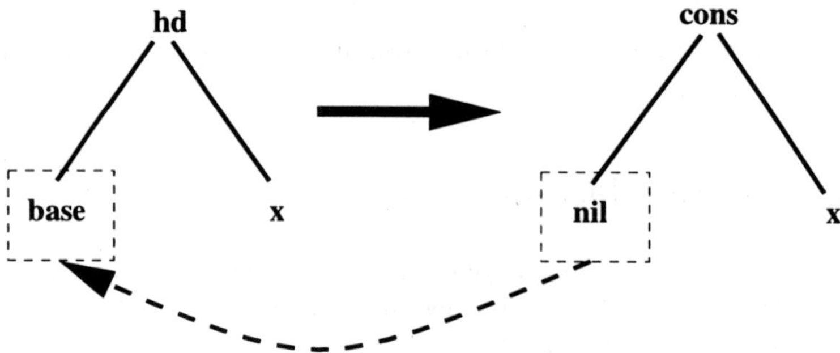

Fig. 4. Example of term matching

The algorithm for matching formulae is very similar to that given for matching terms. Briefly it works by considering the different kinds of patterns. For parametric relations the same sort of strategy is applied as that for parametric functions described above. The noteworthy case is quantified formulae, where we wish to consider the case where the signatures of the variable declaration part of the pattern and target are the same, but where the variables names don't necessarily agree. For this case, some preprocessing is done to the pattern and target, whereby the quantified variables are renamed to new variables which do not appear in either of the two formulae.

Example: Fig. 5 shows how the formula $\forall x : Y, y : Y \bullet P(x, y) \lor Q(x, y)$ can be matched with $\forall a : X, b : Y \bullet a < b \lor b < a$ by firstly replacing the bound

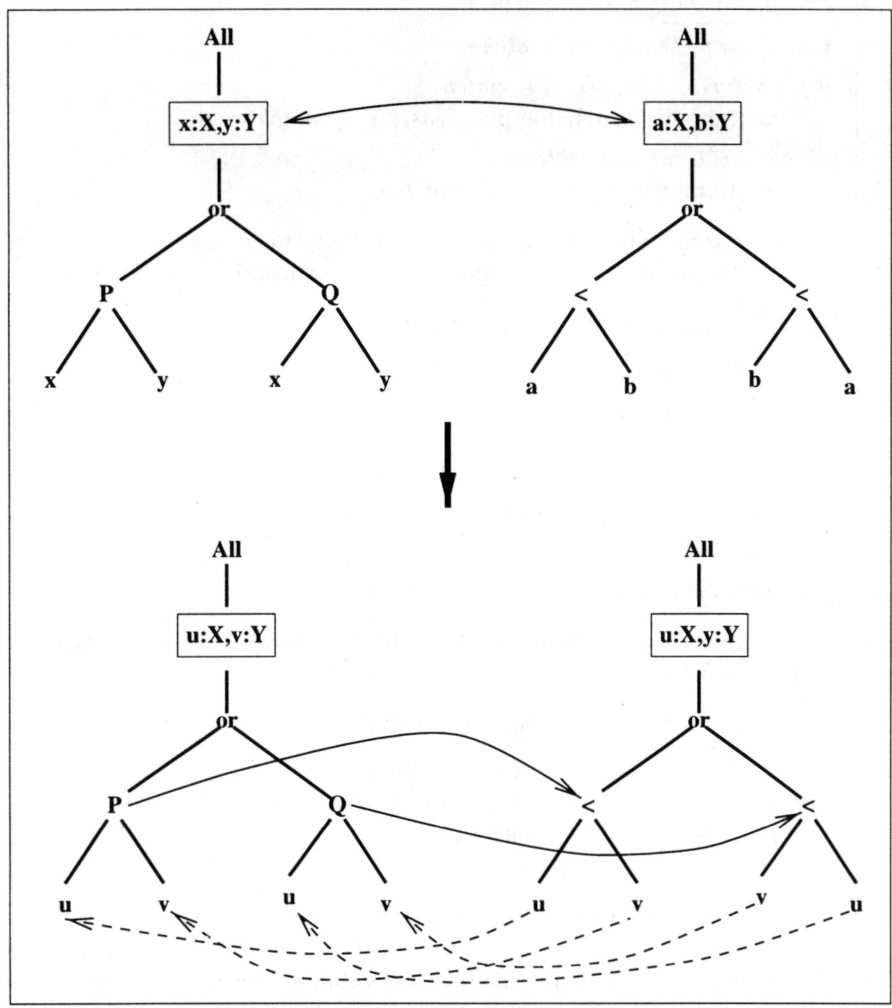

Fig. 5. Example of formula matching

variables and then matching the resulting formulae. The resulting instantiation is $P(a, b) \rightsquigarrow a < b, Q(a, b) \rightsquigarrow b < a$.

The algorithm for matching mathematical sets is straightforward.

5.2 Terms

The predicate *match* is specified by considering the different kinds of patterns: placeholders, variables, non-parameteric functions and parametric functions.

$$
\begin{array}{|l}
\hline
match : \mathbb{P}(Term \times Term \times Inst) \\
\hline
(\mathrm{var}\,x, \mathrm{var}\,x, TrivInst) \in match \\
\#as = \#as_1 \wedge (as, as_1, i) \in match \\
\qquad \Rightarrow (\mathrm{fnapplic}(f, as), \mathrm{fnapplic}(f, as_1), i) \in match \\
(f, as, e, i) \in matchFunction \\
\qquad \Rightarrow (\mathrm{termparam}(f, as), e, i) \in match \\
\end{array}
$$

Two lists of terms of equal length match if the terms in corresponding positions in the lists match, and the resulting set of instantiations are mergeable.

$$
\begin{array}{|l}
\hline
match : \mathbb{P}(\mathrm{seq}\,Term \times \mathrm{seq}\,Term \times Inst) \\
\hline
(\langle\rangle, \langle\rangle, TrivInst) \in match \\
(h_1, h_2, i_1) \in match \\
\qquad \wedge (t_1, t_2, i_2) \in match \\
\qquad\qquad \wedge (i_1, i_2, i) \in areMergeableInsts \\
\qquad\qquad\qquad \Rightarrow (\langle h_1 \rangle \frown t_1, \langle h_2 \rangle \frown t_2, i) \in match \\
\end{array}
$$

5.3   Parametric functions

The following functions model the third case of the algorithm for matching terms given in Section 5.1 above.

$$
\begin{array}{|l}
\hline
matchFunction : \mathbb{P}(FunctionParam \times \mathrm{seq}\,Term \times Term \times Inst) \\
\hline
(e, es, e', i_1) \in isReplacement \wedge freeVars(e') = \varnothing \\
\qquad \wedge i_2.finsts = \{f \mapsto e'\} \wedge i_2.rinsts = \varnothing \wedge i_2.sinsts = \varnothing \\
\qquad \wedge (i_1, i_2, i) \in areMergeableInsts \\
\qquad\qquad \Rightarrow (f, es, e, i) \in matchFunction \\
\end{array}
$$

The functions *isReplacement* and *isReplacementList* model the part of the algorithm where disjoint sets of subterms of a term are replaced by placeholders. *isReplacement* considers the different ways that subterms of a term can be replaced. For a term e, we can either:

1. leave e unchanged, i.e. replace no subterms
2. replace e by a placeholder, or
3. if e is a function call, then replace zero or more of the arguments of e by placeholders

$$
\begin{array}{|l}
\hline
isReplacement : \mathbb{P}(Term \times \mathrm{seq}\,Term \times Term \times Inst) \\
\hline
\neg\, containsPlaceHolders(e) \Rightarrow (e, m, e, TrivInst) \in isReplacement \\
\exists j \in \mathrm{dom}\,m \bullet (m(j), e_1, i) \in match \Rightarrow \\
\qquad (e_1, m, \mathrm{ph}\,j, i) \in isReplacement \\
(as, m, nas, i) \in isReplacementList \Rightarrow \\
\qquad (\mathrm{fnapplic}(g, as), m, \mathrm{fnapplic}(g, nas), i) \in isReplacement \\
\end{array}
$$

A list of terms, *nes*, is a valid replacement for the list *es* if the two lists are of equal length and a replacement $nes(j)$ exists for each $es(j)$.

$$\begin{array}{l}
isReplacementList : \mathbb{P}(\mathrm{seq}\ Term \times \mathrm{seq}\ Term \times \mathrm{seq}\ Term \times Inst) \\[2ex]
\hline
(\langle\rangle, m, \langle\rangle, TrivInst) \in isReplacementList \\[1ex]
(e, m, ne, i_1) \in isReplacement \\
\quad \wedge\ (es, m, nes, i_2) \in isReplacementList \\
\quad\quad \wedge\ (i_1, i_2, i) \in areMergeableInsts \\
\quad\quad\quad \Rightarrow (\langle e\rangle \frown es, m, \langle ne\rangle \frown nes, i) \in isReplacementList
\end{array}$$

5.4 Formulae

The predicate *match* for formulae is defined in terms of the different kinds of patterns.

$$\begin{array}{l}
match : \mathbb{P}(Fmla \times Fmla \times Inst) \\[2ex]
\hline
\#es = \#es_1 \wedge (es, es_1, i) \in match \\
\quad \Rightarrow (\mathrm{relation}(r, es), \mathrm{relation}(r, es_1), i) \in match \\[1ex]
(f_1, f_2, i) \in match \Rightarrow (\mathrm{not}\, f_1, \mathrm{not}\, f_2, i) \in match \\[1ex]
(f_1, f_3, i_1) \in match \wedge (f_2, f_4, i_2) \in match \wedge (i_1, i_2, i) \in areMergeableInst \\
\quad \Rightarrow (\mathrm{binConn}(f_1, f_2), \mathrm{binConn}(f_3, f_4), i) \in match \\[1ex]
(\mathrm{quant}(q, mvs, f), f_1, i) \in matchQuant \\
\quad \Rightarrow (\mathrm{quant}(q, mvs, f), f_1, i) \in match \\[1ex]
(\mathrm{paramrelation}(r, es), f, i) \in matchRelation \\
\quad \Rightarrow (\mathrm{paramrelation}(r, es), f, i) \in match
\end{array}$$

5.5 Parametric relations

The predicate *matchRelation* is similar to *matchFunction*. That is the target is replaced by substituting certain subterms in the formula with placeholders, for which the subterm matched the appropriate argument of the pattern.

$$\begin{array}{l}
matchRelation : \mathbb{P}(RelationParam \times \mathrm{seq}\ Term \times Fmla \times Inst) \\[2ex]
\hline
(f, es, f', i_1) \in isReplacementFmla \wedge freeVars(f') = \varnothing \\
\quad i_2.finsts = \varnothing \wedge i_2.rinsts = \{r \mapsto f'\} \wedge i_2.sinsts = \varnothing \\
\quad\quad \wedge\ (i_1, i_2, i) \in areMergeableInsts \\
\quad\quad\quad \Rightarrow (r, es, f, i) \in matchRelation
\end{array}$$

isReplacementFmla replaces zero or more disjoint subterms of a formula by placeholders; the function is defined by considering the different kinds of formulae.

$$isReplacementFmla : \mathbb{P}(Fmla \times \text{seq } Term \times Fmla \times Inst)$$

$(f, m, f, TrivInst) \in isReplacementFmla$

$(es, m, nes, i) \in isReplacementList$
$\quad\quad \Rightarrow (\text{relation}(r, es), m, \text{relation}(r, nes), i) \in isReplacementFmla$

$(f_1, m, f_2, i) \in isReplacementFmla \Rightarrow$
$\quad\quad (\text{not } f_1, m, \text{not } f_2, i) \in isReplacementFmla$

$(f_1, m, f_3, i_1) \in isReplacementFmla$
$\quad\quad \wedge (f_2, m, f_4, i_2) \in isReplacementFmla$
$\quad\quad\quad (i_1, i_2, i) \in areMergeableInsts$
$\Rightarrow (\text{binaryConn}(b, f_1, f_2), m, \text{binaryConn}(b, f_3, f_4), i) \in isReplacementFmla$

$(f_1, m, f_2, i) \in isReplacementFmla \Rightarrow$
$\quad\quad (\text{quant}(b, vs, f_1), m, \text{quant}(b, vs, f_2), i) \in isReplacementFmla$

5.6   Quantified formulae

Two quantified formulae match if the variable declaration lists have the same
signature and the two formulae are the same up to α-equivalence.

$$matchQuant : \mathbb{P}(Fmla \times Fmla \times Inst)$$

$(vs_q, vs_t, i_1) \in match$
$\wedge \exists vs : \text{iseq } Var \bullet \#vs = \#vs_q$
$\wedge vs \cap (freeVars(f_q) \setminus varSet(vs_q) \cup freeVars(f_t) \setminus varSet(vs_t)) = \varnothing$
$\wedge f_q' = renameVars(f_q, \{k : 1 \mathinner{..} \#vs_q \bullet \text{first } vs_q(k) \mapsto vs(k)\})$
$\wedge f_t' = renameVars(f_t, \{k : 1 \mathinner{..} \#vs_t \bullet \text{first } vs_t(k) \mapsto vs(k)\})$
$\wedge (f_q', f_t', i_2) \in match \wedge (i_1, i_2, i) \in areMergeableInsts$
$\Rightarrow (\text{quant}(b, vs_q, f_q), \text{quant}(b, vs_t, f_t), i) \in matchQuant$

Two variable declaration lists match if the signatures of the two lists match.

$$match : \mathbb{P}(VarDeclars \times VarDeclars \times Inst)$$

5.7   Sets

A given type matches only itself. Two constructed types match if they have the
same type constructor, and if their arguments match. A parameteric set matches
any other set.

$$match : \mathbb{P}(Set \times Set \times Inst)$$

6   Discussion

The paper by Hemer and Lindsay [3] shows how this approach can be extended
to component-wise matching in CARE. Components in CARE include fragments,

types and lemmas. Each component has a specification part and a body. For fragments (which are roughly analogous to functions in an imperative programming language) the specification contains a declaration of input and output variables, as well as pre- and post-conditions. Matching of fragments involves firstly matching the signatures of the inputs and outputs, and then matching the pre- and post-conditions. The specification of a type is a mathematical set, describing what values objects of this type may take - matching of type specifications is done by matching of sets. Similarly, since lemmas are formulae, matching of lemmas is equivalent to matching formulae.

The approach can then be extended further to matching component collections - referred to in CARE as *templates*. Templates may also have applicability conditions, which are conditions stating the subset of valid values that parameters can be instantiated to. Applicability conditions are chosen in such a way that their correctness ensures that the correctness of the template is preserved. In general establishing the correctness of applicability conditions is far easier than re-establishing the correctness of an instantiated template.

By considering associative-commutative functions and relations as a special case in matching we can produce a more effective pattern-matching algorithm. For example, suppose that the function ω is AC, then the pattern $x\omega g(y, z)$ matches $(z\omega y)\omega x$ with one possible instantiation being $g(a, b) \rightsquigarrow a\omega b$. AC pattern-matching is currently being added to the algorithm, based on the paper by Lincoln and Christian [6].

By allowing the software engineer to put parameters in their target specifications, and using unification we enable the software engineer to omit certain details of their desired component(s). For example the software engineer may want a component that adds two numbers, but is not concerned about the exact representation of the numbers (whether they might be naturals, integers, rational etc), so they might leave the set representing the type of numbers as a parameter. An algorithm for unification, based on the pattern-matching algorithm will be designed and implemented in the near future.

A prototype library search tool for CARE, based on the matching algorithm described in this paper has already been developed. The user inputs the specifications of one or more desired components (fragments or types) referred to as the *search query*, and tool searches for templates which contain components matching one or more of the desired components. A signature matching mechanism has been incorporated, which will eliminate any matches that have incompatible types. The tool is currently being extended to allow parameters in query components, allowing the user to under-specify certain properties of their desired component. To enable this, a unification-based search mechanism is being implemented, based on the higher-order unification described by Huet [4].

7 Conclusions

In this paper we have presented an algorithm for pattern-matching Z-like mathematical expressions which may contain parameters. This algorithm can be used as the basis for searching a library of reusable components by giving the specification of the desired component.

A justification that the algorithm satisfies the top-level specification has been given. The top-level specification states that given a pattern p and a target t which match with instantiation i, then instantiating p with i yields t. A search mechanism based on the pattern-matching algorithm given here should be relatively effective, and since it doesn't rely on formal reasoning support like the search mechanisms used by similar systems it should be more efficient. We noted that the algorithm can be strengthened by considering associative-commutative operators as a special case.

Acknowledgements I would like to sincerely thank my supervisor Peter Lindsay whose input, guidance and helpful advice have been invaluable in the writing of this paper, as well as in my other research endeavours.

References

1. B. Fischer, F. Kievernagel, and W. Struckman. VCR: A VDM-based software component retrieval tool. Technical report, Technical University of Braunschwieg, Germany, November 1994.
2. D. Hemer and P.A. Lindsay. The CARE toolset for developing verified programs from formal specifications. In O. Frieder and J. Wigglesworth, editors, *Proceeding of the Fourth International Symposium on Assessment of Software Tools*, pages 24–35. IEEE Computer Society Press, May 1996.
3. D. Hemer and P.A. Lindsay. Reuse of verified design templates. To appear Formal Methods Europe '97, January 1997.
4. G.P. Huet. A unification algorithm for typed λ-calculus. *Theoretical Computer Science*, 1:27–57, 1975.
5. C. B. Jones, K. D. Jones, P. A. Lindsay, and R. Moore. *mural: A Formal Development Support System*. Springer-Verlag, 1991.
6. P. Lincoln and J. Christian. Adventures in associative-commutative unification. In C. Kirchner, editor, *Unification*, pages 393–416. Academic Press, 1990.
7. M.D. McIlroy. Mass produced software components. *Software Engineering Concepts and Techniques*, pages 88–98, 1969.
8. J.M. Spivey. *The Z Notation: A Reference Manual*. Prentice Hall, 1989.
9. A. Moormann Zaremski and J.M. Wing. Specification matching of software components. In *Third ACM SIGSOFT Symposium on the Foundations of Software Engineering*, 1996.

A   Justification

In the following section we wish to give a justification that the algorithm for matching expressions given in §5 satisfies the top-level specification for matching in §4. That is we prove the following theorem.

Theorem 1. *For all patterns p and targets t, if p matches t with instantiation i, then instantiating p with i yields t. i.e.* $\forall (p, t, i) \in match \bullet instantiate(p, i) = t.$

We shall prove this theorem by considering the different kinds of patterns p. Here we shall only consider lists of terms, variables, non-parametric and parametric functions and quantified formulae. The other cases can be proven in a similar manner.

A.1   Preliminaries

Lemma 2. *Given any instantiations i_1 and i_2 which are mergeable (yielding the instantiation i), if an expression h can be instantiated by i_1, then h can also be instantiated by i, giving the same result in both cases; i.e.*

$$(h, i_1) \in \text{dom } instantiate \land (i_1, i_2, i) \in areMergeableInsts$$
$$\Rightarrow (h, i) \in \text{dom } instantiate \land instantiate(h, i) = instantiate(h, i_1)$$

Lemma 3. *For instantiations i_1, i_2 and i*

$$(i_1, i_2, i) \in areMergeableInsts \Leftrightarrow (i_2, i_1, i) \in areMergeableInsts$$

The following corollary follows from lemmas 2 and 3.

Corollary 4. *Given instantiations i_1 and i_2, then*

$$(e, i_2) \in \text{dom } instantiate \land (i_1, i_2, i) \in areMergeableInst$$
$$(e, i) \in \text{dom } instantiate \land instantiate(h, i) =_\alpha instantiate(h, i_2)$$

A.2   Lists

To show that two lists of terms as_1 and as_2 satisfy the theorem, we do a proof by induction on the lists as_1 and as_2 by firstly letting the induction hypothesis be:

$$(as_1, as_2, i) \in match \land \#as_1 = \#as_2 \Rightarrow instantiate(as_1, i) = as_2$$

For the initial case $as_1 = \langle \rangle = as_2$, and $i = TrivInst$. Focussing on the left-hand side of the consequent:

$$instantiate(as_1, i)$$
$$= instantiate(\langle \rangle, TrivInst)$$
$$= \langle \rangle$$

For the general induction step we assume that the hypothesis holds for $as_1 = t_1$ and $as_2 = t_2$, when $(t_1, t_2, i) \in match$ and $\#t_1 = \#t_2$.

We are required to show that the hypothesis holds for $as_1 = \langle h_1 \rangle ^\frown t_2$ and $as_2 = \langle h_2 \rangle ^\frown t_2$. From the definition of $match$ we can infer:

$$(h_1, h_2, i_1) \in match \tag{1}$$
$$(t_1, t_2, i_2) \in matchList \tag{2}$$
$$(i_1, i_2, i) \in areMergeableInsts \tag{3}$$

$$
\begin{aligned}
instantiate(as_1, i) \\
&= instantiate(\langle h_1 \rangle \frown t_1, i) \\
&= \langle instantiate(h_1, i) \rangle \frown instantiate(t_1, i) \\
&= \langle instantiate(h_1, i_1) \rangle \frown instantiate(t_1, i_2) \\
&= \langle h_2 \rangle \frown t_2 = as_2
\end{aligned}
$$

A.3 Variables

For an arbitrary variable x, $(var\, x, var\, x, TrivInst)$ is the only tuple in *match*. It follows from the definition of instantiate that:

$$instantiate(var\, x, TrivInst) = var\, x$$

A.4 Non-parametric functions

Non-parametric function call patterns of the form $f(as_1)$ match targets of the form $f(as_2)$, where $\#as_1 = \#as_2$ and the argument lists as_1 and as_2 match. We are required to prove

$$instantiate(\mathrm{fnapplic}(f, as_1), i) = \mathrm{fnapplic}(f, as_2)$$

but from the definition of instantiate it is sufficient to show that

$$instantiate(as_1, i) = as_2 \tag{4}$$

which follows by induction.

A.5 Parametric functions

A parametric function of the form $f(as)$ matches a term e, with instantiation i, if the tuple (f, as, e, i) is in *matchFunction*.

Before proving the theorem for this case we first prove a number of useful lemmas.

Lemma 5. *Given a parametric function application $f(as)$ and an instantiation i that maps f to a term e containing no placeholders, then $f(as)$ is instantiated to e by i. i.e.*

$$\neg\, containsPlaceHolders(e) \land f \mapsto e \in i.finsts$$
$$\Rightarrow instantiate(\mathrm{termparam}(f, as), i) = e$$

Proof. To prove the above lemma we firstly use the definition of *instantiate*, then make use of the fact that substituting placeholders in an expression e containing no placeholders will return e.

$$
\begin{aligned}
instantiate(\mathrm{termparam}(f, as), i) &= substPlaceHolders(i.finsts(f), m) \\
&= substPlaceHolders(e, m) \\
&= e \quad \square
\end{aligned}
$$

Lemma 6. *Given terms e and e', a sequence of terms as and an instantiation i, then*

$$(e, as, e', i) \in isReplacement$$
$$\Rightarrow instantiate(substPlaceHolders(e', as), i) = e$$

Proof. Consider the three different cases in the definition of *isReplacement*:

1. $e' = e$, $i = TrivInst$ and $\neg\, containPlaceHolders(e)$. Focusing on the left-hand side of the consequent:

$$instantiate(substPlaceHolders(e', as), i)$$
$$= instantiate(substPlaceHolders(e, as), TrivInst)$$
$$= instantiate(e, TrivInst) = e$$

2. $\exists j \in \mathrm{dom}\ as \bullet (as(j), e, i) \in match$ and $e' = \mathrm{ph}\ j$. Focusing on the left-hand side of the consequent:

$$LHS = instantiate(substPlaceHolders(\mathrm{ph}\ j, as), i)$$
$$= instantiate(as(j), i)$$
$$= e$$

3. $e = \mathrm{fnapplic}(g, es)$, $e' = \mathrm{fnapplic}(g, nes)$ and $(es, as, nes, i) \in isReplacementList$. We do a proof by induction on es and nes. For the initial case $es = nes = \langle\rangle$ and $i = TrivInst$;

$$LHS = instantiate(substPlaceHolders(\mathrm{fnapplic}(g, \langle\rangle), as), TrivInst)$$
$$= \mathrm{fnapplic}(g, substPlaceHolders(\langle\rangle, as))$$
$$= e$$

Assuming that the hypothesis is true for $es = t_1$ and $nes = t_2$ with instantiation i_1, show that it holds for $es = \langle h_1 \rangle \frown t_1$ and $nes = \langle h_2 \rangle \frown t_2$. From the definition of *isReplacementList* we can assume:

$$(i_1, i_2, i) \in areMergeableInsts \tag{5}$$
$$(t_1, as, t_2, i_2) \in isReplacementList \tag{6}$$
$$(h_1, as, h_2, i_1) \in isReplacement \tag{7}$$

Focussing on the left-hand side of the consequent:

$$LHS = instantiate(substPlaceHolders(\mathrm{fnapplic}(g, \langle h_2 \rangle \frown t_2), as), i)$$
$$= instantiate(\mathrm{fnapplic}(g, \langle substPlaceHolders(h_2, as) \rangle$$
$$\frown substPlaceHolders(t_2, as)), i)$$
$$= \mathrm{fnapplic}(g, \langle instantiate(substPlaceHolders(h_2, as), i) \rangle$$
$$\frown instantiate(substPlaceHolders(t_2, as), i))$$
$$= \mathrm{fnapplic}(g, \langle instantiate(substPlaceHolders(h_2, as), i_1) \rangle$$
$$\frown instantiate(substPlaceHolders(t_2, as), i_2))$$
$$= \mathrm{fnapplic}(g, \langle h_1 \rangle \frown t_1) \qquad \square$$

The proof of the following lemma is left as an exercise to the reader. It states that result of replacing all placeholders ph j by $instantiate(as(j), i)$ is the same replacing the placeholders ph j by $as(j)$ and then applying the instantiation i to the result.

Lemma 7. *Given a term e, a list of terms as and an instantiation i, then:*

$$substPlaceHolders(e, \{j : 1 \mathinner{.\,.} \#as \bullet j \mapsto instantiate(as(j), i)\})$$
$$= instantiate(substPlaceHolders(e, as), i)$$

Returning to the proof of the theorem for parametric functions, we are required to show that:

$$(e, as, e', i_1) \in isReplacement \wedge i_2 = (\{f \mapsto e'\}, \varnothing, \varnothing)$$
$$\wedge (i_1, i_2, i) \in areMergeableInst$$
$$\Rightarrow instantiate(\mathrm{termparam}(f, as), i) = e$$

Focussing of the left-hand side of the consequent:

$$LHS = instantiate(\mathrm{termparam}(f, as), i)$$
$$= substPlaceHolders(e', \{k : 1 \mathinner{.\,.} \#as \bullet k \mapsto instantiate(as(k), i)\})$$
$$= instantiate(substPlaceHolders(e, as), i)$$
$$= instantiate(substPlaceHolders(e, as), i_1)$$
$$= e$$

A.6   Quantified formulae

Lemma 8. *For an expression e, an instantiation i and a variable mapping vm:*

$$(freeVars(i) \cap \mathrm{dom}\, vm) = \varnothing$$
$$\Rightarrow instantiate(rename(e, vm), i) =_\alpha rename(instantiate(e, i), vm)$$

Proof. The above lemma can be proven by considering the different kinds of expressions. $\square$

Corollary 9. *For expressions f_1 and f_2, variable mapping vm and instantiation i:*

$$f_1 = renameVars(f_2, vm) \wedge (freeVars(i) \cap \mathrm{dom}\, vm) = \varnothing$$
$$\Rightarrow instantiate(f_1, i) =_\alpha instantiate(f_2, i)$$

We will use the above results to show that the refinement for quantified formulae is valid; that is:

$$(vs_q, vs_t, i_1) \in match$$
$$\wedge \exists vs : \mathrm{iseq}\, Var \bullet \#vs = \#vs_q$$
$$\wedge vs \cap (freeVars(f_q) \setminus varSet(vs_q) \cup freeVars(f_t) \setminus varSet(vs_t)) = \varnothing$$
$$\wedge f_q' = renameVars(f_q, \{k : 1 \mathinner{.\,.} \#vs_q \bullet vs_q(k) \mapsto vs(k)\})$$
$$\wedge f_t' = renameVars(f_t, \{k : 1 \mathinner{.\,.} \#vs_t \bullet vs_t(k) \mapsto vs(k)\})$$
$$\wedge (f_q', f_t', i_2) \in match \wedge (i_1, i_2, i) \in areMergeableInsts$$
$$\Rightarrow instantiate(\mathrm{quant}(b, vs_q, f_q), i) =_\alpha \mathrm{quant}(b, vs_t, f_t)$$

From the definition of *instantiate* we know:

$$instantiate(\mathrm{quant}(b, vs_q, f_q), i) =_\alpha \mathrm{quant}(b, vs_1, f_1)$$

where

$$vs_1 =_\alpha instantiate(vs_q, i) \wedge$$
$$(varSet(vs_1) \cap free\,Vars(i)) = \varnothing \wedge$$
$$f_1 = instantiate(rename\,Vars(f_q, \{j : 1 \ldots \# vs_q \bullet \mathrm{first}\ vs_q(j) \mapsto vs_1(j)\}), i)$$

Firstly we show that $vs_1 =_\alpha vs_t$:

$$vs_1 =_\alpha instantiate(vs_q, i)$$
$$= instantiate(vs_q, i_1)$$
$$=_\alpha vs_t$$

Secondly we show that $f_1 =_\alpha f_t$:

$$f_1 = instantiate(rename\,Vars(f_q, \{j : 1 \ldots \# vs_q \bullet \mathrm{first}\ vs_q(j) \mapsto vs_1(j)\}), i)$$
$$= rename\,Vars(instantiate(f_q, i), \{j : 1 \ldots \# vs_q \bullet \mathrm{first}\ vs_q(j) \mapsto vs_1(j)\})$$
$$=_\alpha instantiate(f_q, i)$$
$$=_\alpha instantiate(rename\,Vars(f_q, \{k : 1 \ldots \# vs_q \bullet vs_q(k) \mapsto vs(k)\})$$
$$= instantiate(f_q', i)$$
$$= instantiate(f_q', i_2)$$
$$=_\alpha f_t'$$
$$=_\alpha f_t$$

Towards a Formal Semantic Base for the Type Models of the Unified Modeling Language

María M. Larrondo-Petrie, Robert B. France*,
Monika Saksena and Malcolm Shroff

Department of Computer Science & Engineering
Florida Atlantic University
Boca Raton, FL-33431-0991, USA
{maria,robert,msaksena,mshroff}@cse.fau.edu

Abstract. In order to rigorously analyze the structure and behavior of object-oriented (OO) models, a firm semantic base needs to be developed for the OO concepts. The Unified Modeling Language (UML) is fast becoming the *de facto* OO modeling language, thus it is important that it have a formally defined semantic base.

In this paper we discuss some early results of our efforts to formalize the UML using the Z formal notation. This paper illustrates the application of our recently developed rules to non-trivial UML type structures. The non-trivial type structures considered include combinations of generalization, aggregation, and recursive structures.

Keywords: Formal Specification Techniques, Integrated Methods, Object-Oriented Analysis and Modeling, Unified Modeling Language, Z.

1   Introduction

During requirements analysis, the software engineer needs to systematically transform requirements statements that are often ambiguous and incomplete into precise models that specify required structure and behavior. Rigorous analysis of the resulting models can validate that the models capture the required behavior and verify that implementations comply with the models. Object-oriented (OO) methods provide support for organizing application concepts in a highly-structured manner and abstracting details not pertinent to the early stages of software development. Their graphical models can be easily understood (although the understanding can be more apparent than real) and are very useful when communicating the structure and behavior of the models to clients and modelers. However, the lack of firm semantic bases for the modeling notations used in most OO methods makes it unlikely that the models produced will be precise and rigorously analyzable.

The Unified Modeling Language (UML) [3] is a common OO modeling notation that is becoming the *de facto* standard for specifying, documenting and visualizing OO models. It is based on the Booch Method [2], Rumbaugh's Object

---

* Work partially funded by NSF grant CCR-9410396.

Modeling Technique (OMT) [17], the Fusion Method [8] and Jacobson's Use Case models [14]. In UML Version 1.0 the semantics are specified informally using OO graphical metamodels and natural language. We have found that these specifications are not rigorous enough to determine the precise semantics of certain non-trivial constructs.

Our research on integrating formal and nonformal methods [1, 7, 11] indicates that rigorously formalizing the semantics of OO notations and concepts can greatly benefit the development process. These include uncovering ambiguities and incompleteness in the semantics of the OO notation, and facilitating rigorous analysis of the structure and behavior captured in object models. Our previous research on formalizing Fusion Object Models resulted in a CASE tool, FuZE^2, that supported the formalization process, and provided access to Z analysis tools [5, 6, 11].

Recently, we have developed rules for transforming UML type structures to Z (e.g., see [18]). In this paper we illustrate the application of these rules to non-trivial UML analysis-level Class Diagrams involving only types. In section 2 we present an overview of formalization approaches and of the UML notation. We assume the reader is familiar with the Z notation (for details on the Z notation see [19]). In section 3, the rules for transforming UML type structures to Z are outlined. In section 4 we illustrate the application of the rules on non-trivial type structures. Our conclusions and future work on the formalization of the semantics of UML notation are discussed in section 5.

2   Background

In this section we present the different approaches that have been used to formalize OO concepts and justify the approach we used. We also provide an overview of UML analysis-level Class Diagrams.

2.1   Approaches to Formalizing OO Concepts

Three general approaches have been used to formalize OO modeling concepts:

- *The Supplemental Approach*: In this approach, the OO modeling notation is supplemented with formal notations. Syntropy [9], for example, annotates OMT-like models with mathematical expressions. This approach provides only partial formalization of models, and the formal annotations can clutter the diagrams, obscuring essential aspects of the application being modeled.
- *The OO-Formal Language Approach*: A new formal notation is developed that is essentially an object-oriented version of a mostly textual formal notation. Several object-based formal notations have been proposed in the literature (e.g., FOOM [20] and Object-Z [10]). The disadvantage of creating a new notation is that, at least initially, there will be a lack of available analysis tools that support the new notation.

---

^{2} See http://www.cse.fau.edu/research/MIRG/FUZE

- *The Methods Integration Approach*: An OO modeling technique is made more precise and amenable to rigorous analysis by integrating it with a suitable formal specification notation (e.g., [11, 12, 13]). Some of these approaches have the disadvantage that they introduce artificial notations that are not supported by the underlying formal notation, and thus are not supported by the available analysis tool.

Our research on formalizing OO modeling techniques uses the Methods Integration approach. We integrate the informal and formal methods by providing transformation rules which bridge the graphical UML OO modeling concepts and the formal Z notation. No artificial notations are introduced into the formal expressions, allowing the use of analysis tools available for the formal notation.

We chose to formalize the UML OO notation because it is fast becoming the *de facto* standard, and incorporates concepts from the best OO methods used today. We chose the Z notation because:

- Z is based on mathematical concepts that are well-known and understood (set theory, and predicate logic).
- Z has reached a high level of maturity as a formal specification notation and has been used in industry.
- Z constructs (*schemas*) can be mapped directly to the OO concepts of types, classes, and class structures.
- There are numerous tools available that support Z, such as typecheckers (e.g., ZTC [16]) and animators (e.g., ZANS [15]).

In this paper we discuss some early results of our ongoing efforts to formalize the UML using the Z notation. The intent is to develop a set of rules that supports the mechanical generation of Z specifications from augmented Class Diagrams.

2.2 UML Analysis-Level Class Diagrams

The static structure of a system is modeled in UML as a Class Diagram that can consist of classes, types, objects (class instances) and their associations. UML *types* are specifications of more concrete UML classes. We refer to Class Diagrams that consist solely of types as *Type Diagrams*. Type Diagrams provide appropriate abstractions for constructing requirement analysis level models.

In our formalization, a Type Diagram characterizes a set of instance structures. A *valid instance structure*, also called a *configuration*, exhibits the properties expressed in the Type Diagram, and can be viewed as a snapshot of a system's structure at a given point in time.

A type in UML is represented as a rectangular box separated into three compartments, respectively containing the class name, a list of attributes (with optional types and initial values), and a list of operations (with optional argument lists and return types). In this paper, only the static aspects of classes are formalized; operations are ignored.

An *association* represents a structured relationship between types, and is depicted as a line connecting the type boxes. An association may have a name with an optional directional arrow showing the direction of the named relationship. Each end of an association may have a role name, showing how the class is viewed by the other class in the association, and a multiplicity annotation. Multiplicity of an association indicates how many instances of the type can be associated with one instance of the other type. A range is denoted by $m..n$, where m and n are integers. A single integer is used when the range is $m..m$. The symbol '*' indicates 'many', i.e. an unlimited number of objects. By itself, the symbol '*' is equivalent to '0..*' , i.e. any number including none.

Aggregation is a special form of association that represents the whole-part relationship between types. It is indicated by placing a diamond on the end of the association attached to the whole type of the whole-part relationship. Strong aggregation, also called *composition*, is shown by a filled diamond and requires that there be a life-time binding between the aggregate and its components.

Generalization/specialization is indicated by placing a triangle on the end of the association attached to the superclass. All attributes, operations and associations of the superclass are inherited by the subclasses.

Examples of the UML type structure constructs are given throughout the paper.

3   Formalizing UML Type Models

In this section we outline the rules for transforming UML Type Diagrams to Z specifications. Specifically, we illustrate how types, associations, aggregations and generalization hierarchies can be transformed to Z specifications. The transformations can be carried out in two phases. In the first phase, Z specifications defining the properties of individual types are produced. The specifications produced in this phase are called *type schemas*. In the second phase, the type schemas are used to produce Z specifications, called *configuration schemas*, characterizing the configurations of the models.

3.1   Representing Types in Z

A type is represented by a Z state schema. The attributes of a type and the object identifier of a type are represented by state variables. Any invariants associated with the type, expressed as annotations in the Type Diagram, are stated formally in the predicate part of the Z schema. The type name of an attribute in the Z schema corresponds to the type name of the attribute in the Type Diagram. The type must also be defined in Z as either a basic type or a schema. If no type is associated with an attribute then the capitalized name of the attribute is used as the type name and declared as a Z basic type. The resulting Z schema is called a *type schema*.

Instances of the type are represented as bindings of values to variables declared in the Z type schema. The object identifier of an instance is the value bound to the variable corresponding to the object identifier in the type schema.

Consider the example of a type *Student* with attributes *name, social security number, address* and *phone number.*

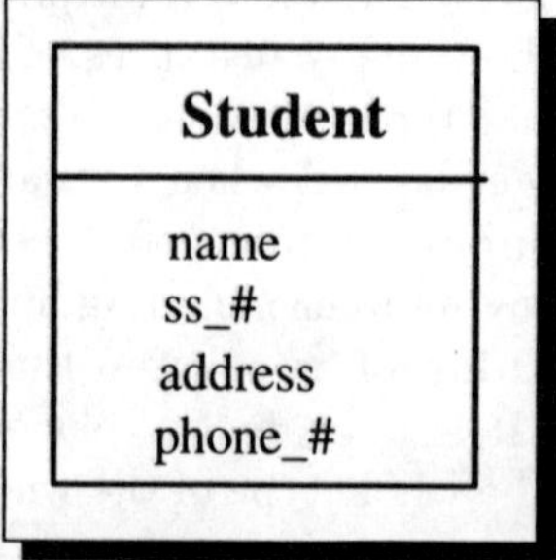

Fig. 1. The type Student

The type schema for *Student* is given below:

[*Student_ID*, *NAME*, *SS_#*, *ADDRESS*, *PHONE_#*]

```
__ Student______________________________________________
  ident : Student_ID
  name : NAME
  ss_# : SS_#
  address : ADDRESS
  phone_# : PHONE_#
```

Since the attribute types are not defined in the Type Diagram, the capitalized attribute names are declared as basic Z types. An instance of *Student* is identified by the object identifier *ident*.

The set of type instances in a configuration is referred to as the type's state. Restrictions on the type state are specified in the type's *configuration schema*. The configuration schema for the *Student* type, *StudentConfig*, given below, specifies that instances in a configuration must have unique identifiers. This constraint must be expressed for every type in a configuration schema.

```
__ StudentConfig________________________________________
  students : ℙ Student
_________________________________________________________
  ∀ i,j : students | i.ident = j.ident • i = j
```

3.2   Representing Associations in Z

An association is a structural relationship between types. An instance of an association is called a link, which can be represented as a tuple of object references.

In a binary association only two object are involved. Ternary and higher-order associations are difficult to implement and are generally avoided.

Transforming associations to Z specifications involves the following activities:

- Create type schemas for each type involved in the association.
- Create a configuration schema for the association. In the declaration part, declare a variable for each type participating in the association, representing the set of type instances in a configuration, and a relational variable representing the association. For n-order associations, where n > 2, the variable representing the association has a n-tuple set type. In the predicate part, state that instances are unique, and that the domain and range of the association variable are restricted to the declared sets of instances. The multiplicity constraints must also be stated in the predicate part. We have developed a set of rules supporting the generation of predicates, given various forms of multiplicity constraints for binary associations. The complex nature of multiplicity constraints for ternary and higher-order associations makes it difficult to develop a similar set of general rules for higher-order associations. Role names are simply names for the domain and range of the association, and can be declared and specified as such in the configuration schema.

Consider the example of a binary association, *Subscribes*, between two types, *Customer* and *Service*, shown in Fig. 2.

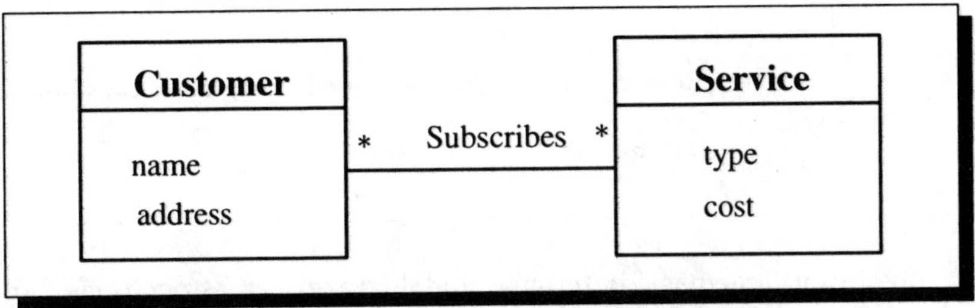

Fig. 2. A binary association

The type schemas for *Customer* and *Service* are given below:

$$[Customer\_ID, Service\_ID, NAME]$$

$$[ADDRESS, TYPE, COST]$$

```
_ Customer ___________________
 ident : Customer_ID
 name : NAME
 address : ADDRESS
```

```
_ Service ___________________
 ident : Service_ID
 cost : COST
 type : TYPE
```

The configuration schema for the association is given below:

$$
\begin{array}{|l}
\hline
\_SubscribesConfig\_\_\_\_\_\_\_\_\_\_ \\
\quad customers : \mathbb{P}\ Customers \qquad\qquad \text{[customer instances in a configuration]} \\
\quad services : \mathbb{P}\ Services \qquad\qquad\quad \text{[service instances in a configuration]} \\
\quad subscribes : Customer \nrightarrow Service \qquad \text{[the association variable]} \\
\hline
\qquad\qquad\qquad \text{[Restrict association to declared configuration instances]} \\
\quad \mathrm{dom}\ subscribes \subseteq customers \\
\quad \mathrm{ran}\ subscribes \subseteq services \\
\qquad\qquad\qquad\qquad \text{[Configuration instances must be unique]} \\
\quad \forall\ i,j : services \bullet (i.ident = j.ident \Leftrightarrow i = j) \\
\quad \forall\ i,j : customers \bullet (i.ident = j.ident \Leftrightarrow i = j) \\
\hline
\end{array}
$$

The domain and range of *subscribes* are subsets of *customers* and *services*, respectively, indicating that not all customers in a configuration need be related to a service, and vice-versa, as indicated by the multiplicity constraint, '*' (0 or more). If a multiplicity indicated a mandatory link then the domain and/or range must be equal to the respective set of instances (indicating that all configuration instances take part in the association).

A partial example of a configuration schema for a ternary association, involving the types A, B, and C, is given below (a multiplicity of '*' is assumed for all the types involved):

$$
\begin{array}{|l}
\hline
\_TernaryConfig\_\_\_\_\_\_\_\_\_\_ \\
\quad as : \mathbb{P}\ A \\
\quad bs : \mathbb{P}\ B \\
\quad cs : \mathbb{P}\ C \\
\quad rel : \mathbb{P}((A \times B) \times C) \\
\hline
\qquad\qquad \text{[Restrict association to declared configuration instances]} \\
\quad \mathrm{ran}\ rel \subseteq cs \\
\quad \forall\ a : A;\ b : B \mid (a,b) \in \mathrm{dom}\ rel \bullet a \in as \wedge b \in bs \\
\quad \ldots \\
\hline
\end{array}
$$

The configuration schemas for ternary and higher-order associations can get quite large if complex multiplicity constraints are given. Currently our rules generate what they can, and leave it up to the modeler to add additional constraints.

3.3　Representing Aggregation in Z

UML supports two forms of aggregation: a weak form, simply called *aggregation*, and a strong form called *composition*. The UML manual defines *composition* as a *whole-part* relationship with strong ownership and coincident lifetimes of part with the whole. The multiplicity of an aggregate (whole) in a composition must not exceed one (it is unshared). Parts with a multiplicity greater than one may be created after the aggregate but once created, they live and die with the whole. This strong aggregation is represented as a solid filled diamond on an association.

The weak form of aggregation allows the sharing of components across aggregates. It is a special form of association that is represented as a hollow diamond attached to the aggregate end of the association.

Generating Z specifications for (non-recursive) aggregations involves the following activities:

- Create type schemas for the components of the composition.
- Create a type schema for the aggregate. The declaration part consists of the usual object identifier and attribute variables, as well as variables representing the components. If the variable represents a component with a cardinality that can be greater than one, then the variable is a set, otherwise the variable represents a single element. The predicate part expresses the components' cardinality constraints.
- Create a configuration schema for the composition. The configuration schema for an aggregation specifies the constraints on the aggregation instances in a configuration. The declaration part declares the configuration instances for the whole and component types, and the predicate part expresses multiplicity and uniqueness constraints. The predicate part also restricts components to declared sets of instances.

Consider the composition shown in Fig. 3, in which the aggregate is a window that contains two scrollbars, a header and a panel.

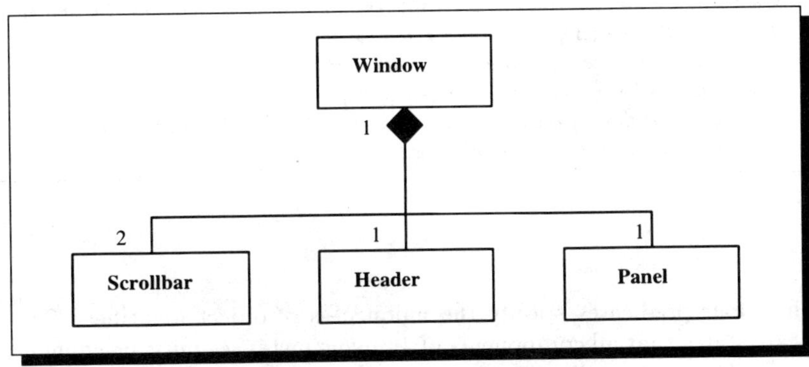

Fig. 3. A UML compostion

The type schemas for the model are given below:

$[Window\_ID, Scrollbar\_ID, Header\_ID, Panel\_ID]$

```
┌─ Scrollbar ─────────────        ┌─ Header ───────────────
│ ident : Scrollbar_ID            │ ident : Header_ID
└──────────────────────────       └────────────────────────
```

$$
\begin{array}{|l}
\hline
\underline{Panel} \\
\;\;ident : Panel\_ID \\
\hline
\end{array}
\qquad
\begin{array}{|l}
\hline
\underline{Window} \\
\;\;ident : Window\_ID \\
\;\;scrollbar : \mathbb{P}\,Scrollbar \\
\;\;header : Header \\
\;\;panel : Panel \\
\hline
\;\;\#scrollbar = 2 \\
\hline
\end{array}
$$

The configuration schema for the composition is given below:

$$
\begin{array}{|l}
\hline
\underline{WindowConfig} \\
\;\;windows : \mathbb{P}\,Window \\
\;\;scrollbars : \mathbb{P}\,Scrollbar \\
\;\;headers : \mathbb{P}\,Header \\
\;\;panels : \mathbb{P}\,Panel \\
\hline
\;\;\forall\, w1, w2 : windows \bullet w1.ident = w2.ident \Leftrightarrow w1 = w2 \\
\;\;\forall\, s1, s2 : scrollbars \bullet s1.ident = s2.ident \Leftrightarrow s1 = s2 \\
\;\;\forall\, h1, h2 : headers \bullet h1.ident = h2.ident \Leftrightarrow h1 = h2 \\
\;\;\forall\, p1, p2 : panels \bullet p1.ident = p2.ident \Leftrightarrow p2 = p2 \\
\;\;\forall\, w : windows \bullet w.scrollbar \subseteq scrollbars \wedge w.header \in headers \wedge w.panel \in panels \\
\;\;\forall\, sc : scrollbars \bullet (\exists\, w : windows \bullet sc \in w.scrollbar) \\
\;\;\forall\, h : headers \bullet (\exists\, w : windows \bullet w.header = h) \\
\;\;\forall\, p : panels \bullet (\exists\, w : windows \bullet w.panel = p) \\
\;\;\forall\, w1, w2 : windows \mid w1.scrollbar = w2.scrollbar \vee w1.header = w2.header \\
\;\;\;\;\;\; \vee\; w1.panel = w2.panel \bullet w1 = w2 \\
\hline
\end{array}
$$

The first four predicates specify the uniqueness of object identifiers. The fifth predicate states that all components of *Window* instances must be configuration instances. The sixth predicate states that all scrollbars in a configuration must be a component of a window in the configuration. The seventh and eighth predicates express similar constraints for the headers and panels in a configuration. The ninth predicate states that no sharing of window components is allowed in a configuration.

A variation of the *Window* aggregate that involves a weak form of aggregation is shown in Fig. 4. In the model, panels and headers can be shared across windows, but not scrollbars.

The type schemas for this variation of the *Window* structure are the same as above. Its configuration schema is given below:

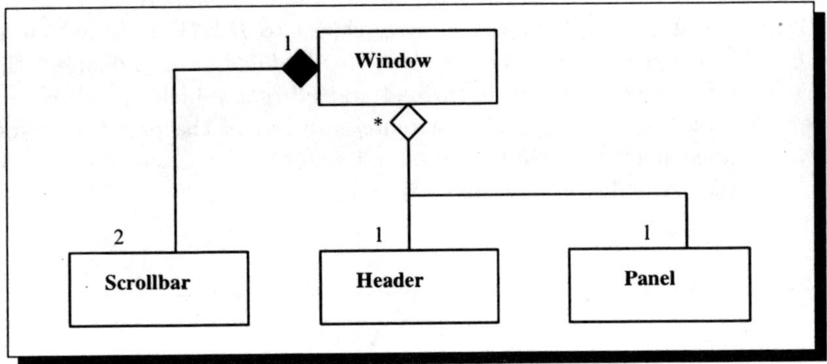

Fig. 4. A structure with a weak form of aggregation

─── *WindowConfig2* ──────────────────────────────────────
$windows : \mathbb{P}\ Window$
$scrollbars : \mathbb{P}\ Scrollbar$
$headers : \mathbb{P}\ Header$
$panels : \mathbb{P}\ Panel$
─────────────────────────────
$\forall\, w1, w2 : windows \bullet w1.ident = w2.ident \Leftrightarrow w1 = w2$
$\forall\, s1, s2 : scrollbars \bullet s1.ident = s2.ident \Leftrightarrow s1 = s2$
$\forall\, h1, h2 : headers \bullet h1.ident = h2.ident \Leftrightarrow h1 = h2$
$\forall\, p1, p2 : panels \bullet p1.ident = p2.ident \Leftrightarrow p2 = p2$
$\forall\, w : windows \bullet w.scrollbar \subseteq scrollbars \wedge w.header \in headers \wedge w.panel \in panels$
$\forall\, sc : scrollbars \bullet (\exists\, w : windows \bullet sc \in w.scrollbar)$
$\forall\, w1, w2 : windows \mid w1.scrollbar = w2.scrollbar \bullet w1 = w2$
──

The *WindowConfig2* schema is obtained from the *WindowConfig* schema by removing the restrictions that all headers and panels must be components, and that they must not be shared.

3.4 Representing Generalization Structures in Z

The generalization/specialization structure in UML defines the relationship between a type and one or more specialized versions of it. The specialist type is called a *subtype* and the type being specialized is called the *supertype*. One can view a generalization hierarchy as expressing desirable subdivisions of the root type's instance space.

For example, in Fig. 5, a generalization/specialization structure in which land vehicles and water vehicles inherit attributes from a *Vehicle* supertype, is shown. Each instance of a vehicle must either be a land or water vehicle, but cannot be both (the types are disjoint). When each instance of the supertype is an instance of one (or more) of its subtype, we say that the supertype is *abstract*[3]. *LandVehicle* is an abstract supertype with respect to *RoadVehicle* and *OffRoad-Vehicle*. The subtypes *RoadVehicle* and *OffRoadVehicle* are not disjoint, that is there can be instances that are both road and offroad vehicles (multiple inheritance). We can view this model as a representation of the partitioning of the *Vehicle* instance space into land vehicles and water vehicles, where land vehicles are further subdivided into road and offroad vehicles.

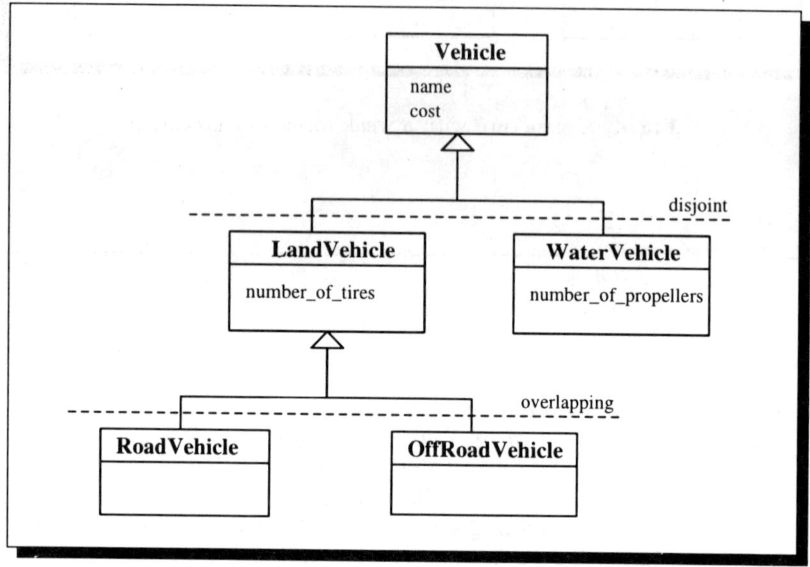

Fig. 5. A UML generalization hierarchy

In our approach, the generation of Z specifications from generalization structures involves the following activities:

- Create a type schema for the root supertype.
- Create a type schema for each subtype. In the declaration part, include a variable representing the supertype (thus inheriting its attributes and constraints), and variables for additional attributes of the subtype. The predicate part includes the additional constraints required by the subtype.
- Create a configuration schema for the structure. In the declaration part include variables representing configuration instances of the types involved. If any of the supertypes are abstract, or any subset of the subtypes are

---

[3] Our notion of abstract type is different than the programming language notion of abstract class

disjoint, then the constraints must be stated in the predicate part of the schema.

The type schemas for the structure shown in Fig. 5 is given below:

$[Vehicle\_ID, NAME, COST]$

$$\begin{array}{|l}\hline Vehicle \\\hline ident : Vehicle\_ID \\ name : NAME \\ cost : COST \\\hline\end{array}$$

$$\begin{array}{|l}\hline LandVehicle \\\hline vehicle : Vehicle \\ number\_of\_tires : \mathbb{N} \\\hline\end{array}$$

$$\begin{array}{|l}\hline WaterVehicle \\\hline vehicle : Vehicle \\ number\_of\_propellers : \mathbb{N} \\\hline\end{array}$$

$$\begin{array}{|l}\hline RoadVehicle \\\hline vehicle : LandVehicle \\\hline\end{array}$$

$$\begin{array}{|l}\hline OffRoadVehicle \\\hline vehicle : LandVehicle \\\hline\end{array}$$

The configuration schema for the structure is given below:

$$\begin{array}{|l}\hline VehicleConfig \\\hline vehicles : \mathbb{P}\,Vehicle \\ lvehicles : \mathbb{P}\,LandVehicle \\ wvehicles : \mathbb{P}\,WaterVehicle \\ rvehicles : \mathbb{P}\,RoadVehicle \\ orvehicles : \mathbb{P}\,OffRoadVehicle \\\hline \forall\,v1, v2 : vehicles \bullet v1.ident = v2.ident \Leftrightarrow v1 = v2 \\ \langle\{l : lvehicles \bullet l.vehicle\}, \{w : wvehicles \bullet w.vehicle\}\rangle \text{ partition } vehicles \\ \forall\,lv : lvehicles \bullet lv \in \{rv : rvehicles \bullet rv.vehicle\} \\ \qquad \lor\ lv \in \{or : orvehicles \bullet or.vehicle\} \\\hline\end{array}$$

4   Formalizing Non-Trivial Structures

In this section we apply the UML formalization rules to small, but non-trivial, UML Type Diagrams. The Type Diagrams considered involve generalization and aggregation structures with recursive properties. Such models are often difficult to understand because the multi-dimensional structure is abstracted out in two-dimensional presentations of the models. This means that additional effort is required to visualize the actual structure of the system.

The difficulty of visualizing models involving multi-dimensional structures often results in their underspecification. Formalizing such structures can help clarify structure, and can lead to a more complete specification of required structure. Formalization can also reveal that seemingly complex models are not as intricate as they appear, and may lead one to consider simpler models that are equivalent. On the other hand, formalization can also reveal that certain complex structures cannot be realized in a finite system.

In the following subsections we formalize two models. The first is an example of a generalization structure in which the general class is an aggregate that consists of components that are also part of (another) generalization hierarchy. The second is an example of recursive composition.

4.1   Formalization Example

In Fig. 6, a model that involves generalization and aggregation is shown. In the model, *Employee* is specialized by *Hourly Employee*, *Salaried Employee*, and *Exempt Employee*. *Employee* is also an aggregate that has one component, an *Employee Pension*, which is specialized as shown. The model seems complex because of the two-dimensional presentation of orthogonal structuring concepts. As will be shown by our formalization, the structure is not as complex as it seems.

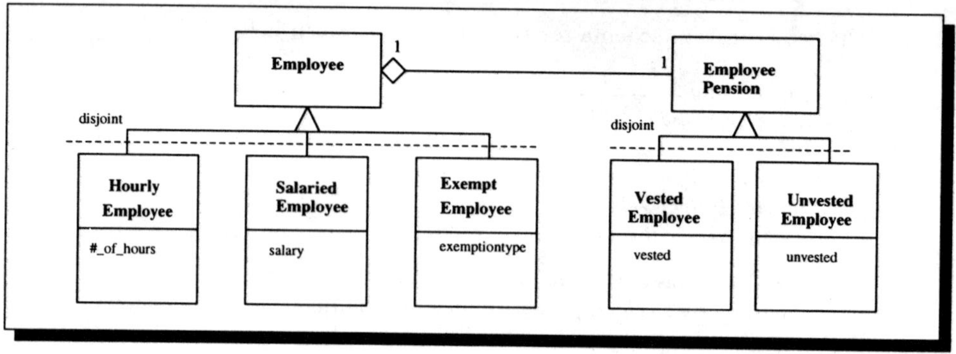

Fig. 6. A generalization structure with aggregation

The type schemas for the aggregate structure of *Employee* are given below:

[*Employee_ID, Employee_Pension_ID*]

```
┌ EmployeePension ─────────────────
│ ident : Employee_Pension_ID
└──────────────────────────────────
```

```
┌ Employee ─────────────────────────
│ ident : Employee_ID
│ emp_pension : EmployeePension
└───────────────────────────────────
```

The type schemas for the two generalization structures are given below:

[*VESTED, UNVESTED*]
[*EXEMPTIONTYPE*]

```
┌ HourlyEmployee ───────────────────
│ employee : Employee
│ #_of_hours : ℕ
```

$$\boxed{\begin{array}{l} \underline{SalariedEmployee} \\ employee : Employee \\ salary : \mathbb{N} \end{array}}$$

$$\boxed{\begin{array}{l} \underline{ExemptEmployee} \\ employee : Employee \\ exemption\_type : EXEMPTIONTYPE \end{array}}$$

$$\boxed{\begin{array}{l} \underline{VestedEmployee} \\ employeepension : EmployeePension \\ vested : VESTED \end{array}}$$

$$\boxed{\begin{array}{l} \underline{UnvestedEmployee} \\ employeepension : EmployeePension \\ unvested : UNVESTED \end{array}}$$

Using the above schemas, the following configuration schema, specifying constraints on the instances of the structure in a configuration, is obtained by applying the rules for creating configuration schemas for aggregation and generalization structures:

$$\boxed{\begin{array}{l} \underline{EmployeeConfig} \\ em : \mathbb{P}\,Employee \\ semp : \mathbb{P}\,SalariedEmployee \\ hemp : \mathbb{P}\,HourlyEmployee \\ eemp : \mathbb{P}\,ExemptEmployee \\ epen : \mathbb{P}\,EmployeePension \\ vemp : \mathbb{P}\,VestedEmployee \\ uemp : \mathbb{P}\,UnvestedEmployee \\ \hline \langle\{h : hemp \bullet h.employee\}, \{s : semp \bullet s.employee\}, \\ \quad \{ee : eemp \bullet ee.employee\}\rangle \text{ partition } em \\ \langle\{v : vemp \bullet v.employeepension\}, \{ue : uemp \bullet ue.employeepension\}\rangle \\ \quad \text{partition } epen \\ \forall\, e : em \bullet e.emp\_pension \in epen \\ \forall\, ep : epen \bullet (\exists\, e : em \bullet e.emp\_pension = ep) \\ \forall\, e1, e2 : em \bullet e1.ident = e2.ident \Leftrightarrow e1 = e2 \\ \forall\, ep1, ep2 : epen \bullet ep1.ident = ep2.ident \Leftrightarrow ep1.ident = ep2.ident \end{array}}$$

The above formalization simply applied the rules for formalizing aggregation and generalization in a straightforward manner. In this case the complexity of the model is more apparent than real, as the formalization process makes clear.

4.2   Formalization of a Recursive Composition

In Fig. 7, a model of glyph structures is shown. Here, a *glyph* is an object that consists of two components, pictures and lines, both of which are also glyphs. In the model, the component classes *Picture* and *Line* are disjoint, and the class *Glyph* is abstract, meaning that a *Glyph* instance is either a *Picture* instance or a *Line* instance, but not both. A *Glyph* instance is composed of zero or more *Picture* instances and zero or more *Line* instances. *Picture* and *Line* instances can be components of zero or more *Glyphs*. The resulting recursive structure is required to be antisymmetric and irreflexive.

Ignoring the aggregate structure, the type schemas for the generalization structure of the model are given below:

$$[Glyph\_ID]$$

$$\boxed{\begin{array}{l} \underline{Glyph} \\ ident : Glyph\_ID \end{array}}$$

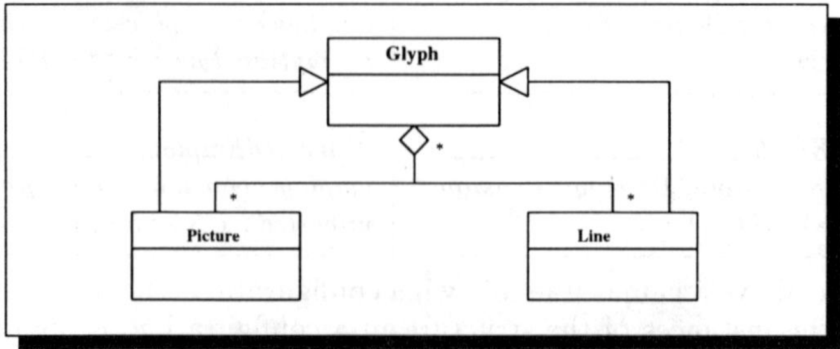

Fig. 7. A recursive aggregation structure

$$\begin{array}{|l|}\hline \textit{Line} \\\hline \textit{glyph} : \textit{Glyph} \\\hline \end{array} \qquad \begin{array}{|l|}\hline \textit{Picture} \\\hline \textit{glyph} : \textit{Glyph} \\\hline \end{array}$$

When aggregation and recursion are taken into consideration, direct application of our rules for formalizing aggregation results in a complex Z specification. Specifically, application of the rules for aggregation given previously will result in the following type schema for *Glyph*:

$$
\begin{array}{|l|}
\hline
\textit{Glyph} \\
\hline
\textit{ident} : \textit{Glyph\_ID} \\
\textit{lines} : \mathbb{P}\,\textit{Line} \\
\textit{pictures} : \mathbb{P}\,\textit{Picture} \\
\hline
\ldots \qquad\qquad\qquad\qquad\qquad\qquad\qquad\qquad \text{[recursion properties]} \\
\hline
\end{array}
$$

Specifying the recursion properties using the above schema requires the definition of auxiliary functions for traversing the implicit glyph tree structure. Rather than use this approach, we used another that utilizes a more explicit representation of the glyph tree structure. One can view the glyph structure as a set of two relations: the relation p_comp, relating glyphs to their picture components, and the relation l_comp, relating glyphs to their line components. The antisymmetric and irreflexive properties can be expressed as conditions on the transitive closure of these relations (p_comp^{+} is the transitive closure of p_comp, and l_comp^{+} is the transitive closure of l_comp). The schema defining the general recursive structure is given below:

$$
\begin{array}{|l|}
\hline
\textit{GlyphComp} \\
\hline
p\_comp, l\_comp : \textit{Glyph} \leftrightarrow \textit{Glyph} \\
\hline
\forall\, a, b, c, d : \textit{Glyph} \mid (a, b) \in p\_comp^{+} \,\wedge \\
\quad (c, d) \in l\_comp^{+} \bullet (b, a) \notin p\_comp^{+} \wedge (d, c) \notin l\_comp^{+} \\
\forall\, i : \textit{Glyph} \bullet (i, i) \notin p\_comp^{+} \wedge (i, i) \notin l\_comp^{+} \\
\hline
\end{array}
$$

The structure characterized by *GlyphComp* is further constrained in the configuration schema for the recursive structure given below:

___ *GlyphConfig* ____

$glyphs : \mathbb{P}\ Glyph$
$pictures : \mathbb{P}\ Picture$
$lines : \mathbb{P}\ Line$
$GlyphComp$

____

$\langle\{p : pictures \bullet p.glyph\}, \{l : lines \bullet l.glyph\}\rangle$ partition $glyphs$
$\mathrm{dom}\ p\_comp \subseteq glyphs$
$\mathrm{dom}\ l\_comp \subseteq glyphs$
$\mathrm{ran}\ p\_comp \subseteq \{p : pictures \bullet p.glyph\}$
$\mathrm{ran}\ l\_comp \subseteq \{l : lines \bullet l.glyph\}$
$\forall g1, g2 : glyphs \bullet g1.ident = g2.ident \Leftrightarrow g1 = g2$

____

Given that lines and pictures are glyphs (and that they partition the glyph type space), the relations l_comp and p_comp can be combined to form a single relation without changing the semantics of the structure. This observation led us to the simpler structure shown in Fig. 8, which, when formalized resulted in an equivalent specification.

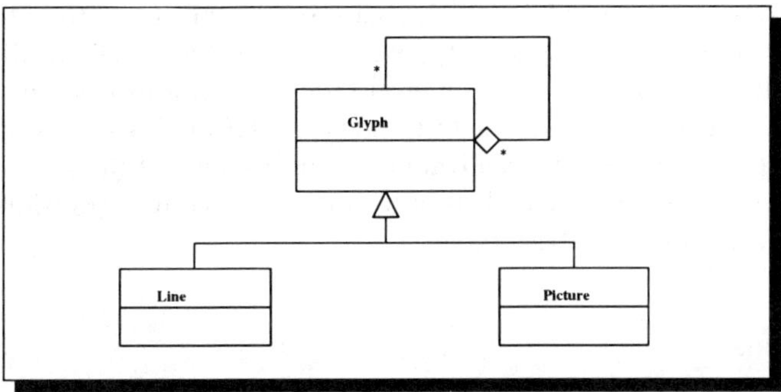

Fig. 8. An equivalent recursive aggregation structure

We also explored other multiplicity constraints with the recursive structure and showed that certain multiplicities resulted in structures that violated the desired properties when finite systems are considered. For example, if a multiplicity of 1 is placed at the whole end (requiring that all pictures and lines be components of a glyph) then the assumption that there is a finite number of glyphs at any time in the system results in a structure that violates the irreflexive property. The same is true if the multiplicity at any of the component ends is 1 or more. In fact, the multiplicity shown is the only one possible for a finite system that satisfies the properties.

5   Conclusions

This paper uses rules resulting from our ongoing research on transforming UML Type Diagrams to Z specifications. We formalized variations of non-trivial structures involving combinations of generalization and aggregation, and recursive composition. By formalizing the required structural constraints, we were able to gain a deeper understanding of the non-trivial structures. Seemingly complex structures turned out to be semantically straightforward, and the insights gained through formalization can lead to the identification of simpler, semantically equivalent models.

Experience gained exercising the transformation rules on non-trivial UML type structures indicates that there are alternative styles of formalization. Some alternatives that appear straightforward can yield very complex formalizations when applied to non-trivial structures, while others that initially appear more complex can produce clear formalizations when applied to complex structures. We plan to compare the different styles in the future and suggest criteria for selecting the most appropriate. Currently we are exercising the styles on larger case studies and extending the transformation rules to other Class Diagram constructs. We plan to gradually cover the entire UML notation.

The process of transforming the UML models to Z formal specifications is a very tedious one when done by hand, making tool support is a necessity. The transformation rules we are developing can and should be automated. Our experience developing FuZE, a prototype CASE tool that mechanically generates Z specification from Fusion analysis-level models, indicates that this can be accomplished. Our prototype successfully integrated Z analysis tools into a Fusion development environment, to facilitate rigorous analysis of the generated formal models. We plan to build a CASE tool to support UML to Z transformation and seamlessly integrate Z analysis tools.

References

1. B. W. Bates, J.-M. Bruel, R. B. France, and M. M. Larrondo-Petrie. Formalizing Fusion Object-Oriented Analysis Models. In E. Najm and J.-B. Stephani, editors, *Proceedings of the First IFIP International Workshop on Formal Methods for Open Object-based Distributed Systems,* Paris, France. Chapman & Hall, London, UK, 4–6 Mar. 1996.

2. G. Booch.   *Object-Oriented Analysis and Design with Applications.*   Benjamin/Cummings, Menlo Park, CA, Second edition, 1994.

3. G. Booch, J. Rumbaugh, and I. Jacobson. Unified Modeling Language. Version 1.0, Rational Software Corporation, Santa Clara, CA-95051, USA, Jan. 1997.

4. J. P. Bowen and J. A. Hall, editors. *Z User Workshop, Cambridge 1994,* Workshops in Computing. Springer-Verlag, New York, 1994.

5. J.-M. Bruel, R. B. France, and A. Benzekri. A Z-based Approach to Specifying and Analyzing Complex Systems. In *Proceedings of the Second IEEE International Conference on Engineering of Complex Computer Systems (ICECCS'96),* Montreal, Canada, 21–25 Oct. 1996.

6. J.-M. Bruel, R. B. France, B. Chintapally, and G. K. Raghavan. A Tool for Rigorous Analysis of Object Models. In *Proceedings of the 20th International Conference on Technology of Object-Oriented Languages and Systems (TOOLS'96)*, Santa Barbara, California, July 29–August 2 1996.

7. J.-M. Bruel, R. B. France, M. M. Larrondo-Petrie, B. Chintapally, and G. K. Raghavan. CASE-based Rigorous Object-Oriented Modeling. In *Proceedings of the Northern Formal Methods Workshop,* Bradford, UK, 23–24 Sept. 1996.

8. D. Coleman, P. Arnold, S. Bodoff, C. Dollin, H. Gilchrist, F. Hayes, and P. Jeremaes. *Object-Oriented Development: The Fusion Method.* Prentice Hall, Englewood Cliffs, NJ, Object-Oriented Series edition, 1994.

9. S. Cook and J. Daniels. Let's get formal. *Journal of Object-Oriented Programming (JOOP)*, pages 22–24 and 64–66, July–Aug. 1994.

10. R. Duke, P. King, G. A. Rose, and G. Smith. The Object-Z specification language. In T. D. Korson, V. K. Vaishnavi, and B. Meyer, editors, *Technology of Object-Oriented Languages and Systems: TOOLS 5*, pages 465–483. Prentice Hall, 1991.

11. R. B. France, J.-M. Bruel, and M. M. Larrondo-Petrie. An Integrated Object-Oriented and Formal Modeling Environment. *To appear in the Journal of Object-Oriented Programming (JOOP)*, 1997.

12. J. A. Hall. Specifying and Interpreting Class Hierarchies in Z. In Bowen and Hall [4], pages 120–138.

13. J. A. R. Hammond. Producing Z Specifications from Object-Oriented Analysis. In Bowen and Hall [4], pages 316–336.

14. I. Jacobson, M. Christerson, P. Jonsson, and G. Overgaard. *Object-Oriented Software Engineering–A Use Case Driven Approach.* Addison-Wesley, Reading, MA, 1992.

15. X. Jia. *An Approach to Animating Z Specifications.* Division of Software Engineering, School of Computer Science, Telecommunication, and Information Systems, DePaul University, Chicago, IL 60604, USA, 1995. ZANS is a Z animator available on Internet via anonymous ftp at `ise.cs.depaul.edu`.

16. X. Jia. *ZTC: A Z Type Checker, User's Guide, version 2.01.* Division of Software Engineering, School of Computer Science, Telecommunication, and Information Systems, DePaul University, Chicago, IL 60604, USA, May 1995. The ZTC tool and documentation are available on Internet via anonymous ftp at `ise.cs.depaul.edu`.

17. J. Rumbaugh, M. R. Blaha, W. Premerlani, F. Eddy, and W. Lorensen. *Object-Oriented Modeling and Design.* Prentice Hall, Englewood Cliffs, NJ, 1991.

18. M. Shroff and R. B. France. Towards a Formalization of UML Class Structures in Z. FAU Technical Report TR-CSE-97-4, Department of Computer Science & Engineering, Florida Atlantic University, Boca Raton, FL–33431, USA, Jan. 1997.

19. J. M. Spivey. *The Z Notation: A Reference Manual.* Prentice Hall, Englewood Cliffs, NJ, Second edition, 1992.

20. P. A. Swatman and P. M. C. Swatman. FOOM - A Method for Acquiring Appropriate Inter-Organisational Systems. In *Proceedings of the 8^{t}h International EDI-IOS Conference,* Bled, Slovenia, 5–8 June 1995.

Compilation as Refinement

Karl Lermer   Colin Fidge

Software Verification Research Centre
School of Information Technology,
The University of Queensland. Australia. 4072.

Abstract. Program refinement usually translates an abstract specification to a high-level language program. However, this process can be taken further by refining a high-level language 'specification' to an assembler code 'implementation'. It is shown how this can be done in the familiar refinement calculus framework. Several derived refinement rules for modelling program compilation are presented.
Keywords: Program refinement; compilation; action systems

1   Introduction

Compilation of high-level language programs to assembler code is among the oldest and most well-explored technologies in computer programming. Nevertheless, stories of production compilers containing bugs abound! Often this is merely an annoyance, but in *safety-critical* applications the danger of unknown compilation errors is unacceptable.

One solution to this is to develop a verified, trustworthy compilation strategy for a simplified programming language. Such a strategy can then be used as a basis for either (directly) proving the correctness of object code against its specification, or (indirectly) proving correctness by verifying the compiler.

In this report we show how such a strategy can be presented as a program 'refinement', with the 'specification' being a high-level language program, and the target 'implementation' being assembler code. To do this we extend the modelling language underlying the refinement calculus, so that it can represent assembler code behaviour, and then develop the compilation strategy as a number of derived refinement rules.

2   Previous Work

The idea of modelling program compilation as a formal development process has surfaced several times [17, 16, 24, 12, 13, 8], but has proven to be significantly challenging. Notable large-scale attempts include the U.K.-based **safemos** [9] and European ProCoS [15] projects, which developed elaborate, multi-layered formalisms for modelling program compilation. Despite making considerable progress, the scale of the problem was so great that the **safemos** group concluded, "in the short term, it seems unlikely that a production compiler will ever be completely verified" [9, p. 146].

Mindful of these experiences, we are keen to pursue very simple models, applicable to trusted safety-critical language subsets. For instance, Norvell [23, 24] showed how an elegant compilation strategy for a small programming language could be derived using formal models of the source and target programming languages. His assembler code model was complicated, however, by the need for assembler instructions to be 'interpreted' by an imaginary abstract machine. Hale [13, §7.2.2], Fränzle and Müller-Olm [12, p. 303], Hoare [16, §4] and He Jifeng [15, §6.2] adopted similar models, with the semantics of machine code instructions given by their interpretation.

This need for an interpreter stems from the potentially unstructured control flow of assembler programs—such code cannot be easily represented in the structured guarded command language used by the refinement calculus. Unfortunately the presence of this interpreter introduces a significant paradigm shift during attempts to model compilation as refinement—the source and target languages are represented in very different ways. Nevertheless, Back [2] demonstrated through a case study that his 'action system' guarded command language subset *is* capable of representing assembler-like programs. One of the motivations of the current study, therefore, is to apply Back's approach to the compilation problem, in order to develop a simpler, more uniform way of representing compilation as refinement.

3 Language Models

Our derived refinement rules are expressed in the familiar refinement calculus of Back and Morgan. We use guarded command language, augmented with specification statements, as our general modelling language. Both our source 'high-level' and target 'assembler' languages are then merely distinguished subsets.

3.1 Modelling language

As usual in the refinement calculus, our wide-spectrum modelling language consists of Dijkstra's guarded command language, augmented with specification statements [21]. Let S be a statement in the language; v a variable name; T a type; c a constant name; X a (statically evaluable) expression; P and Q predicates; E a (dynamically evaluable) expression; B a (dynamically evaluable) boolean expression; i an indexing variable.

$$
\begin{array}{lll}
S & ::= & |[\, \text{var } v : T \bullet S \,]| & \text{– (logical) variable block} \\
  & | & |[\, \text{con } c : T = X \bullet S \,]| & \text{– (logical) constant block} \\
  & | & \text{skip} & \text{– null statement} \\
  & | & v := E & \text{– assignment} \\
  & | & S_1 \,;\, S_2 & \text{– sequential composition} \\
  & | & \text{if}\,(\,[\!]\,i \bullet B_i \to S_i)\,\text{fi} & \text{– conditional composition} \\
  & | & \text{do}\,(\,[\!]\,i \bullet B_i \to S_i)\,\text{od} & \text{– iterative composition} \\
  & | & \{P\} & \text{– assertion} \\
  & | & [Q] & \text{– coercion} \\
  & | & v:[P\,,\,Q] & \text{– specification}
\end{array}
$$

Generalisation of variable and constant blocks to multiple declarations, and assignments to multiple assignments, is assumed implicitly. Typing and initialisation of constants are both optional. Scoping brackets $|[\cdots]|$ can be omitted when programs are displayed vertically [21, pp. 55–6].

We make use of the following well-known refinement rules for manipulating this language.

R1 Strengthen postcondition [20, Law 1.1]. If $Q' \Rrightarrow Q$, then

$$v \colon [P , Q] \sqsubseteq v \colon [P , Q'] \,.$$

R2 Introduce (initialised) local variable [20, Law 6.1; Law 8.5]. If x does not appear in v, P, and Q then

$$v \colon [P , Q] \sqsubseteq |[\, \mathsf{var}\ x : T \bullet x := E \, ; v, x \colon [P , Q]\,]| \,.$$

R3 Introduce logical constant [20, Law 6.2]. If $P \Rrightarrow (\exists\, c : T \bullet P')$ and c does not appear in v, P, Q then

$$v \colon [P , Q] \sqsubseteq |[\, \mathsf{con}\ c : T \bullet v \colon [P' , Q]\,]| \,.$$

To support specification and refinement of parallel programs, Back identified a guarded command language subset capable of representing non-sequential behaviours, the *action system* [6, §2.1][6, §5]. In essence an action system $\mathcal{A}$ is merely a variable block containing an initialised loop, where each alternative in the loop is an 'action' [6, §3.1].

$$
\begin{aligned}
\mathcal{A} = \ &\mathsf{var}\ v : T \bullet \\
&v := E \, ; \\
&\mathsf{do} \\
&\qquad (\,[\!]\, i \bullet B_i \to S_i) \\
&\mathsf{od}
\end{aligned}
$$

This construct is of interest because, although such a loop normally defines sequential behaviour, it can be *interpreted* as representing a parallel system specification. No ordering is imposed on *independent* actions, so an action system description can be used to define an interleaving-semantics representation of concurrent behaviour [2, §3].

Several program transformation rules for manipulating entire action system constructs are defined. Their provisos are somewhat complex, requiring reasoning about how actions 'enable' or 'disable' one another. We rely on the following rules in this paper, but refer the interested reader to the appropriate references for more detail.

R4 Making a loop directly [1, Rule 1]. If $Q \Rightarrow B$ and action $B \to S$ disables itself, then

$$\{Q\} \, ; S \sqsubseteq \mathsf{do}\ B \to S\ \mathsf{od} \,.$$

R5 Changing a sequence to a loop [1, Rule 2]. If $Q \Rightarrow B_1$; action $B_i \rightarrow S_i$ enables action $B_{i+1} \rightarrow S_{i+1}$, for $i \in 1 \mathinner{\ldotp\ldotp} n - 1$; action $B_n \rightarrow S_n$ establishes $\neg (\bigvee i : 1 \mathinner{\ldotp\ldotp} n \bullet B_i)$; and each action $B_i \rightarrow S_i$ excludes $B_j \rightarrow S_j$, where $i \neq j$; then

$$\{Q\} \, ; \, S_1 \, ; \, \ldots \, ; \, S_n \sqsubseteq \mathsf{do}\,(\llbracket i : 1 \mathinner{\ldotp\ldotp} n \bullet B_i \rightarrow S_i)\,\mathsf{od}\,.$$

R6 Merging nested loops. Let I and J be disjoint indexing sets, with $i \in I$ and $j \in J$. If $Q \Rightarrow \neg (\bigvee i : I \bullet B_i)$; no action $B_j \rightarrow S_j$ can enable or disable any action $B_i \rightarrow S_i$; action $B_0 \rightarrow S_0$ is excluded by every action $B_i \rightarrow S_i$; and each $B_j \rightarrow S_j$ is excluded by every $B_i \rightarrow S_i$; then

$$
\begin{aligned}
&\{Q\} \, ; \, \mathsf{do}\; B_0 \rightarrow S_0 \, ; \, \mathsf{do} \\
&\qquad\qquad\qquad (\llbracket i : I \bullet B_i \rightarrow S_i) \\
&\qquad\qquad \mathsf{od} \\
&\qquad \llbracket \;\; (\llbracket j : J \bullet B_j \rightarrow S_j) \\
&\quad\; \mathsf{od} \\
&\sqsubseteq \mathsf{do}\; B_0 \rightarrow S_0 \\
&\qquad \llbracket \;\; (\llbracket i : I \bullet B_i \rightarrow S_i) \\
&\qquad \llbracket \;\; (\llbracket j : J \bullet B_j \rightarrow S_j) \\
&\quad\; \mathsf{od}\,.
\end{aligned}
$$

This is a more restrictive version of Back and Sere's rule [1, Rule 6]. The last proviso is stronger than theirs, but suffices for our application.

R7 Introduce stuttering actions. Let I and J be disjoint indexing sets, with $i \in I$ and $j \in J$. If there is an abstraction relation R linking existing variables and new variables v such that R is established by the initial assignment $v := E$; each new action $B_i' \rightarrow S_i'$ has the same effect as an old action $B_i \rightarrow S_i$ on old variables and re-establishes R; each auxiliary action $B_j \rightarrow S_j$ has no effect on old variables and re-establishes R; repeatedly executing auxiliary actions $B_j \rightarrow S_j$ will eventually lead to termination; and whenever R and some B_i is true then some new B_i' or auxiliary B_j guard is true, then

$$
\begin{aligned}
\mathsf{do}\,(\llbracket i : I \bullet B_i \rightarrow S_i)\,\mathsf{od} \sqsubseteq\; &\mathsf{var}\; v : T \; \bullet \\
&v := E \, ; \\
&\mathsf{do}\,(\llbracket i : I \bullet B_i' \rightarrow S_i') \\
&\llbracket \;\; (\llbracket j : J \bullet B_j \rightarrow S_j) \\
&\mathsf{od}\,.
\end{aligned}
$$

Refer to Back's general rule for more detail [2, p. 7][6, §4.1]. Since we are interested in refining loops, rather than action systems per se, we have omitted the distinction between global and 'old' local variables.

3.2 High-level language

Our source language is a generic sequential high-level language (HLL), featuring the usual structured programming statements. Let H be a statement in the high-level language.

$$
\begin{array}{llll}
H & ::= & \textbf{var } v : T \textbf{ begin } H \textbf{ end} & \text{– variable declaration} \\
& | & \textbf{con } c = X \textbf{ begin } H \textbf{ end} & \text{– constant declaration} \\
& | & v := E & \text{– assignment} \\
& | & H_1 \; ; \; H_2 & \text{– sequence} \\
& | & \textbf{if } B \textbf{ then } H \textbf{ end} & \text{– condition} \\
& | & \textbf{if } B \textbf{ then } H_1 \textbf{ else } H_2 \textbf{ end} & \text{– choice} \\
& | & \textbf{while } B \textbf{ do } H \textbf{ end} & \text{– iteration}
\end{array}
$$

Semantically, however, these constructs merely denote a distinguished subset of our modelling language. The following table shows the equivalences.

$$
\begin{aligned}
\textbf{var } v : T \textbf{ begin } H \textbf{ end} &= |[\, \textsf{var } v : T \bullet H \,]| \\
\textbf{con } c = X \textbf{ begin } H \textbf{ end} &= |[\, \textsf{con } c = X \bullet H \,]| \\
v := E &= v := E \\
H_1 \; ; \; H_2 &= H_1 \; ; \; H_2 \\
\textbf{if } B \textbf{ then } H \textbf{ end} &= \textsf{if } B \rightarrow H \textsf{ fi} \\
\textbf{if } B \textbf{ then } H_1 \textbf{ else } H_2 \textbf{ end} &= \textsf{if } B \rightarrow H_1 \, [\!] \, \neg B \rightarrow H_2 \textsf{ fi} \\
\textbf{while } B \textbf{ do } H \textbf{ end} &= \textsf{do } B \rightarrow H \textsf{ od}
\end{aligned}
$$

3.3   Machine-level language

Our target, machine-level language is also a distinguished subset of the modelling notation.

The target language includes the following primitive statements. Let M be an instruction in our target machine language; reg a register; c a (statically evaluable) constant; loc the address of a memory location.

$$
\begin{array}{llll}
M & ::= & \textbf{load } reg, loc & \text{– load register } reg \text{ from memory location } loc \\
& | & \textbf{loadi } reg, c & \text{– load register } reg \text{ with constant } c \\
& | & \textbf{store } loc, reg & \text{– store memory location } loc \text{ from register } reg \\
& | & \textbf{add } reg_1, reg_2 & \text{– store } reg_2 + reg_1 \text{ in } reg_1 \\
& | & \textbf{addi } reg, i & \text{– store } reg + i \text{ in } reg, \text{ for } i \in \mathbb{N} \\
& | & \textbf{jump } loc & \text{– jump to location } loc \\
& | & \textbf{bgt } reg_1, reg_2, loc & \text{– branch to location } loc \text{ if } reg_1 > reg_2 \\
& | & \textbf{blez } reg, loc & \text{– branch to location } loc \text{ if } reg \leqslant 0
\end{array}
$$

To be as flexible as possible we assume that instruction addresses range over some set Σ, inductively ordered by relation $\prec$, with successor function $succ_\Sigma$. See Section 4.1.

$$
succ_\Sigma(s) = \begin{cases} min\{t : \Sigma \bullet s \prec t\}, & \exists\, t : \Sigma \bullet s \prec t \\ s\,, & \text{otherwise} \end{cases}
$$

Following the approach of Norvell [23, §6.1], Hale [13, p. 133] and He Jifeng [15, §6.1.2], we model the effect of executing *individual* instructions on the state of the processor via multiple assignment statements. We assume a simple target processor whose behaviour is observable through a program counter pc, several temporary registers reg_1 to reg_8, and a location-indexed array of memory values mem. The instructions above are thus equivalent to the following statements in our modelling language.

$$\begin{aligned}
\textbf{load } reg, loc &= pc, reg := succ_\Sigma(pc), mem(loc) \\
\textbf{loadi } reg, c &= pc, reg := succ_\Sigma(pc), c \\
\textbf{store } loc, reg &= pc, mem(loc) := succ_\Sigma(pc), reg \\
\textbf{add } reg_1, reg_2 &= pc, reg_1 := succ_\Sigma(pc), reg_1 + reg_2 \\
\textbf{addi } reg_1, i &= pc, reg := succ_\Sigma(pc), reg + i \\
\textbf{jump } loc &= pc := loc \\
\textbf{bgt } reg_1, reg_2, loc &= \textsf{if } reg_1 > reg_2 \rightarrow pc := loc \\
&\quad [\!]\ reg_1 \leqslant reg_2 \rightarrow pc := succ_\Sigma(pc) \\
&\quad \textsf{fi} \\
\textbf{blez } reg, loc &= \textsf{if } reg \leqslant 0 \rightarrow pc := loc \\
&\quad [\!]\ reg > 0 \rightarrow pc := succ_\Sigma(pc) \\
&\quad \textsf{fi}
\end{aligned}$$

Unfortunately, defining the effect of several machine-level instructions is harder. A 'sequence' of assembler instructions is not necessarily executed sequentially, due to the presence of jump and branch instructions. In essence, machine code is not bound by the principles of structured programming, and is thus awkward to model in the guarded command language. Furthermore, we need to consider *where* the instructions reside in memory to know when they will be executed [23, §6.3][15, §6.2]!

Although action systems were originally devised for modelling parallel systems, Back showed that their usefulness extends to other non-sequential behaviours. In particular, in a brief case study illustrating refinement of a mutual exclusion algorithm [2, §7], he showed how execution of assembler-like primitive actions could be modelled using an action system controlled by a program counter variable. We adopt this model here. Let $\langle M_1 ; \ldots ; M_n \rangle @\, loc$ denote a list of consecutive instructions M_1 to M_n, starting at memory location $loc \in \Sigma$. Then this is equivalent to the following iterative statement, within the scope of appropriate declarations for pc, reg_1 to reg_8 and mem. Let $succ_\Sigma^0 = id$ and $succ_\Sigma^{i+1} = succ_\Sigma \circ succ_\Sigma^i$.

$$\langle M_1 ; \ldots ; M_n \rangle @\, loc = \textsf{do}$$
$$([\!]\, i : 1 \ldots n \bullet pc = succ_\Sigma^{i-1}(loc) \rightarrow M_i)$$
$$\textsf{od}$$

In effect, we use the **do**...**od** statement as the 'machine' to interpret our assembler-level instructions.

This simple approach proves to be surprisingly effective. It allows the instructions to occur in any order, as controlled by the program counter pc. The first action executed depends on the initial value of pc when the whole construct begins. As long as pc is in the range $loc \ldots succ_\Sigma(loc)^{n-1}$ the appropriate instruction is selected and executed. The construct terminates as soon as pc leaves this range. The model is even robust enough to correctly represent pathological examples: an instruction that jumps to itself defines an infinite loop, as we would expect:

$$\langle \mathbf{jump}\ loc \rangle\ @\ loc\ =\ \mathsf{do}\ pc = loc\ \rightarrow\ pc := loc\ \mathsf{od}\,.$$

4 Derived Refinement Rules for Compilation

In this section we derive several refinement rules, representing simple compilation strategies for our high-level language statements. Each rule has the form

$$\{pc = \alpha\}\,;\,P\,;\,[pc = \omega] \sqsubseteq \mathsf{con}\ \Sigma = \{\alpha, \nu_\omega^\alpha.2, \ldots, \nu_\omega^\alpha.n, \omega\}\ \bullet$$
$$\mathsf{do}\ pc = \alpha \rightarrow P_1$$
$$\|\quad pc = \nu_\omega^\alpha.2 \rightarrow P_2$$
$$\vdots$$
$$\|\quad pc = \nu_\omega^\alpha.n \rightarrow P_n$$
$$\mathsf{od}$$

where α and ω are instruction address constants. For the assembler code generated corresponding to program P, α denotes its starting address, and ω its 'exit' address. Constant Σ is the set of instruction addresses over which this particular program fragment ranges. As explained in Section 4.1, these addresses obey a well-defined ordering,

$$\alpha \preccurlyeq \nu_\omega^\alpha.2 \preccurlyeq \cdots \preccurlyeq \nu_\omega^\alpha.n \preccurlyeq \omega\,.$$

Thus each refinement rule divides the program into its components P_i, instantiates them with different program counter values (see below), and encloses the resulting program in an execution loop.

Program 'compilation' then consists of repeated application of such rules. This results in a model consisting of nested loops, which are then unfolded and optimised to obtain the final assembler program.

4.1 Address management

Unique addresses must be introduced for each assembler instruction generated, but this proves to be quite awkward without knowing in advance how many instructions each HLL statement will produce. Previous formal compilation approaches have represented addresses via natural numbers, with symbolic constants or expressions representing as-yet-unknown values [23]. A 'second pass' is then required to instantiate these values once the compilation process is complete. Here we avoid this by adopting a hierarchical numbering system that generates unique addresses as our refinement rules 'parse' the HLL code.

Addresses $\mathbb{A}$ are represented as values of type

$$\mathbb{A} \stackrel{\mathrm{def}}{=} \mathsf{seq}\ \mathbb{N}\,,$$

over which is defined a well-ordering $\preccurlyeq$, and successor function $succ_\Sigma$ for a given inductively ordered set of addresses Σ. For conciseness when using sequences of natural numbers as addresses, we omit the sequence brackets $\langle \cdots \rangle$, and separate elements by '.', i.e., "$\langle 1, 2, 1 \rangle$" is written "1.2.1", and so on.

Note that $\mathbb{A}$ bears a natural well-ordering,

$$a \preccurlyeq b :\Leftrightarrow M \in \mathrm{dom}(a) \Rightarrow (M \in \mathrm{dom}(b) \wedge a(M) < b(M)),$$

where $M = max\{i \in \mathbb{N} \bullet \forall j < i \bullet j \in \mathrm{dom}(a) \cap \mathrm{dom}(b) \wedge a(j) = b(j)\}$. For example $1.1 \preccurlyeq 2$ and $2.15.4 \preccurlyeq 2.16.1.1$. Strict ordering is defined easily:

$$a \prec b :\Leftrightarrow a \preccurlyeq b \wedge a \neq b.$$

Starting from two values $a, b \in \mathbb{A}$, the *new address* function

$$\nu_b^a = \begin{cases} a, & \text{if } \#a \geqslant \#b \\ front\, b \frown \langle 1 \rangle, & \text{otherwise} \end{cases},$$

is used to construct addresses. For instance:

$$\nu_2^1 = 1$$
$$\nu_{2.3}^1 = 2.1.$$

Note that $\#\nu_b^a = max\{\#a, \#b\}$. Furthermore, for $a \neq b$ and $n \in \mathbb{N}$, all elements of $\{a, \nu_b^a \frown \langle 1 \rangle, \nu_b^a \frown \langle 2 \rangle, \ldots, \nu_b^a \frown \langle n \rangle, b\}$ are pairwise distinct.

Significantly, given two addresses $\alpha, \omega \in \mathbb{A}$, such that $\alpha \prec \omega$, this function can always introduce intermediate addresses 'between' these values, as new assembler instructions are generated.

Each of the following refinement rules begins with a program fragment of the form $\{pc = \alpha\}\,;\mathcal{A}\,;[pc = \omega]$ with entry value α and exit value ω where we assume the following properties,

$$\alpha \prec \omega, \tag{1}$$
$$last\, \omega \geqslant 2. \tag{2}$$

With the notational convention,

$$\nu_b^a.n = \begin{cases} \nu_b^a \frown \langle n \rangle, & n > 1 \\ a, & n = 1 \end{cases},$$

we state our method for computing fresh instruction addresses between α and ω.

Lemma 1. *If α and ω satisfy (1) and (2), then, for any $i \geqslant 2$,*

$$\alpha \prec \nu_\omega^\alpha.i \prec \nu_\omega^\alpha.i + 1 \prec \omega.$$

Proof. Consider the following two cases.

1. $\#\alpha \geqslant \#\omega \Rightarrow \nu_\omega^\alpha = \alpha \Rightarrow \alpha \prec \alpha.i \prec \alpha.i + 1 \prec \omega$
2. $\#\alpha < \#\omega \Rightarrow \nu_\omega^\alpha = (front\, \omega).1 \Rightarrow \nu_\omega^\alpha.i \prec \omega$, because of property (2).

Combined with property (1), we get $\alpha \prec \nu_\omega^\alpha$, as required. $\qquad\qquad\square$

4.2   Introduce assembler-level variables

Compilation of an entire HLL specification $\mathcal{A}$ begins by introducing those variables needed for the assembler model, and enclosing the whole system within an execution loop [2]. The most prominent of these variables are the program counter pc and the local memory mem, which we model as a function from addresses to integers.

C1 Compile program.

$$
\begin{aligned}
\mathcal{A} \sqsubseteq\ & \text{var } pc : \mathbb{A};\ mem : \mathbb{A} \to \mathbb{Z} \bullet \\
& \quad \text{con } \Sigma = \{1,2\} \bullet \\
& \quad pc := 1\ ; \\
& \quad \text{do} \\
& \qquad\qquad pc = 1 \to \mathcal{A}\ ; [pc = 2] \\
& \quad \text{od}
\end{aligned}
$$

The initial assignment establishes the starting address for the program (and allows us to readily introduce assertion $\{pc = 1\}$ before program $\mathcal{A}$). The coercion $[pc = 2]$ ensures that the right-hand side program terminates whenever $\mathcal{A}$ does. The set $\Sigma \subseteq \mathbb{A}$ defines the domain of pc for this loop.

Proof.

$$
\begin{aligned}
\mathcal{A} \sqsubseteq\ & \text{``by } \mathbf{R2}\text{''} \\
& \quad \text{var } pc : \mathbb{A};\ mem : \mathbb{A} \to \mathbb{Z} \bullet \\
& \quad pc := 1\ ; \mathcal{A} \\
\sqsubseteq\ & \text{``by } \mathbf{R3}\text{''} \\
& \quad \text{var } pc : \mathbb{A};\ mem : \mathbb{A} \to \mathbb{Z} \bullet \text{con } \Sigma = \{1,2\} \bullet \\
& \quad pc := 1\ ; \mathcal{A} \\
\sqsubseteq\ & \text{``by } \mathbf{R1}\text{''} \\
& \quad \text{var } pc : \mathbb{A};\ mem : \mathbb{A} \to \mathbb{Z} \bullet \text{con } \Sigma = \{1,2\} \bullet \\
& \quad pc := 1\ ; \mathcal{A}\ ; [pc = 2] \\
\sqsubseteq\ & \text{``by } \mathbf{R4}\text{''} \\
& \quad \text{var } pc : \mathbb{A};\ mem : \mathbb{A} \to \mathbb{Z} \bullet \text{con } \Sigma = \{1,2\} \bullet \\
& \quad pc := 1\ ; \text{do } pc = 1 \to \mathcal{A}\ ; [pc = 2] \text{ od}
\end{aligned}
$$

□

A correspondence must also be established between HLL program variables and particular *mem* locations. Registers could also be introduced at this stage but, because their role is often temporary, we use a different rule below to introduce them locally as needed.

4.3   Compile sequential composition

Let $\mathcal{A}$ be a HLL program fragment of the form

$$
\{pc = \alpha\}\ ; S_1\ ; S_2\ ; [pc \in \mathcal{M}],
$$

where S_1 leaves the value of pc unspecified, α denotes an element of $\mathbb{A}$, $\mathcal{M}$ is a member of $\mathbb{P}\mathbb{A}$ and $a \notin \mathcal{M}$. Assume further that $\omega = min\{m \in \mathcal{M} \bullet m \succ a\}$ exists and α and ω satisfy obligations (1) and (2). Then the following law can be used to instantiate each component S_i with different program counter values.

C2 Compile sequence.

$$
\begin{aligned}
\mathcal{A} &\sqsubseteq \mathsf{con}\ \Sigma = \{\alpha, \nu_\omega^\alpha.2\} \cup \mathcal{M}\ \bullet \\
&\quad \mathsf{do}\ pc = \alpha \to S_1\ ; [pc = \nu_\omega^\alpha.2] \\
&\quad [\!]\quad pc = \nu_\omega^\alpha.2 \to S_2\ ; [pc \in \mathcal{M}] \\
&\quad \mathsf{od}
\end{aligned}
$$

Note that $a \prec \nu_\omega^\alpha.2 \prec \omega$ and, α and $\nu_\omega^\alpha.2$ define the starting addresses for program components S_1 and S_2, respectively. Addresses x, such that $\alpha \preccurlyeq x \prec \nu_\omega^\alpha.2$ are available for instructions implementing S_1, and addresses y, such that $\nu_\omega^\alpha.2 \preccurlyeq y \prec \omega$, are available for instructions implementing S_2. Obligations (1) and (2) are maintained and each component S_i can now be refined independently.

Proof. To prove this law we must change a sequence of actions into a loop [1].

$$
\begin{aligned}
\mathcal{A} &\sqsubseteq \text{``by } \mathbf{R3}\text{''} \\
&\quad \mathsf{con}\ \Sigma = \{\alpha, \nu_\omega^\alpha.2\} \cup \mathcal{M}\ \bullet \\
&\quad \{pc = \alpha\}\ ; S_1\ ; S_2\ ; [pc \in \mathcal{M}] \\
&\sqsubseteq \text{``by } \mathbf{R1}\text{''} \\
&\quad \mathsf{con}\ \Sigma = \{\alpha, \nu_\omega^\alpha.2\} \cup \mathcal{M}\ \bullet \\
&\quad \{pc = \alpha\}\ ; S_1\ ; [pc = \nu_\omega^\alpha.2]\ ; S_2\ ; [pc \in \mathcal{M}] \\
&\sqsubseteq \text{``by } \mathbf{R5}\text{''} \\
&\quad \mathsf{con}\ \Sigma = \{\alpha, \nu_\omega^\alpha.2\} \cup \mathcal{M}\ \bullet \\
&\quad \{pc = \alpha\}\ ; \\
&\quad \mathsf{do}\ pc = \alpha \to S_1\ ; [pc = \nu_\omega^\alpha.2] \\
&\quad [\!]\quad pc = \nu_\omega^\alpha.2 \to S_2\ ; [pc \in \mathcal{M}] \\
&\quad \mathsf{od}
\end{aligned}
$$

$\square$

Rule **C2** allows the final pc address to be a value in set $\mathcal{M}$, rather than just a single exit value, even though any sequential HLL construct always has a single exit point. This generalisation was introduced so that the rule may also be used for manipulating assembler-level specifications, which include branch instructions with more than one possible final pc value. See Section 4.8.

4.4   Compile iteration

We present two refinement rules for iteration, representing two different compilation strategies. Let HLL specification $\mathcal{A}$ be of the form

$$\{pc = \alpha\}\ ; \mathbf{while}\ B\ \mathbf{do}\ H\ \mathbf{end}\ ; [pc = \omega]\,,$$

where pc does not occur in B or H, and α and ω are defined in the enclosing program frame and denote different elements of $\mathbb{A}$ which satisfy obligations (1) and (2) above.

The first rule places evaluation of the loop expression before the loop body.

C3 Compile iteration 1.

$$\mathcal{A} \sqsubseteq \mathbf{con}\ \Sigma = \{\alpha, \nu_\omega^\alpha.2, \nu_\omega^\alpha.3, \omega\} \bullet$$
$$\mathbf{do}\ pc = \alpha \to \mathbf{if}\ B \to pc := succ_\Sigma(pc)$$
$$[\!]\ \neg B \to pc := \omega$$
$$\mathbf{fi}$$
$$[\!]\quad pc = \nu_\omega^\alpha.2 \to H\ ; [pc = \nu_\omega^\alpha.3]$$
$$[\!]\quad pc = \nu_\omega^\alpha.3 \to \mathbf{jump}\ \alpha$$
$$\mathbf{od}$$

The resulting specification consists of three actions. The first is effectively a branch instruction to location ω if B is false (although we cannot directly represent it as such until code is generated for evaluating expression B—see Section 4.8). The second is the loop body. The third is an unconditional jump back to re-evaluate the loop condition, which occurs each time the loop body ends.

The new addresses introduced all satisfy obligations (1) and (2). For the branch 'instruction' we have entry address α and for exit address set $\mathcal{M} = \{\nu_\omega^\alpha.2, \omega\}$ with $min\{m \in \mathcal{M} \bullet m \succ \alpha\} = \nu_\omega^\alpha.2$.

Proof. We use the stuttering refinement rule to prove the refinement in one step. For brevity let c_1 to c_4 denote the distinct address constants.

$$\{pc = c_1\}\ ; \mathbf{do}\ B \to H\ \mathbf{od}$$
$$\sqsubseteq\ \text{``by } \mathbf{R7}\text{''}$$
$$\mathbf{con}\ \ c_2, c_3, c_4 : \mathbb{A} \bullet$$
$$\{pc = c_1\}\ ;$$
$$\mathbf{do}\ pc = c_1 \to \mathbf{if}\ B \to pc := c_2$$
$$[\!]\ \neg B \to pc := c_4$$
$$\mathbf{fi}$$
$$[\!]\quad pc = c_2 \to H\ ; pc := c_3$$
$$[\!]\quad pc = c_3 \to pc := c_1$$
$$\mathbf{od}$$

Here the second action, containing loop body H, is the 'main action' and the first and third actions are the new 'auxiliary actions'.

Firstly we identify a suitable abstraction relation to link the old and new program counters pc and pc', and the other observable variables.

$$R \stackrel{\text{def}}{=} (pc = c_1 \wedge ((pc' = c_1) \vee (pc' = c_2 \wedge B) \vee (pc' = c_3)))$$
$$\vee\ (pc' = c_4 \wedge \neg B)$$

In this case the program counter is the 'local' variable, and all other variables are 'global'.

In order to apply the data refinement rule we have to verify the following obligations [2].

Initialisation The abstraction relation holds initially given assertions $\{pc = c_1\}$ and $\{pc' = c_1\}$.

Main action To show that the main action is currectly refined under relation R we must prove

$$B \rightarrow H \sqsubseteq_R pc' = c_2 \rightarrow H \; ; pc' := c_3 \, .$$

1. The action guard is correctly refined under relation R because $R \wedge pc' = c_2 \Rightarrow B$.

2. To show that the statement part is correctly refined we must demonstrate that the action bodies refine under R. This is true because when $R \wedge pc' = c_2$ is true then H has the same effect on the observables as $H \wedge pc' = c_3$ and R still holds. The effect is the same and the invariant is preserved.

Exit condition $R \wedge B \Rightarrow pc = c_1 \wedge ((pc' = c_1) \vee (pc' = c_2 \wedge B) \vee (pc' = c_3)))$ and so either the main action or one of the auxiliary actions is enabled.

Auxiliary actions For each new action, we must show it is a refinement of skip (on the observables) with respect to R.

$$\mathsf{skip} \sqsubseteq_R pc' = c_2 \rightarrow \mathsf{if}\ B \rightarrow pc' := c_2 \ [\!]\ \neg B \rightarrow pc' := c_4\ \mathsf{fi}$$
$$\mathsf{skip} \sqsubseteq_R pc' = c_3 \rightarrow pc' := c_1$$

In both cases there is no change to global variables, so the invariant is preserved.

Internal convergence Executing only auxiliary actions will eventually terminate because they leave $pc' \in \{c_2, c_4\}$ which disables both actions. $\qquad\square$

The second possible compilation strategy for loops puts evaluation of the loop expression after the loop body. This introduces an initial jump before the expression is evaluated, but avoids the need for the final jump back to the 'top' at every iteration.

C4 Compile iteration 2.

$$
\begin{aligned}
\mathcal{A} \sqsubseteq\ &\mathsf{con}\ \varSigma = \{\alpha, \nu_\omega^\alpha.2, \nu_\omega^\alpha.3, \omega\} \bullet \\
&\mathsf{do}\ pc = \alpha \rightarrow \mathbf{jump}\ \nu_\omega^\alpha.3 \\
&\quad [\!]\quad pc = \nu_\omega^\alpha.2 \rightarrow H \; ; [pc = \nu_\omega^\alpha.3] \\
&\quad [\!]\quad pc = \nu_\omega^\alpha.3 \rightarrow \mathsf{if}\ B \rightarrow pc := \nu_\omega^\alpha.2 \\
&\qquad\qquad\qquad\qquad\quad [\!]\ \neg B \rightarrow pc := succ_\varSigma(pc) \\
&\qquad\qquad\qquad\ \mathsf{fi} \\
&\mathsf{od}
\end{aligned}
$$

Note that in the final 'conditional branch' action, address $succ_\varSigma(pc)$ is equivalent to exit address ω.

4.5   Compile choice

For conditional statements we use the following refinement rules which can be proved with similar computations as above, again using the stuttering refinement rule **R7** and lemma 1.

Let $\mathcal{A}$ be an **if** statement without an **else** alternative,

$$\{pc = \alpha\} \,;\, \textbf{if } B \textbf{ then } H \textbf{ end} \,;\, [pc = \omega]\,,$$

where α and ω are elements of $\mathbb{A}$ which satisfy obligations (1) and (2).

C5 Compile simple conditional.

$$
\begin{aligned}
\mathcal{A} \sqsubseteq \textsf{con } \varSigma = \{\alpha, \nu_\omega^\alpha.2, \omega\} \,\bullet \\
\textsf{do } pc = \alpha \to \textsf{if } B \to pc := succ_\varSigma(pc) \\
[\!] \;\neg\, B \to pc := \omega \\
\textsf{fi} \\
[\!] \quad pc = \nu_\omega^\alpha.2 \to H \,;\, [pc = \omega] \\
\textsf{od}
\end{aligned}
$$

The first action represents evaluation of the boolean expression and conditional flow control to address $\nu_\omega^\alpha.2$ (i.e., $succ_\varSigma(\alpha)$) if B is true. Otherwise the entire construct exits.

Statements with two alternatives can be treated similarly. Let $\mathcal{A}$ be

$$\{pc = \alpha\} \,;\, \textbf{if } B \textbf{ then } H_1 \textbf{ else } H_2 \textbf{ end} \,;\, [pc = \omega]\,.$$

C6 Compile alternatives.

$$
\begin{aligned}
\mathcal{A} \sqsubseteq \textsf{con } \varSigma = \{\alpha, \nu_\omega^\alpha.2, \nu_\omega^\alpha.3, \nu_\omega^\alpha.4, \omega\} \,\bullet \\
\textsf{do } pc = \alpha \to \textsf{if } B \to pc := succ_\varSigma(pc) \\
[\!] \;\neg\, B \to pc := \nu_\omega^\alpha.4 \\
\textsf{fi} \\
[\!] \quad pc = \nu_\omega^\alpha.2 \to H_1 \,;\, [pc = \nu_\omega^\alpha.3] \\
[\!] \quad pc = \nu_\omega^\alpha.3 \to \textbf{jump } \omega \\
[\!] \quad pc = \nu_\omega^\alpha.4 \to H_2 \,;\, [pc = \omega] \\
\textsf{od}
\end{aligned}
$$

4.6   Loop optimisation laws

After refining a **while** statement using the above rules we may generate an assembler program where a load from memory instruction occurs at the beginning. If the corresponding memory location is not changed during the subsequent iterations there is no need to load this value again. The following optimisation law allows us to efficiently change the control flow in such cases.

We assume a program $\mathcal{A}$ of the form

$$
\begin{aligned}
\{pc = \alpha\} \,;\, \textsf{do } pc = \alpha \to S \,;\, [pc = a_{i_0}] \\
[\!] \quad ([\!]\, i : 1 \,.\,.\, n \,\bullet\, pc = \alpha_i \to S_i) \\
\textsf{od}
\end{aligned}
$$

with distinct addresses $\alpha, \alpha_1, \ldots, \alpha_n$ and a distinguished element $i_0 \in 1 \ldots n$. If statement S, after it has been executed for the first time, becomes equivalent to skip, we can ignore it for the remainder of this loop's execution. In this situation the following refinement can be undertaken.

C7 Loop optimisation 1.

$$A \sqsubseteq \{pc = \alpha\} ; \textbf{do } pc = \alpha \to S ; [pc = \alpha_{i_0}]$$
$$[\!] \quad ([\!] \ i : 1 \ldots n \bullet pc = \alpha_i \to S_i[\alpha_{i_0}/\alpha])$$
$$\textbf{od}$$

where $S_i[\alpha_{i_0}/\alpha]$ denotes statement S_i with free occurrences of 'α' replaced by 'α_{i_0}'.

Proof sketch If $\neg (B_1 \wedge B_2)$ is always true then

$$\textbf{do } B_1 \to S_1 [\!] B_2 \to S_2 \textbf{ od} = \textbf{do } B_1 \to S_1$$
$$[\!] \quad B_2 \to S_2 ; \textbf{if } B_1 \to S_1 [\!] \neg B_1 \to \textbf{skip fi}$$
$$\textbf{od} .$$

This rule can be applied to derive

$$A = \{pc = \alpha\} ;$$
$$\textbf{do } pc = \alpha \to S ; [pc = \alpha_{i_0}]$$
$$[\!] \quad ([\!] \ i : 1 \ldots n \bullet pc = \alpha_i \to S_i ; \textbf{if } pc = \alpha \to S ; [pc = \alpha_{i_0}]$$
$$[\!] \ pc \neq \alpha \to \textbf{skip}$$
$$\textbf{fi})$$
$$\textbf{od} .$$

Remember that after its first execution S equals **skip**. Hence, instruction

$$S_i ; \textbf{if } pc = \alpha \to S ; [pc = \alpha_{i_0}]$$
$$[\!] \ pc \neq \alpha \to \textbf{skip}$$
$$\textbf{fi}$$

is the same as $S_i[\alpha_{i_0}/\alpha]$. $\qquad\qquad\Box$

If $S ; [pc = \alpha_{i_0}]$ is equivalent to a load instruction **load** reg, z, for a register reg and a HLL variable z (treating 'z' as the address in *mem* corresponding to this variable), and if $S_1, \ldots, S_n$ do not change z or reg, we are allowed to substitute reg for z in any S_i.

C8 Loop optimisation 2.

$$A \sqsubseteq \{pc = \alpha\} ; \textbf{do } pc = \alpha \to S ; [pc = \alpha_{i_0}]$$
$$[\!] \quad ([\!] \ i : 1 \ldots n \bullet pc = \alpha_i \to S_i[reg/z])$$
$$\textbf{od}$$

4.7   Introduction of registers

As mentioned before we distinguish between local (hidden) and global (observable) variables. Wherever possible our compilation strategy associates local variables with registers, rather than their memory location. In effect, local variable names are substituted by register names at the end of the refinement process.

On the other hand, global variables are always associated with locations in memory, and must be loaded into a register to be manipulated. Assume a program $\mathcal{A}$ and a global HLL variable z of type T. Again we use 'z' below to denote the address in array *mem* associated with this variable. We assume further that α and ω fulfill obligations (1) and (2).

C9 Introduce temporary register.

$$\{pc = \alpha\} \; ; \mathcal{A} \; ; [pc = \omega] \sqsubseteq \mathbf{var} \; reg : T \; \bullet$$
$$\mathbf{con} \; \Sigma = \{\alpha, \nu_\omega^\alpha.2, \nu_\omega^\alpha.3, \omega\} \; \bullet$$
$$\mathbf{do} \; pc = \alpha \to \mathbf{load} \; reg, z$$
$$[\!] \quad pc = \nu_\omega^\alpha.2 \to \mathcal{A}[reg/z] \; ; [pc = \nu_\omega^\alpha.3]$$
$$[\!] \quad pc = \nu_\omega^\alpha.3 \to \mathbf{store} \; z, reg$$
$$\mathbf{od}$$

C10 Compile assignment of constant to global variable. Let C be a compile-time constant of type T.

$$\{pc = \alpha\} \; ; z := C \; ; [pc = \omega] \sqsubseteq \mathbf{var} \; reg : T \; \bullet$$
$$\mathbf{con} \; \Sigma = \{\alpha, \nu_\omega^\alpha.2, \omega\} \; \bullet$$
$$\mathbf{do} \; pc = \alpha \to \mathbf{loadi} \; reg, C$$
$$[\!] \quad pc = \nu_\omega^\alpha.2 \to \mathbf{store} \; z, reg$$
$$\mathbf{od}$$

Proof of both refinement laws is a straightforward application of rule **C2**.

4.8   Boolean expression evaluation

We state a rule which allows us to evaluate boolean expressions in conditional statements. This rule is commonly needed after application of rules **C3**, **C4**, **C5** or **C6**.

Let $\mathcal{A}$ be of the form

$$\{pc = \alpha\} \; ; \mathbf{if} \; B \to pc := \psi \; [\!] \; \neg \, B \to pc := \omega \; \mathbf{fi}$$

where $\alpha \notin \mathcal{M}$, $\mathcal{M} = \{\psi, \omega\}$ and $\phi = min\{m \in \mathcal{M} \bullet m \succ \alpha\}$ exists. Observe that this branching construct can terminate with pc equal to either ψ or ω. We assume further B is equivalent to expression $R(E_1, E_2)$, for a relation R and sub-expressions E_1 and E_2.

C11 Compile boolean expression.

$$
\begin{aligned}
&\mathcal{A} \sqsubseteq \textsf{var } reg : T \bullet \\
&\qquad \textsf{con } \Sigma = \{\alpha, \nu_\phi^\alpha.2, \psi, \omega\} \bullet \\
&\qquad \textsf{do } pc = \alpha \rightarrow reg := E_2 \,;\, [pc = \nu_\phi^\alpha.2] \\
&\qquad \llbracket \;\; pc = \nu_\phi^\alpha.2 \rightarrow \textsf{if } R(E_1, reg) \rightarrow pc := \psi \\
&\qquad\qquad\qquad\qquad\qquad\;\; \llbracket \;\neg\, R(E_1, reg) \rightarrow pc := \omega \\
&\qquad\qquad\qquad\qquad\;\; \textsf{fi} \\
&\qquad \textsf{od}
\end{aligned}
$$

The proof is a straightforward application of rule **C2** with $\mathcal{M}$ as defined above.

4.9 Nested loops

Application of the refinement rules introduced above to a high-level language program leads to a series of nested loops, whose control flow is directed by the program counter variable pc. Significantly, the way pc addresses are created in the refinement process by the ν operator, coupled with the use of the $succ_\Sigma$ operator, makes it possible to 'flatten' this nested structure without changing any of the addresses, guards or statements. The successor ordering on generated instruction addresses in a nested loop remains stable under unfolding! (Interestingly, the size of the pc-base address for a particular loop is proportional to the number of refinement laws which have been applied in order to produce this statement.)

Assume $\mathcal{A}$ is of the form

$$
\begin{aligned}
&\textsf{con } \Sigma_1 \bullet \\
&\{pc = \alpha_1\}\,; \\
&\textsf{do} \\
&\qquad (\llbracket \, i : 1 .. n \bullet pc = \alpha_i \rightarrow S_i) \\
&\textsf{od}\,,
\end{aligned}
$$

where $succ_{\Sigma_1}(\alpha_i) = \alpha_{i+1}$, for $i \in 1..n$, with $\alpha_{n+1} = \omega$. Assume further that component S_{i_0} has the form

$$
\begin{aligned}
&\textsf{con } \Sigma_2 \bullet \\
&\{pc = \alpha_{i_0}\}\,; \\
&\textsf{do } pc = \alpha_{i_0} \rightarrow S' \\
&\llbracket \;\; (\llbracket \, j : 1 .. m \bullet pc = \beta_j \rightarrow S'_j) \\
&\textsf{od}
\end{aligned}
$$

with addresses $succ_{\Sigma_2}(\beta_j) = \beta_{j+1}$, for $j \in 0..m$, $\beta_{m+1} = \alpha_{i_0+1}$, and $\beta_0 = \alpha_{i_0}$.

In $\mathcal{A}$ and S_{i_0} we permit assignments of the form

$$
pc := succ_\Sigma(pc)
$$

with respect to the corresponding successor function.

We then have distinct addresses

$$\alpha_1 \prec \alpha_2 \prec \cdots \prec \alpha_{i_0} \prec \beta_1 \prec \cdots \prec \beta_m \prec \alpha_{i_0+1} \prec \cdots \prec \alpha_n \prec \beta$$

and it is possible to merge the nested loops without changing any addresses or instructions. (Nevertheless, it may be necessary to rename variables local to S_{i_0} when expanding their scope.)

C12 Flatten nested loops.

$$\begin{aligned}
&\mathcal{A} \sqsubseteq \mathsf{con}\ \Sigma = \Sigma_1 \cup \Sigma_2 \bullet \\
&\quad \mathsf{do}\ (\![\ i : 1 \mathinner{\ldotp\ldotp} n \setminus \{i_0\} \bullet pc = \alpha_i \to S_i) \\
&\quad [\!]\quad pc = \alpha_{i_0} \to S' \\
&\quad [\!]\quad (\![\ j : 1 \mathinner{\ldotp\ldotp} m \bullet pc = \beta_j \to S'_j) \\
&\quad \mathsf{od}
\end{aligned}$$

Proof sketch We prove first that it is possible to replace S_{i_0} in $\mathcal{A}$ by

$$S'\ ;\mathsf{do}\ (\![\ j : 1 \mathinner{\ldotp\ldotp} m \bullet pc = \beta_j \to S'_j)\ \mathsf{od}\ .$$

For the remaining proof step we use rule **R6** for merging loops and our knowledge of the nested structure. □

Finally, we can state the main tool for any recursive application of the above introduced rules.

Theorem 2. *Every nested loop structure that is generated by recursive application of rules* **C1** *to* **C12** *on a high level language program* $\mathcal{A}$ *can be unfolded by rule* **C12**.

Proof. For the proof we use Lemma 1 and the following result in order to verify the proof obligations of rule **C12**:

> Given any high level language program $\mathcal{A}$ and any recursive application of the rules **C1** to **C12**, then each of the rule applications was performed on a program of the form $\{pc = \alpha\}\ ;\ P\ ;\ [pc \in \mathcal{M}]$ with $succ_\Sigma(\alpha) = min\{m \in \mathcal{M} \bullet m \succ \alpha\}$.

This can be seen with a straightforward induction over the rules **C1** to **C12**. □

5　Example

Consider the following program fragment $\mathcal{A}$ that sums all non-negative numbers in an array.

$$\mathbf{var}\ size : \mathbb{N};\ numbers : \mathbf{array}\ 1 \mathinner{\ldotp\ldotp} size\ \mathbf{of}\ \mathbb{Z};\ total : \mathbb{Z}$$

$$\vdots$$

$$\mathbf{var}\ count : \mathbb{N}$$
$$\mathbf{begin}$$

$$total := 0\,;$$
$$count := 1\,;$$
$$\textbf{while } count \leqslant size \textbf{ do}$$
$$\quad \textbf{if } numbers(count) > 0$$
$$\quad \textbf{then } total := total + numbers(count)$$
$$\quad \textbf{end}\,;$$
$$\quad count := count + 1$$
$$\textbf{end}$$
$$\textbf{end}$$

Here we treat *count* as a local variable and all remaining variables as global.

For conciseness during refinement we abbreviate the code fragments as follows.

$$\mathcal{A} = P_1\,;\,P_2$$
$$P_1 = P_1^1\,;\,P_1^2$$
$$P_1^1 = total := 0$$
$$P_1^2 = count := 1$$
$$P_2 = \textbf{while } count \leqslant size \textbf{ do } P_2^1\,;\,P_2^2 \textbf{ end}$$
$$P_2^1 = \textbf{if } numbers(count) > 0$$
$$\qquad\quad \textbf{then } total := total + numbers(count) \textbf{ end}$$
$$P_2^2 = count := count + 1$$

'Compilation' begins by introducing the machine-level variables.

$$\mathcal{A} \sqsubseteq \text{``by } \mathbf{C1}; \text{ defn. } \mathcal{A}\text{''}$$
$$\quad \textbf{var } pc : \mathbb{A};\ mem : \mathbb{A} \to \mathbb{Z} \bullet$$
$$\quad \textbf{con } \Sigma = \{1, 2\} \bullet$$
$$\quad pc := 1\,;$$
$$\quad \textbf{do}$$
$$\qquad pc = 1 \to P_1\,;\,P_2\,;\,[pc = 2]$$
$$\quad \textbf{od}$$

Next we compile the sequential components of the loop above, making use of the guard as the initial assertion, so that the program counter controls the order in which they are performed, rather than the HLL ';' operator.

$$\{pc = 1\}\,;\,P_1\,;\,P_2\,;\,[pc = 2] \sqsubseteq \text{``by } \mathbf{C2}\text{''}$$
$$\qquad\qquad \textbf{con } \Sigma = \{1, 2, 1.2\} \bullet$$
$$\qquad\qquad \textbf{do } pc = 1 \to P_1\,;\,[pc = 1.2]$$
$$\qquad\qquad \llbracket \quad pc = 1.2 \to P_2\,;\,[pc = 2]$$
$$\qquad\qquad \textbf{od}$$

The first of these sequential components can then be compiled similarly.

$$\{pc = 1\}\,;\,P_1\,;\,[pc = 1.2] \sqsubseteq \text{``by } \mathbf{C2}\text{''}$$
$$\qquad\qquad \textbf{con } \Sigma = \{1, 1.2, 1.1.2\} \bullet$$
$$\qquad\qquad \textbf{do } pc = 1 \to P_1^1\,;\,[pc = 1.1.2]$$
$$\qquad\qquad \llbracket \quad pc = 1.1.2 \to P_1^2\,;\,[pc = 1.2]$$
$$\qquad\qquad \textbf{od}$$

Since the first of the two actions introduced in the previous step is an assignment to a global variable, it can be immediately compiled to a load/store combination. Let loc_{total} be the memory address holding the value of global HLL variable $total$.

$$\{pc = 1\} \, ; P_1^1 \, ; [pc = 1.1.2] \sqsubseteq \text{``by C10''}$$

$$\begin{aligned}
&\textbf{con } \Sigma = \{1, 1.1.2, 1.1.1.2\} \bullet \\
&\textbf{var } reg_1 : \mathbb{N} \bullet \\
&\textbf{do } pc = 1 \to \textbf{loadi } reg_1, 0 \\
&\quad[\!] \quad pc = 1.1.1.2 \to \textbf{store } loc_{total}, reg_1 \\
&\textbf{od}
\end{aligned}$$

The **while** loop is compiled as follows. Firstly the loop compilation rule is used to divide the HLL statement into three parts, evaluation of the boolean expression, the statement body and an unconditional jump. Then the rule for compiling expression evaluation is applied to the first 'branching' action and the resulting model is 'flattened'. Let reg_{count} be the register used to hold the value of local HLL variable $count$, and let mem_{size} be the memory value corresponding to HLL variable $size$, i.e., $mem(loc_{size})$.

$$\{pc = 1.2\} \, ; P_2 \, ; [pc = 2]$$
$$\sqsubseteq \text{``by C3''}$$
$$\begin{aligned}
&\textbf{con } \Sigma = \{1.2, 2, 1.1.2, 1.2.3\} \bullet \\
&\textbf{do } pc = 1.2 \to \textbf{if } reg_{count} \leqslant mem_{size} \to pc := succ_\Sigma(pc) \\
&\qquad\qquad\qquad\quad [\!] \ reg_{count} > mem_{size} \to pc := 2 \\
&\qquad\qquad\quad \textbf{fi} \\
&\quad [\!] \quad pc = 1.2.2 \to P_2^1 \, ; P_2^2 \, ; [pc = 1.2.3] \\
&\quad [\!] \quad pc = 1.2.3 \to \textbf{jump } 1.2 \\
&\textbf{od}
\end{aligned}$$
$$\sqsubseteq \text{``by C11, C12, C7''}$$
$$\begin{aligned}
&\textbf{var } reg_1 : \mathbb{N} \bullet \\
&\textbf{con } \Sigma = \{1.2, 2, 1.2.1.2, 1.2.2, 1.2.3\} \bullet \\
&\textbf{do } pc = 1.2 \to \textbf{load } reg_1, loc_{size} \\
&\quad [\!] \quad pc = 1.2.1.2 \to \textbf{if } reg_{count} \leqslant reg_1 \to pc := succ_\Sigma(pc) \\
&\qquad\qquad\qquad\qquad\quad [\!] \ reg_{count} > reg_1 \to pc := 2 \\
&\qquad\qquad\qquad\quad \textbf{fi} \\
&\quad [\!] \quad pc = 1.2.2 \to P_2^1 \, ; P_2^2 \, ; [pc = 1.2.3] \\
&\quad [\!] \quad pc = 1.2.3 \to \textbf{jump } 1.2.1.2 \\
&\textbf{od}
\end{aligned}$$

We now compile the body of the **while** loop, i.e., the second action in the **do** loop defined above, whose sequential components can be compiled as follows,

$$\{pc = 1.2.2\} \, ; P_2^1 \, ; P_2^2 \, ; [pc = 1.2.3]$$
$$\sqsubseteq \text{``by C2''}$$
$$\begin{aligned}
&\textbf{con } \Sigma = \{1.2.2, 1.2.3, 1.2.2.2\} \bullet \\
&\textbf{do } pc = 1.2.2 \to P_2^1 \, ; [pc = 1.2.2.2] \\
&\quad [\!] \quad pc = 1.2.2.2 \to P_2^2 \, ; [pc = 1.2.3] \\
&\textbf{od}
\end{aligned}$$

The first of these components, the **if** statement, is then compiled as follows. Firstly the boolean expression and body are made into separate actions. Then evaluation of the boolean guard is compiled by introducing register variables to hold the values of *numbers*(*count*) and *total*, and the resulting structure is flattened and optimised. Let mem_{total} be shorthand for $mem(loc_{total})$.

$$\{pc = 1.2.2\}\,;\, P_2^1\,;\, [pc = 1.2.2.2]$$

$\sqsubseteq$ "by **C5**"

    con $\Sigma = \{1.2.2, 1.2.2.1.2, 1.2.2.2\} \bullet$

    do $pc = 1.2.2 \rightarrow$ **if** $mem(loc_{numbers} + reg_{count}) > 0 \rightarrow pc := succ_{\Sigma}(pc)$

                      $[\!]$ $mem(loc_{numbers} + reg_{count}) \leqslant 0 \rightarrow pc := 1.2.2.2$

          fi

    $[\!]$ $pc = 1.2.2.1.2 \rightarrow mem_{total} := mem_{total} + mem(loc_{numbers} + reg_{count})\,;$

                   $[pc = 1.2.2.2]$

    od

$\sqsubseteq$ "by **C11, C9, C12, C8**"

    var $reg_1, reg_2 : \mathbb{Z} \bullet$

    con $\Sigma = \{1.2.2, 1.2.2.1.1.2, 1.2.2.1.2,$

               $1.2.2.1.2.2, 1.2.2.1.2.3, 1.2.2.2\} \bullet$

    do $pc = 1.2.2 \rightarrow reg_1 := mem(loc_{numbers} + reg_{count})\,;\, [pc = 1.2.2.1.1.2]$

    $[\!]$ $pc = 1.2.2.1.1.2 \rightarrow$ **if** $reg_1 > 0 \rightarrow pc := succ_{\Sigma}(pc)$

                      $[\!]$ $reg_1 \leqslant 0 \rightarrow pc := 1.2.2.2$

          fi

    $[\!]$ $pc = 1.2.2.1.2 \rightarrow$ **load** reg_2, loc_{total}

    $[\!]$ $pc = 1.2.2.1.2.2 \rightarrow$ **add** reg_2, reg_1

    $[\!]$ $pc = 1.2.2.1.2.3 \rightarrow$ **store** reg_2, loc_{total}

    od

Finally, we collect all the above steps together and flatten the entire nested structure to create a model of the final assembler program. In doing so we must be careful to rename register variables reg_x, introduced locally, when their scope is expanded, to avoid name clashes. The remaining abstract assignments at locations 1.2.2 and 1.2.2.2 and the branch instructions at 1.2.1.2 and 1.2.2.1.1.2 can be replaced using our machine language definitions.

$\sqsubseteq$ "by **C12**, defn. **load**, **addi**, **bgt**, **blez**"

    var $pc : \mathbb{A};\ mem : \mathbb{A} \rightarrow \mathbb{Z};\ reg_1, reg_3 : \mathbb{N};\ reg_2 : \mathbb{Z} \bullet$

    con $\Sigma = \{1, 2, 1.1.1.2, 1.1.2, 1.2, 1.2.1.2, 1.2.2, 1.2.2.1.1.2,$

               $1.2.2.1.2, 1.2.2.1.2.2, 1.2.2.1.2.3, 1.2.2.2, 1.2.3\} \bullet$

    $pc := 1\,;$

$$
\begin{aligned}
\textbf{do } & pc = 1 \rightarrow \textbf{loadi } reg_1, 0 \\
[\!] \quad & pc = 1.1.1.2 \rightarrow \textbf{store } loc_{total}, reg_1 \\
[\!] \quad & pc = 1.1.2 \rightarrow \textbf{loadi } reg_{count}, 1 \\
[\!] \quad & pc = 1.2 \rightarrow \textbf{load } reg_1, loc_{size} \\
[\!] \quad & pc = 1.2.1.2 \rightarrow \textbf{bgt } reg_{count}, reg_1 \\
[\!] \quad & pc = 1.2.2 \rightarrow \textbf{load } reg_2, loc_{numbers} + reg_{count} \\
[\!] \quad & pc = 1.2.2.1.1.2 \rightarrow \textbf{blez } reg_2, 1.2.2.2 \\
[\!] \quad & pc = 1.2.2.1.2 \rightarrow \textbf{store } loc_{total}, reg_3 \\
[\!] \quad & pc = 1.2.2.1.2.2 \rightarrow \textbf{add } reg_3, reg_2 \\
[\!] \quad & pc = 1.2.2.1.2.3 \rightarrow \textbf{store } loc_{total}, reg_3 \\
[\!] \quad & pc = 1.2.2.2 \rightarrow \textbf{addi } reg_{count}, 1 \\
[\!] \quad & pc = 1.2.3 \rightarrow \textbf{jump } 1.2.1.2 \\
\textbf{od} &
\end{aligned}
$$

It then remains to allocate a free register, in this case reg_4 to local variable *count*, and use the syntactic shorthand introduced in Section 3.3 to express the final assembler code in a more familiar form.

$$
\begin{aligned}
&\textbf{var } pc : \mathbb{A};\ mem : \mathbb{A} \rightarrow \mathbb{Z};\ reg_1, reg_3, reg_4 : \mathbb{N};\ reg_2 : \mathbb{Z} \bullet \\
&\textbf{con } \Sigma \bullet \\
&pc := 1; \\
&\langle \textbf{loadi } reg_1, 0\ ;\ \textbf{store } loc_{total}, reg_1\ ;\ \textbf{loadi } reg_4, 1\ ;\ \textbf{load } reg_1, loc_{size}; \\
&\quad \textbf{bgt } reg_4, reg_1, 2\ ;\ \textbf{load } reg_2, loc_{numbers} + reg_4\ ;\ \textbf{blez } reg_2, 1.2.2.2; \\
&\quad \textbf{store } loc_{total}, reg_3\ ;\ \textbf{add } reg_3, reg_2\ ;\ \textbf{store } loc_{total}, reg_3\ ;\ \textbf{addi } reg_4, 1; \\
&\quad \textbf{jump } 1.2.1.2 \rangle\, @\, 1
\end{aligned}
$$

Now it is only a small data refinement step to define the program counter over the natural numbers instead of the abstract inductively ordered set Σ. In this case we assume the final assembler code fragment will be loaded at memory location 100.

$$
\begin{aligned}
&\textbf{var } pc : \mathbb{N};\ mem : \mathbb{A} \rightarrow \mathbb{Z};\ reg_1, reg_3, reg_4 : \mathbb{N};\ reg_2 : \mathbb{Z} \bullet \\
&pc := 100; \\
&\langle \textbf{loadi } reg_1, 0\ ;\ \textbf{store } loc_{total}, reg_1\ ;\ \textbf{loadi } reg_4, 1\ ;\ \textbf{load } reg_1, loc_{size}; \\
&\quad \textbf{bgt } reg_4, reg_1, 112\ ;\ \textbf{load } reg_2, loc_{numbers} + reg_4\ ;\ \textbf{blez } reg_2, 110; \\
&\quad \textbf{store } loc_{total}, reg_3\ ;\ \textbf{add } reg_3, reg_2\ ;\ \textbf{store } loc_{total}, reg_3\ ;\ \textbf{addi } reg_4, 1; \\
&\quad \textbf{jump } 104 \rangle\, @\, 100
\end{aligned}
$$

6   Conclusion

The notion of treating program compilation as a formal development process is by no means new [16], and our specific methods have borrowed from Norvell [23], Back [2] and the **safemos** [9] and ProCoS [15] projects.

What *is* new, however, is our determination to use models and rules that are founded in the familiar refinement calculus of Morgan, Back, et al, rather than devising largely new formalisms, as most previous attempts have done. This gives us a greater chance of achieving acceptance from the (quite justifiably!) conservative and sceptical safety-critical programming community. In addition, we can easily extend the presented model by defining and verifying more refinement rules or by incorporating a more sophisticated variable and register treatment.

A particularly interesting problem solved in this paper was the challenge of 'numbering' assembler code segments as they are introduced during refinement. The inductively-ordered numbering system adopted above avoided the need to guess how many instructions would be generated for a particular action in order to 'reserve' sufficient memory locations for them. Indeed, it can be seen that the numbering system mimics the structure of an actual compiler's parse tree.

Acknowledgements This research is funded by Australian Research Council grant A49600176, *Verified compilation rules for real-time programs via program refinement*.

References

1. R.-J. Back and K. Sere. Stepwise refinement of action systems. *Structured Programming*, 12(1):17–30, 1991.
2. R.-J. R. Back. Refinement of parallel and reactive programs. Technical Report Caltech-CS-TR-92-23, California Institute of Technology, 1992.
3. R.-J. R. Back. Atomicity refinement in a refinement calculus framework. Technical Report 141, Åbo Akademi, Dept. of Computer Science, March 1993.
4. R.-J. R. Back and K. Sere. Deriving an occam implementation of action systems. In C. Morgan and J. Woodcock, editors, *3rd Refinement Workshop*, pages 9–30. Springer-Verlag, 1990.
5. R.-J. R. Back and K. Sere. From action systems to modular systems. *Software— Concepts and Tools*, 17(1):26–39, 1996.
6. R.-J. R. Back and K. Sere. Superposition refinement of reactive systems. *Formal Aspects of Computing*, 8:324–346, 1996.
7. R.-J. R. Back and J. von Wright. Trace refinement of action systems. In B. Jonsson and J. Parrow, editors, *CONCUR'94: Concurrency Theory*, volume 836 of *Lecture Notes in Computer Science*, pages 367–384. Springer-Verlag, 1994.
8. E. Börger and I. Durdanović. Correctness of compiling occam to transputer code. *The Computer Journal*, 39(1):52–92, 1996.
9. J. Bowen, editor. *Towards Verified Systems*, volume 2 of *Real-Time Safety Critical Systems*. Elsevier, 1994.
10. M. J. Butler. Refinement and decomposition of value-passing action systems. In E. Best, editor, *Concur'93*, volume 715 of *Lecture Notes in Computer Science*, pages 217–232. Springer-Verlag, 1993.
11. C. J. Fidge and A. J. Wellings. An action-based formal model for concurrent, real-time systems. *Formal Aspects of Computing*, 1997. To appear.
12. M. Fränzle and M. Müller-Olm. Towards provably correct code generation for a hard real-time programming language. In P. Fritzson, editor, *Compiler Construction*, volume 786 of *Lecture Notes in Computer Science*, pages 294–308. Springer-Verlag, 1994.

13. R. W. S. Hale. Program compilation. In J. Bowen, editor, *Towards Verified Systems*, volume 2 of *Real-Time Safety Critical Systems*, chapter 7, pages 131–146. Elsevier, 1994.

14. I. J. Hayes and M. Utting. Coercing real-time refinement: A transmitter. In *Proc. Northern Formal Methods Workshop*. Springer-Verlag, 1997. To appear.

15. He Jifeng. *Provably Correct Systems*. McGraw-Hill, 1995.

16. C. A. R. Hoare. Refinement algebra proves correctness of compiling specifications. In C. Morgan and J. Woodcock, editors, *3rd Refinement Workshop*, pages 33–48. Springer-Verlag, 1990.

17. C. A. R. Hoare and He Jifeng. Refinement algebra proves correctness of a compiler. Draft, June 1990.

18. R. Kurki-Suonio. Stepwise design of real-time systems. *IEEE Trans. Software Engineering*, 19(1):56–69, January 1993.

19. B. Mahony. Using the refinement calculus for dataflow processes. Technical Report 94-32, Software Verification Research Centre, October 1994.

20. C. Morgan. *Programming from Specifications*. Prentice-Hall, second edition, 1994.

21. C. Morgan. The specification statement. In C. Morgan and T. Vickers, editors, *On the Refinement Calculus*, pages 1–21. Springer-Verlag, 1994.

22. C. Morgan and T. Vickers. *On the Refinement Calculus*. Springer-Verlag, 1994.

23. T. S. Norvell. Machine code programs are predicates too. In D. Till, editor, *Sixth Refinement Workshop*, pages 188–204. Springer-Verlag, 1994.

24. T. S. Norvell. *A Predicative Theory of Machine Languages and its Application to Compiler Correctness*. PhD thesis, Graduate Department of Computer Science, University of Toronto, 1994.

Version and Configuration Management of Formal Theories

Peter A. Lindsay and Owen Traynor

Software Verification Research Centre
School of Information Technology,
The University of Queensland, Brisbane 4072, Australia

Abstract. This paper reports the results of an investigation of the problems of version management of systems of objects which themselves are under version control, and where there are complex consistency and completeness requirements. The issues are illustrated on a case study concerning management of a formal theory which consists of theorems and their proofs. Conclusions are drawn about a framework for fine-grained configuration management for formal methods of software development.

1 Introduction

1.1 Software Configuration Management (SCM)

Managing and controlling change is a critical part of software engineering. Software components typically pass through many different versions during both the initial development of a system and the ongoing maintenance of the system once deployed. In its most general form, *Software Configuration Management* concerns the control of all development artifacts throughout the system life-cycle, to preserve the definition of versions of components and the relationships among them [8].

As well as providing the framework within which developers work to construct consistent system "builds", SCM provides the mechanisms needed to demonstrate traceability between the built system, the design, the requirements documents, and other tools and artifacts of a development (such as compilers and test reports). SCM provides the means for recording and controlling the "configuration" of versions of documents associated with software development, including inter- and intra-document dependencies. Regulatory and standards authorities have long recognised the importance of reliable SCM mechanisms, especially in the development of high integrity software systems [5, 13].

An important supporting technology for SCM is *version control*, which concerns storage and retrieval of different versions of development components. Most version control systems attempt to maintain a record of the changes ("deltas") between different versions cof components. This provides the basis for tracing the evolution of a system through its lifetime. The majority of software development companies currently use version control facilities such as RCS [20] or SCCS [16] to manage their documents and software, but such facilities operate at an inadequately coarse level of granularity (typically, whole documents or whole modules) and fall far short of users' desires.

1.2   SCM for Formal Methods

Formal Methods of software development have particular needs in relation to SCM. Formal Methods are based on the use of mathematically precise definitions of development components and their relationships, together with the use of mathematical analysis techniques – including theorem proving – for establishing correctness. The fact that individual development components have mathematical meaning makes it possible to formally verify that desired relationships hold within and between development components. In contrast to traditional development methods, cross-development configuration consistency can be defined precisely and at fine levels of granularity [18].

One of the main practical problems currently hindering the uptake of Formal Methods, however, is the sheer number and complexity of its artifacts. For example, proofs are typically orders of magnitude larger than the programs they prove. By separating management of dependencies between formal entities from their construction, fine-grained SCM has much to offer Formal Methods, especially in the area of management of change. Most *Formal Development Support Environments* (FDSEs) currently do not provide specialised support for configuration management.

Direct adoption of standard approaches to SCM are not sufficient, however. Software configuration and version management for traditional software engineering focus on the requirements associated with the management of things like source code, object code, etc. In the context of formal development, the artifacts are more complex and more numerous and there is wider variety of dependencies between these artifacts. The concept of an automated "build" management system (upon which these traditional systems rely) is inappropriate for a FDSE, since the activities that bring configurations into consistent states cannot be fully automated.

The ARC-funded *Fine-Grained Configuration Management (FGCM)* project at the SVRC is establishing a framework for fine-grained configuration management. The framework builds on a programme of work carried out by PhD student Kelvin Ross under the supervision of the first author, investigating the application of SCM techniques to formal development [17, 18]. The aim of the framework is to allow developers to support their correctness claims with evidence that, not only have the individual components of a system been developed correctly, but that the combination and integration of these components has been done in a consistent manner and that the final result is derived from consistent, complete and up-to-date development components. The framework is intended to apply not only to specifications, designs and programs, but also to fine-grained development components such as the specification components, reviews, change requests, refinements, design decisions, test sets, theories and proofs that are generated as part of the development process.

Early versions of the framework have concentrated on traceability and version management [12] for a number of reasons. First, we noted that it is often desirable to be able to store multiple versions of development artifacts – for example, in order to support team working, where different components of a system are developed largely independently and then brought back together [15]. Another reason is because of regulatory requirements: for example, one of the companies with which we have collaborated in the past is required to be able to reconstruct all development artifacts (including system software, analysis and test results) associated with the release of every one of its implantable medical devices.

As well as being able to trace the development of systems, we would like to be able to trace (in isolation as far as possible) the development history of individual fine-grained artifacts. For example, in critical system development it is important to be able to trace the evolution of individual safety requirements right through the design to final implementation [13]. Another example is the ability to trace (and reconstruct) artifacts which have been reused in different developments, such as components from standard libraries.

Other projects are looking at version management of structured objects (e.g. Goodstep [19], Merlin [14], UQ* [10, 21]) and have proposed generic approaches, with a good deal of success. However, the configuration consistency and completeness conditions they consider are not sophisticated enough to meet the above requirements.

1.3　High integrity software engineering

We consider the definition of a coherent framework, within which configuration and version management can be carried out, as an important prerequisite in the development of trusted and cost-effective environments for the development of critical software. The processes that define the development activities in such trusted environments must be based on sound underlying technology and models that allow the impact of any development step (in terms of the consistency of the relationships between the underlying artifacts) to be accurately assessed. Existing FDSEs use relatively untrusted standard version-management technology in the development of critical systems; this is clearly a weak link since these technologies have no coherent formal basis for consistency checking.

The need for careful control of the development process must be balanced against the need for flexibility. Users will not accept a development process that is overly constraining. Similarly, it is vitally important for encouraging industrial uptake that Formal Methods be adaptable to different situations, project structures and so on, without sacrificing the trustworthiness of the environments.

The approach taken here is to define configuration consistency models (or *configuration models*, for short) which define the key configuration items, the relationships between them, and the consistency and completeness conditions desired for the configuration. In our approach, configuration models would form the core part of FDSEs, with development processes defined relative to the core models. This means that the consistency and completeness of a development could be established largely independently of the development process applied, giving flexibility and trustworthiness in the one framework.

It is worth noting at this point that we consider version management as an integral part of the configuration management activity. Whereas it is common to have some form of configuration management without a version management system, version management systems rely critically on a configuration management framework for defining the collections of versions which denote a specific configuration.

1.4 The problem and its motivation

The approach will be illustrated here on version management of a formal theory, consisting of theorems and their proofs. This is an area of particular concern to FDSEs such as the B-Tool [1], Cogito [3] and `mural` [9], in which formal specifications have corresponding formal theories in which consequences of the specification are deduced. Version management is of critical importance in such systems because specifications – and hence the associated theories – often change, yet the soundness of deductions depends on the specification and associated theory being in step with one another.

For some time, version control (at the theory level) has been used in theorem proving; when a change is made to a given theory, the old version is stored and the system is rebuilt with the new version. At the SVRC, we use CVS [4] for this task in the context of the Ergo theorem prover used with the Cogito system.

Similarly, most interactive theorem proving systems have some notion of configuration management at fine levels of granularity. For example, the Ergo theorem prover will warn the user if a circularity has been introduced during proof construction. This form of dependency management is a good illustration of the (simple) use of fine-grained configuration management principles. However, we are not aware of any theorem-proving systems which use notions of version control at this level of granularity.

The motivation for considering fine-grained version control of formal theories has come from a number of sources. Our experience in deploying complex theory structures for use in actual formal development, and our use of theorem provers (without fine-grained version control) to support the formal development activity, have illustrated the need for finer levels of configuration and version control within the proof activity. In an industrial pilot project of Cogito [7] for example, the following scenario occurred a number of times.

During validation activities, from time to time the application developers would uncover a deficiency in a core theory. The maintainers of the theorem provers would then release a theory update which corrected the deficiency. Typically the deficiency would be in a small portion of a single theory, and an update would only be required for the current proof in that theory, since no other proofs had used that part of the theory so far. Since configuration control (and by implication version control) was at the theory level, however, no advantage could be taken of the fact that the theorem or definition being updated was used only in a single proof. Instead, the complete set of proofs for the development, and all consistency proofs for the core theories, had to be redone!

The same problem arose again at the application level. When a portion of a specification (say a definition) was modified, then all other specifications that relied on the changed specification (and by implication all the associated theories and proofs) required updating. Even for the relatively small specification used in the particular development, this often meant days of extra work – work which would not be necessary if effective configuration management support had been in place.

1.5   This paper

This paper reports preliminary conclusions from a case study in adding fine-grained versioning to a simple theory store. As it happens, the case study is a good one for illustrating our general approach, because it is relatively concise, yet has complex consistency and completeness requirements. A key focus is the impact of the versioning policy on the consistency requirements of the theories, theorems and proofs that are the subject of version control. In conducting the analysis, we found that a variety of versioning policies were possible, all with different implications for the consistency of the theory store.

For simplicity, we have presented a simple model of theories and a very straightforward version control policy. However, even such a simple model has rather surprising implications for both the theory developer and user, which are discussed in more detail later.

Section 2 presents a basic model for version management of individual objects, covering notions of derivation history and freezing of version contents. The paper uses a variant of Object-Z [6] as the modelling language, for precision and concision. Section 3 defines the requirements for the case study. Section 4 models versions of theorems in terms of versions of statements and proofs. Section 5 models theories as collections of theorems. Section 6 defines the consistency and completeness conditions for the case study. Finally, the different models are brought together in Section 7, which treats version management of theories.

2   Version Management

The model of version management considered here uses *version trees* to trace the development history of objects: see Fig. 1. Branching corresponds to derivation of variants of the *parent* version. (It is possible to add version merging to the model, but we have not done so here for simplicity.) The content of a version can be *frozen*, meaning the content cannot subsequently be changed. In particular, a version must be frozen before (or when) a new version is derived from it, since to allow subsequent changes would violate the traceability requirement.

The rest of this section defines a generic class *VersionTree*[*Content*] of version trees, where *Content* is a generic set representing the kind of objects that are being put under version control.

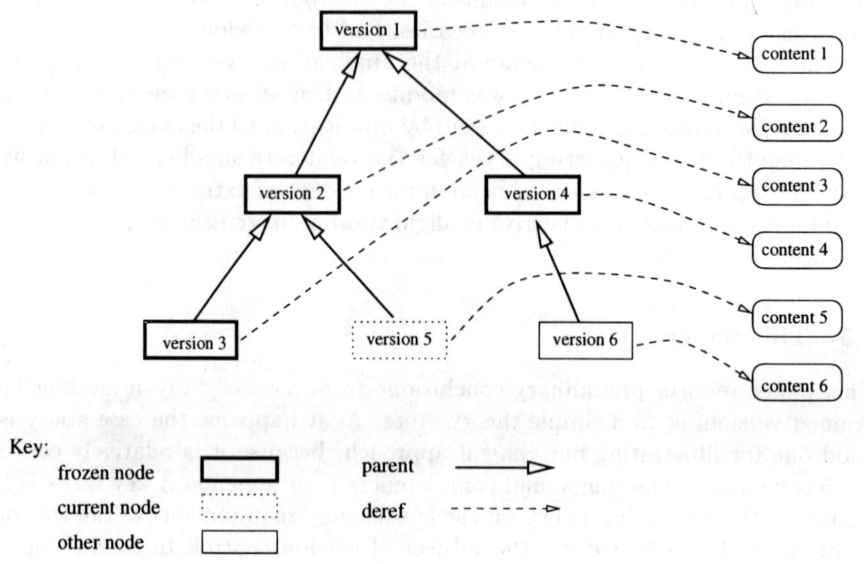

Fig. 1. An example version tree.

2.1 Object state

The attributes of version trees are defined as follows:

- Each node in the tree is labelled with a unique *version label*, from the set *VLabel*.
- The mapping *deref* 'dereferences' version labels, yielding the content of a particular version.
- The set *frozen* contains the labels of the frozen nodes.
- The mapping *parent* returns the (unique) parent of any given non-root node in the tree.
- One node is nominated as the *current* version.

$$
\begin{array}{|l}
\hline
\mathit{root} : \mathit{VLabel} \\
\mathit{deref} : \mathit{VLabel} \nrightarrow \mathit{Content} \\
\mathit{frozen} : \mathbb{P}\, \mathit{VLabel} \\
\mathit{parent} : \mathit{VLabel} \nrightarrow \mathit{VLabel} \\
\mathit{current} : \mathit{VLabel} \\
\hline
\mathit{nodes} : \mathbb{P}\, \mathit{VLabel} \\
\mathit{currentcontent} : \mathit{Content} \\
\hline
\mathit{nodes} = \mathrm{dom}\, \mathit{deref} \\
\mathit{currentcontent} = \mathit{deref}(\mathit{current}) \\
\mathit{frozen} \cup \{\mathit{root}, \mathit{current}\} \subseteq \mathit{nodes} \\
\mathrm{dom}\, \mathit{parent} = \mathit{nodes} \setminus \{\mathit{root}\} \\
\mathrm{ran}\, \mathit{parent} \subseteq \mathit{frozen} \\
\hline
\end{array}
$$

Note that *nodes* and *currentcontent* are derived attributes: i.e., their values can be calculated from the other attributes. Such attributes do not need to be mentioned explicitly in operations' delta sets.

The following predicate defines the initial values of version trees. The predicate takes the tree's initial content *content?* as a parameter.

$$
\begin{array}{|l}
\hline
\_\mathit{INIT} \\
\hline
\mathit{content?} : \mathit{Content} \\
\hline
\mathit{deref} = \{\mathit{root} \mapsto \mathit{content?}\} \\
\mathit{frozen} = \varnothing \\
\hline
\end{array}
$$

Note that the values of all the other attributes can be deduced from the above (e.g. *current* = *root*).

2.2 Operations

This section defines the basic operations that can be performed on a version tree. Note that each operation preserves the state invariant.

$$
\begin{array}{|l}
\hline
\_\mathit{FreezeNode} \\
\hline
\Delta(\mathit{frozen}) \\
\mathit{v?} : \mathit{VLabel} \\
\hline
\mathit{v?} \in \mathit{nodes} \\
\mathit{frozen'} = \mathit{frozen} \cup \{\mathit{v?}\} \\
\hline
\end{array}
$$

Freeze a given node.

```
┌─ DeriveNewVersion ──────────────────
│ Δ(nodes, parent, deref, frozen, current)
│ newcontent? : Content
├──────────────────────────────────────
│ FreezeNode[current/v?]
│ ∃ new : VLabel •
│   new ∉ nodes
│   parent' = parent ∪ {new ↦ current}
│   deref' = deref ∪ {new ↦ newcontent?}
│   frozen' = frozen ∪ {current}
│   current' = new
└──────────────────────────────────────
```

Freeze the current node and create a new node with this as its parent and with the given content, making it the new current node.

```
┌─ DeriveNewCopy ─────────────────────
│ Δ(nodes, parent, deref, frozen, current)
├──────────────────────────────────────
│ DeriveNewVersion
│   [currentcontent/newcontent?]
└──────────────────────────────────────
```

Similar to DeriveNewVersion but takes a copy of the current content.

```
┌─ ChangeCurrent ─────────────────────
│ Δ(current)
│ v? : VLabel
├──────────────────────────────────────
│ v? ∈ nodes
│ current' = v?
└──────────────────────────────────────
```

Change which node is considered the current node.

```
┌─ EditCurrent ───────────────────────
│ Δ(deref)
│ content? : Content
├──────────────────────────────────────
│ current ∉ frozen
│ deref' = deref ⊕ {current ↦ content?}
└──────────────────────────────────────
```

Change the content of the given node, provided it is not frozen.

3   The Case Study: Requirements

The case study is based on a (hypothetical) theory manager, in which a *formal theory*, consisting of theorems and their proofs, is put under version control. This section defines the basic requirements for the case study.

3.1   The objects and relationships to be managed

Theorems have *statements* and *proofs*. (In this simplified model, axioms are theorems with 'axiomatic' proofs.) We shall assume that proofs and statements can be edited separately, and so may get out of step (i.e., may not necessarily agree with one another). The internal structure of statements and proofs will not be considered here, other than to say that a proof may make reference to ('use') other theorems. This leads to a transitive *dependency relation* between theorems: see Fig. 2.

The system we consider is a flexible yet powerful one in which theorems can be proposed and used before their proofs are complete: Ergo and `mural` [9] are examples of such systems. For flexibility, we even allow the possibility that dependencies between theorems may be circular, although this is obviously a situation users would prefer to avoid (cf. Fig. 2). Making the model sufficiently general to support circularities means that a greater range of interaction models is possible, and more flexible user interfaces; however, these issues will not be addressed here.

To define the configuration consistency and completeness requirements we first make some definitions. A *theory* is a set of theorems. A theory is *closed* if all proofs use only theorems from within the theory (i.e., there are no dangling references). A theory is *circular* if there are loops in the dependency graph. A theory is *complete* if it is closed, non-circular, and all its theorems have complete and valid proofs.

We shall assume tools are available for:

- checking whether a proof is complete;
- extracting the theorems used in a proof; and
- editing statements and proofs.

Building on these, the paper specifies requirements for establishing and maintaining configuration consistency and completeness conditions. (Note however that *logical* consistency of the theory is not being considered here, just *structural* consistency.)

3.2   Versioning requirements

We want to be able to store multiple versions of statements and proofs of a given theorem, and store versions of proofs separately from statements. For the purposes of the investigation we shall assume that there may be more than one proof corresponding to each statement, but that each proof has a unique corresponding statement.

A *theorem instance* is a particular version of a statement and a particular version of a proof of a theorem. Thus, what we said above was not quite accurate: proofs use theorem instances rather than simply theorems, and a theory is a set of theorem instances. The other definitions carry through *mutatis mutandis*.

To use the model of version management given in Section 2, we first need to come up with a *versioning policy* which includes an interpretation of when things can be frozen and what implications can be drawn (if any) from the fact that something is frozen. For the purposes of the investigation we adopt the following versioning policy:

- When a theorem instance is frozen, all of the theorem instances on which it depends should also be frozen. (Intuitively, the full justification of a theorem consists not only of its statement and proof, but also of the statement and proof of all the theorems on which it depends.)

Example theorem statement
and proof

Theorem A: $1 + 1 = 3$	
Proof:	
1. 1+1+1=3	definition
2. 1=0	Theorem B
3. 1+1+0=3	substitution(1,2)
4. 1+1+0=1+1	Theorem C
5. 1+1=3	substitution(3,4)

Example theorem dependency graph

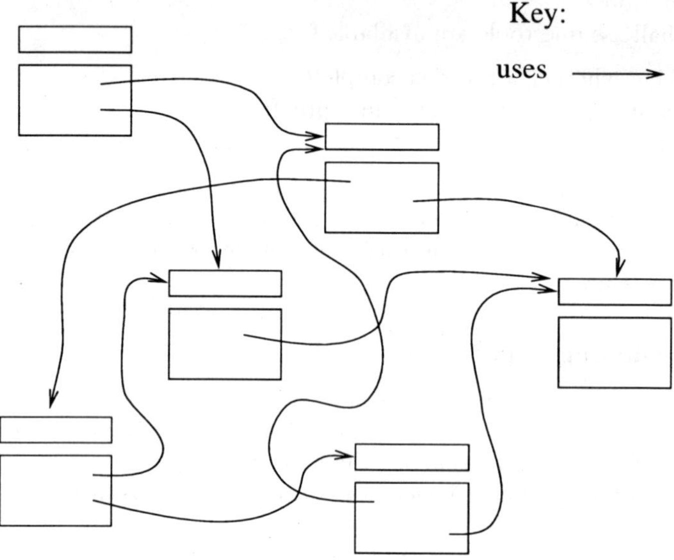

Fig. 2. (a) Example theorem, showing its statement and proof. (b) Example theory, showing dependency relation between theorems.

– A version of a theory can be frozen only if it is complete.

This policy attempts to preserve the "semantic integrity" of theorems and theories. Of course, other versioning policies are possible for theory management, and this policy may be too stringent in practice — the point is that versioning policies are application-dependent.

3.3 Formal modelling

The following definitions are needed for the models in later sections of this paper. Let *Stmt* be the set of all possible statements of theorems, and let *Proof* be the set of all (possibly incomplete) proofs of theorems. *ThmInst* is the set of theorem instances (to be defined further below).

The basic tool-established relationships will be modelled as follows:

- the function *uses* : *Proof* $\rightarrow$ $\mathbb{F}$ *ThmInst* extracts the theorem instances which are used in a given proof;
- the relation *proves* : *Proof* $\leftrightarrow$ *Stmt* checks whether a given proof is a complete and valid proof of a given statement.

The **mural** system [9] has such tools, for example.

4 The 'Theorem Version History' Object Class

This section defines a class *ThmVHistory* for the version histories of particular theorems.

4.1 Object state

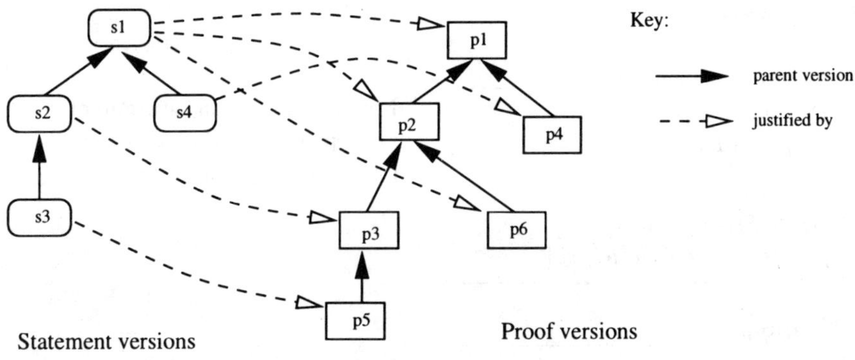

Fig. 3. Example version trees for a theorem.

A *theorem version history* is a tree of versions of statements and proofs of a particular theorem, together with an indication of the correspondence between proofs and statements (called 'justifies' here): see Fig. 3. Note that the 'justifies' relation simply notes which statement corresponds to which proof in the configuration: it is a structural relation only, and nothing can be inferred from it about whether the given proof actually 'proves' the given statement.

$$
\begin{array}{|l}
\hline
stmts : VersionTree[Stmt] \\
proofs : VersionTree[Proof] \\
justif : VLabel \leftrightarrow VLabel \\
\hline
\forall\, p : proofs.nodes \bullet \exists_1\, s : stmts.nodes \bullet justif(p, s) \\
\mathrm{dom}\, justif = proofs.nodes \\
\mathrm{ran}\, justif = stmts.nodes \\
\hline
\end{array}
$$

In accordance with the requirements defined in Section 3, the invariant says that 'justifies' is a many-one correspondence: i.e., each proof has exactly one corresponding statement, and each statement has at least one corresponding proof, with no dangling references.

A new theorem version history is created by supplying an initial statement and proof:

$$
\begin{array}{|l}
\hline
\_\,INIT \\
\hline
stmt? : Stmt \\
proof? : Proof \\
\hline
stmts.INIT[stmt?/content?] \\
proofs.INIT[proof?/content?] \\
justif = \{proofs.root \mapsto stmts.root\} \\
\hline
\end{array}
$$

4.2 Operations

$$
\begin{array}{|l}
\hline
\_\,FreezeThmVersion \\
\hline
\Delta(stmts, proofs) \\
s?, p? : VLabel \\
\hline
justif(p?, s?) \\
stmts.FreezeNode[s?/v?] \\
proofs.FreezeNode[p?/v?] \\
\hline
\end{array}
$$

Freeze the given statement and proof of the theorem, provided they correspond.

$$
\begin{array}{|l}
\hline
\_\,EditStmt \\
\hline
\Delta(stmts) \\
new? : Stmt \\
\hline
stmts.EditCurrent[new?/content?] \\
\hline
\end{array}
$$

Edit the current statement, provided it is not frozen. (The precondition $stmts.current \notin stmts.frozen$ is carried over from $EditCurrent$.)

$$
\begin{array}{|l}
\hline
\_\,EditProof \\
\hline
\Delta(proofs) \\
new? : Proof \\
\hline
proofs.EditCurrent[new?/content?] \\
\hline
\end{array}
$$

Edit the current proof, provided it is not frozen.

$$
\begin{array}{|l}
\hline
\_\,ChangeCurrentThmVersion \_\_\_\_ \\
\Delta(stmts, proofs) \\
s?, p? : VLabel \\
\hline
justif(p?, s?) \\
stmts.ChangeCurrent[s?/v?] \\
proofs.ChangeCurrent[p?/v?] \\
\hline
\end{array}
$$

Change which statement and proof are considered the current versions for the theorem, provided they correspond.

5 The 'Theory' Object Class

This section defines a class *Theory* for collections of theorem instances.

5.1 Preliminaries

A *theorem instance* consists of the name and statement of a theorem and two version labels, for selecting the particular statement and proof from the theorem's version trees, respectively.

$$
\begin{array}{|l}
\hline
\_\,ThmInst \_\_\_\_\_\_\_\_\_\_\_\_\_\_\_\_\_\_ \\
name : ThmName \\
sversion : VLabel \\
pversion : VLabel \\
\hline
\end{array}
$$

5.2 Object state

A *theory* is a set of theorem instances such that no theorem name appears more than once (i.e., a theory contains *at most one* version of each theorem). For convenience, a derived attribute *workingset* is introduced to stand for the 'working set' of (names of) theorems in the theory.

$$
\begin{array}{|l}
\hline
thms : \mathbb{F}\ ThmInst \\
\hline
workingset : \mathbb{F}\ ThmName \\
\hline
workingset = \{a : thms \bullet a.name\} \\
\forall\, a, b : thms \bullet a.name = b.name \Rightarrow a = b \\
\hline
\end{array}
$$

Initially, a theory is empty:

$$
\begin{array}{|l}
\hline
\_\,INIT \_\_\_\_\_\_\_\_\_\_\_\_\_\_\_\_\_\_ \\
thms = \varnothing \\
\hline
\end{array}
$$

5.3   Operations

$$
\begin{array}{|l}
\hline \text{\textit{AddThm}} \rule[-0.3em]{0pt}{1em}\hline
\Delta(\textit{thms}) \\
a? : \textit{ThmInst} \\
\hline
a?.name \in \textit{workingset} \Rightarrow \\
\quad thms' = (thms \setminus \{a\}) \cup \{a?\} \\
\textbf{where } a = (\mu\, a : thms \mid a.name = a?.name) \\
a?.name \notin \textit{workingset} \Rightarrow \\
\quad thms' = thms \cup \{a?\} \\
\hline
\end{array}
$$

Add the given theorem instance to the theorem, replacing the instance of the same name if necessary.

$$
\begin{array}{|l}
\hline \text{\textit{RemoveThm}} \rule[-0.3em]{0pt}{1em}\hline
\Delta(\textit{thms}) \\
a? : \textit{ThmInst} \\
\hline
thms' = thms \setminus \{a?\} \\
\hline
\end{array}
$$

Remove the given theorem instance from the theory.

6   Theorem Networks

This section defines the configuration consistency and completeness properties for the case study. First we define a *theorem map* to be a collection of theorem names and their corresponding version histories:

$$ ThmMap == ThmName \nrightarrow ThmVHistory $$

Theorem maps contain the contextual information required for the definitions below.

6.1   Auxiliary functions

The following relation represents the theorem instances which belong to ('are valid in') a given theorem map:

$$
\begin{array}{|l}
\hline
isValidInst : ThmInst \leftrightarrow ThmMap \\
\hline
isValidInst(a, lookup) \Leftrightarrow \\
\quad a.name \in \text{dom } lookup \,\wedge \\
\quad a.sversion \in t.stmts.nodes \,\wedge \\
\quad a.pversion \in t.proofs.nodes \,\wedge \\
\quad t.justif(a.pversion, a.sversion) \\
\quad \textbf{where } t = lookup(a.name) \\
\hline
\end{array}
$$

(Where variable declarations in Z definitions are obvious, they are dropped for brevity.) The following function returns the proof corresponding to a (valid) theorem instance:

$$getProof : ThmInst \times ThmMap \nrightarrow Proof$$

$$\text{dom } getProof = isValidInst$$
$$getProof(a, lookup) = (t.proofs.deref)(a.pversion)$$
$$\textbf{where } t = lookup(a.name)$$

The function $getStmt : ThmInst \times ThmMap \nrightarrow Stmt$ is defined similarly.

6.2 Dependency checking functions

We say a theorem instance a *depends directly on* theorem instance b if the proof of a uses b:

$$ddo : \mathbb{P}(ThmInst \times ThmInst \times ThmMap)$$

$$ddo(a, b, lookup) \Leftrightarrow isValidInst(a, lookup) \wedge b \in uses(getProof(a, lookup))$$

A *dependency chain* is a sequence of theorem instances, each of which depends directly on the next one in the chain:

$$isDepChain : (\text{seq } ThmInst) \leftrightarrow ThmMap$$

$$isDepChain(as, lookup) \Leftrightarrow \forall i : 1 \ .. \ \#as - 1 \bullet ddo(as(i), as(i+1), lookup)$$

Theorem instance a *depends* on theorem instance b if there is a non-trivial dependency chain from a to b:

$$dependsOn : \mathbb{P}(ThmInst \times ThmInst \times ThmMap)$$

$$dependsOn(a, b, lookup) \Leftrightarrow$$
$$\exists \, as : \text{seq } ThmInst \bullet$$
$$\#as \geq 2 \wedge isDepChain(as, lookup) \wedge head \ as = a \wedge last \ as = b$$

The *supporters* of theorem instance a is the set of all theorem instances on which a depends:

$$supporters : ThmInst \times ThmMap \rightarrow \mathbb{P} \ ThmInst$$

$$supporters(a, lookup) = \{b : ThmInst \mid dependsOn(a, b, lookup)\}$$

A theory is closed if it has no dangling references (i.e., it contains all supporters of its theorem instances):

$$isClosed : Theory \leftrightarrow ThmMap$$

$$isClosed(T, lookup) \Leftrightarrow \forall \, a : T.thms \bullet supporters(a, lookup) \subseteq T.thms$$

A theory is *circular* if it contains a theorem instance which depends on itself:

$$isCircular : Theory \leftrightarrow ThmMap$$

$$isCircular(T, lookup) \Leftrightarrow \exists \, a : T.thms \bullet dependsOn(a, a, lookup)$$

6.3   Theorem checking functions

The following checks whether a particular theorem instance is *proven*:

$$isProven : ThmInst \leftrightarrow ThmMap$$

$$isProven(a, lookup) \Leftrightarrow$$
$$isValidInst(a, lookup) \wedge proves(getProof(a, lookup), getStmt(a, lookup))$$

Note that, in this model, being 'proven' just means that the immediate proof is complete. Since the proof may however depend on proofs which are not themselves complete, being proven in this sense is not sufficient to fully justify a theorem.

A theorem is *established* if it is proven and everything on which it depends is proven:

$$isEstablished : ThmInst \leftrightarrow ThmMap$$

$$isEstablished(a, lookup) \Leftrightarrow$$
$$isProven(a, lookup) \wedge \forall b : supporters(a, lookup) \bullet isProven(b, lookup)$$

Again, a theorem may be 'established' and yet not be valid: e.g. if reasoning is circular.

A theory is *compete* if it is closed and non-circular, and all its theorems are established:

$$isComplete : Theory \leftrightarrow ThmMap$$

$$isComplete(T, lookup) \Leftrightarrow$$
$$isClosed(T, lookup) \wedge \neg\ isCircular(T, lookup)$$
$$\wedge\ \forall a : T.thms \bullet isEstablished(a, lookup)$$

7   The 'Theory Configuration' Object Class

This section describes a class *TheoryConfig* for version management of a theory. This establishes two levels of version: versions of theorems within the theory, and versions of the theory itself.

7.1   Object state

A *theory configuration* (or configuration, for short) consists of a theorem network, together with versions of a theory.

$$lookup : ThmMap$$
$$theory : VersionTree[Theory]$$

$$currentTheory : Theory$$

$$currentTheory = theory.currentcontent$$

Note that the 'working' version of a theory and the 'current' version of a theory may get out of step: i.e., theorem instances in the current version are not necessarilly the current versions of the theorems in the working set.

Initially the configuration is empty:

$$
\begin{array}{|l}
\_INIT \_\_\_\_\_\_ \\
\hline
lookup = \varnothing \\
theory.INIT[T/content?] \textbf{ where } T = Theory.INIT \\
\end{array}
$$

7.2   Operations

The following are some of the operations that would be available to users:

$$
\begin{array}{|l}
\_AddNewThm \_\_\_\_\_ \\
\hline
\Delta(lookup) \\
name? : ThmName \\
stmt? : Stmt \\
proof? : Proof \\
\hline
name? \notin \mathrm{dom}\ lookup \\
lookup' = lookup \oplus \{name? \mapsto t\} \\
\textbf{where } t = ThmVHistory.INIT \\
\end{array}
$$

Add a new theorem to the configuration.

$$
\begin{array}{|l}
\_AddThmToTheory \_\_\_\_\_ \\
\hline
\Delta(theory) \\
a? : ThmInst \\
\hline
isValidInst(a?, lookup) \\
theory.current \notin theory.frozen \\
currentTheory.AddThm \\
\end{array}
$$

Add the given theorem instance to the current theory, provided it is not frozen. (If there is a theorem instance with the same name in the theory, it will be removed.)

$$
\begin{array}{|l}
\_RemoveThm \_\_\_\_\_ \\
\hline
\Delta(theory) \\
a? : ThmInst \\
\hline
isValidInst(a?, lookup) \\
theory.current \notin theory.frozen \\
currentTheory.RemoveThm \\
\end{array}
$$

Remove the given theorem from the current theory, provided it is not frozen.

$$
\begin{array}{|l}
\_FreezeThm \_\_\_\_\_ \\
\hline
\Delta(lookup) \\
a? : ThmInst \\
\hline
isValidInst(a?, lookup) \\
\bigwedge a : supporters(a?, lookup) \cup \{a?\} \bullet \\
\quad lookup(a.name).FreezeThmVersion[a.sversion, a.pversion/s?, p?] \\
\end{array}
$$

Freeze the given theorem instance and all its supporters.

```
┌─ FreezeTheory ────────────────┐
│  Δ(lookup, theory)            │
│ ─────────────────────────     │
│  isComplete(T, lookup)        │
│  ⋀ a : T.thms • FreezeThm[a/a?] │
│  T.FreezeCurrent              │
│  where T = currentTheory      │
└───────────────────────────────┘
```

Freeze all theorem instances in the current theory, provided it is complete.

```
┌─ DeriveNewTheory ─────────────┐
│  Δ(theory)                    │
│ ─────────────────────────     │
│  theory.current ∈ theory.frozen │
│  theory.DeriveNewCopy         │
└───────────────────────────────┘
```

Derive a new theory by taking a copy of the current theory; as a precondition we require the current theory to be frozen.

Other operations would include:

- changing what is considered the current version of a theory or theorem;
- editing (non-frozen) statements and proofs;
- deriving new versions of statements and proofs;
- changing the version of the theorem referenced in a proof.

7.3   Adherence to versioning policy

It is easy to show that the following properties are "behavioural invariants" of the class (i.e., they are true initially and are preserved by all enabled operations):

```
┌───────────────────────────────────────────────────────────────────┐
│  ∀ t : ran lookup • justif(t.proofs.current, t.stmts.current)      │
│  ∀ t : ran lookup • ∀ p : ran(t.proofs.deref) • ∀ a : uses(p) •    │
│      isValidInst(a, lookup)                                        │
│  ∀ T : ran(theory.deref) • ∀ a : T.thms • isValidInst(a, lookup)  │
│  ∀ i : theory.frozen • let T = deref(i) in                        │
│      isComplete(T, lookup) ∧ ∀ a : T.thms • let t = lookup(a.name) in │
│          a.sversion ∈ t.stmts.frozen ∧ a.pversion ∈ t.proofs.frozen │
└───────────────────────────────────────────────────────────────────┘
```

In words, the behavioural invariant says:

1. the current statement and proof of any theorem in the configuration are always in correspondence;
2. all proofs in the configuration use only theorem instances which are valid in the configuration (i.e., there are no dangling references in proofs);
3. all theorem instances in all theory versions are valid in the configuration (i.e., no dangling references in theories); and
4. all frozen versions of the theory are complete and all of their theorems are frozen.

Properties 1–3 make explicit the configuration consistency conditions implicit in Section 3. Property 4 shows that our model satisfies the versioning policy defined in Section 3.

8 Conclusions

8.1 Summary

This paper described a case study in fine-grained version and configuration management. The subject of the case study was the management of a formal theory consisting of theorems and proofs. We demonstrated that fine-grained versioning of versioned objects provides substantially more flexibility than traditional approaches that manage only high-level coarse-grained objects.

There are two dimensions to the benefits accrued by the use of fine-grained versioning models in our case study. The first is due to the fact that we consider the individual components of a theory as first class citizens in the context of configurations. This allows substantial flexibility in the way in which consistency of an overall theory store is determined, as well as focusing attention on the specific objects undergoing change. The second benefit comes from the actual versioning of the theory components themselves. It allows us to define consistency criteria in terms of the conditions that must be satisfied by the individual components. We can then show that the chosen versioning model actually meets the criteria.

A key issue arising in the case study concerned requirements associated with the definition of a versioning policy. We presented a generic versioning class which was used as the basis for versioning of all objects within configurations. As we have shown, the definition of the basic versioning framework is insufficient in itself to guarantee the consistency of configurations. Thus the versioning policy must be constructed with some care, to ensure that consistency criteria of the overall system are met. In our example, this required that careful consideration be given to the operations that implement the versioning policy model. We were required to define additional constraints in the pre-conditions of the versioning operations that used primitive versioning mechanisms, to ensure that these operations preserved the configuration consistency criteria.

The simple version control model illustrated in this paper is adequate for simple versioning policies (and does solve the problems we have encountered and outlined in section 1.2). However, due to the consistency constraints present in the model, it has deficiencies which would make the versioning model outlined here impractical for large development environments:

- Some principles of abstraction and independence are violated. For example, freezing a theorem results in all supporters being frozen, even those whose proofs are still under construction. When the latter are subsequently modified (by deriving and editing new versions), it will be necessary to relink all proofs which refer to the old version. We have identified a solution to this problem, however it does complicate the overall model.
- The versioning offered in our model provides a opportunity to identify the minimal amount of rework required to accommodate a change in a theory. However, this does require strategic decisions to be made, such as whether or not to update the version of a theorem referenced in some other proof when a new version of the theorem in created. This may in turn complicate the

overall theory structure: for example, it may be desirable to allow different versions of theorems to be used simultaneously in a single version of a theory store. (Our model currently precludes this.) The desire for such flexibility has to be balanced against the complication it adds to the process of theory maintenance (e.g. when freezing a theory).

Using versioning in the way proposed in this paper offers developers the opportunity to make decisions as to how to react to a given change and to control the immediate extent to which a change (new version) will impact on the rest of the system. Without the versioning, such options would be extremely difficult (or impossible) to provide.

8.2 Further Work

During our work a number of other issues arose, particularly with respect to possible extensions to our framework which would allow developers to reduce the impact of change (as opposed to accurately assessing the actual work required to react to a change). The case study considered a single theory. It's clear that, by supporting more sophisticated theory structuring mechanisms (such as theory interpretation and instantiation), the impact of change could be localised better. We intend to test our hypothesis that the versioning model outlined here will scale to more complex theory structures, to provide a basis for the flexible theory construction and maintenance facilities suitable for large-scale formal development.

The model we have presented is a core model. In addition to this one would typically define high (user)-level processes based on these models. Whereas consistency constraints for objects within a theory store are defined by the model, the process which evolves and uses these underlying concepts need not be fixed. Our framework can be used as a basis upon which more sophisticated process models can be developed. Such models can offer context-sensitive guidance to the users of systems, as well as the opportunity to further constrain the way in which a system is used. It is possible to define and reason about intermediate states of configuration consistency, and to offer guidance on how to bring the system back into a consistent state, for example. We have illustrated these ideas on theory management (but without versioning) [11], and have extended them to the case study presented in this paper.

Finally, we are prototyping tools to support our fine-grained configuration framework using object-oriented database technology [2].

Acknowledgements The authors gratefully acknowledge the useful contributions of Yaowei Liu and Sabine Sachweh to the work presented here.

References

1. J-R. Abrial. *The B Book: Assigning Programs to Meanings.* Cambridge University Press, 1996.

2. F. Bancilhon, C. Delobel, and G. Harrus. *Building an Object-Oriented Database System: the Story of O2*. Morgan Kaufman, 1992.

3. A. Bloesch, E. Kazmierczak, P. Kearney, and O. Traynor. Cogito: A Methodology and System for Formal Software Development. *International Journal of Software Engineering and Knowledge Engineering*, 5(4), December 1995.

4. P. Cederqvist. CVS – concurrent versions system. http://uther1.phy.ornl.gov/offline/manual/cvs/cvs.html.

5. International Electrotechnical Commission. Functional safety: safety-related systems. Draft International Standard IEC 1508, June 1995.

6. R. Duke, P. King, G. Rose, and G. Smith. The Object-Z specification language. SVRC TR 91-1, 1991.

7. T. Hart et al. Formal methods pilot project. In *Proc. Asia-Pacific Software Engineering Conference (APSEC'96), Seoul*, December 1996. Also SVRC TR 96-17.

8. IEEE. IEEE Std 1042: IEEE guide to software configuration management, 1987.

9. C. B. Jones, K. D. Jones, P. A. Lindsay, and R. Moore. *mural: A Formal Development Support System*. Springer-Verlag, 1991.

10. T. Jones and J. Welsh. Requirements for a generic, language-based diagram editor. In Malti Patel, editor, *Proc. 20th Australian Computer Science Conference*, pages 316–325, Sydney, Australia, 1997. Also SVRC TR 96-10.

11. P.A. Lindsay. A formal basis for modelling process and task management aspects of user interface design. In *Proc. Formal Aspects of The Human Computer Interface*, Sheffield, UK, Sept 1996. Also SVRC TR 96-07.

12. P.A. Lindsay, Y. Liu, and O. Traynor. Managing document conformance: a case study in fine-grained configuration management. *AustComp. Sci. Communications*, 19(1):373–382, 1997. Also SVRC TR 96-20.

13. U.K. Ministry of Defence. Safety Management Requirements for Defence Systems Containing Programmable Electronics. Second Draft Defence Standard 00-56.

14. B. Peuschel. Merlin user guide. Technical Report 364, ESPRIT project 415, FB Informatik, University of Dortmund, 1992.

15. B. Peuschel, W. Schäfer, and S. Wolf. A knowledge-based software development environment supporting cooperative work. *International Journal of Software Engineering and Knowledge Engineering*, 2(1):79–106, 1992.

16. M. J. Rochkind. The source code control system. *IEEE Transactions on Software Engineering*, 1:24–36, 1975.

17. K. Ross. Models for configuration management of refinement calculus developments. In *Proc. 7th Refinement Workshop*, pages 1–26. BCS-FACS, July 1996.

18. K.J. Ross and P.A. Lindsay. Maintaining consistency under changes to formal specifications. In *Proc. 1st Int. Symp. of Formal Methods Europe (FME'93)*, LNCS 670, pages 558–577. Springer Verlag, 1993. Also SVRC TR 93-3.

19. S. Sachweh and W. Schäfer. Version management for tightly integrated software engineering environments. In *Proc. 7th Int. Conf. on Software Eng. Environments*, pages 21–31, The Netherlands, 1995. IEEE Computer Society Press.

20. W. Tichy. RCS - a system for version control. *Software – Practice and Experience*, 15, 1985.

21. J. Welsh, B. Broom, and D. Kiong. A design rationale for a language-based editor. *Software – Practice and Experience*, 21(9):923–948, 1991.

A Tactic Language for Ergo

Andrew Martin,[1] Ray Nickson,[1] and Mark Utting[2]

[1] Software Verification Research Centre
School of Information Technology
The University of Queensland
Brisbane Qld 4072 Australia
{apm,nickson}@it.uq.edu.au
[2] Department of Computer Science
School of Computing and Mathematical Sciences
The University of Waikato
Private Bag 3105
Hamilton New Zealand
marku@cs.waikato.ac.nz

Abstract. A new version of the Ergo theorem prover is under development. It uses a single tactic language, based on *Angel*, for tactic programming, user interface, and proof representation. This paper describes the language as it is used in each of these cases, and explains the details of its implementation in Qu-Prolog. An example from classical propositional calculus is included.

1 Introduction

Ergo is an interactive proof tool that has been designed and implemented at the SVRC over the last ten years. It is implemented in Qu-Prolog (Robinson and Hagen, 1997), and is designed to be extensible, so that users can add new theories, tactics and user interfaces. Ergo 5 is currently under development. Having no inbuilt object logic, it is a generic prover that can be instantiated by providing a collection of axiomatic and/or definitional theories. The core of Ergo 5 provides support for (uninterpreted) sequents with named tuples of arbitrary terms as antecedents and single terms as consequents. The interpretation of sequents, and the axioms and fundamental inference rules that relate them to object-level terms and formulas, are under the control of the theory developer. In this paper, our examples will use the sequents with a single set of antecedents called hypotheses, which are to be understood as the antecedents in Gentzen's sequent calculus (Gentzen, 1969).

Ergo 5 proofs are tree-structured, constructed backwards in a style reminiscent of natural deduction. An inference rule is a meta-level structure of the form **from** $P_1 \& \ldots \& P_n$ **infer** C, in which each term $P_1, \ldots, P_n, C$ is a sequent. The connectives & and **from-infer** are respectively meta-level conjunction and implication, which can be related to object-level connectives by the inference rules of a theory of intuitionistic logic, for example. The meaning of the meta-level implication **from** P **infer** C is that conclusion C can be deduced from premiss P. The meaning of the meta-level conjunction $P_1\&P_2$ is that both P_1 and P_2 are (to be) proved.

A successful application of an inference rule **from** $P_1\&\ldots\&P_n$ **infer** C (where $n \geq 0$) transforms the proof tree. Some identified open node whose term matches C is extended with n child nodes, the terms of which are the corresponding matches for the premisses. We can leave the proof tree implicit, instead viewing a rule as a mapping from some specified open node to a sequence of n open nodes.

Earlier versions of Ergo used Qu-Prolog (the implementation language) as a tactic language (Whitwell, 1992). This allowed efficient and powerful tactics to be written, but created some problems.

- Some users were daunted by the power of unrestricted Prolog, and would prefer a simpler dedicated language.
- Full Prolog is a complex language and is not amenable to reasoning.
- The interface presented to the tactic programmer was not precisely defined (Qu-Prolog lacks a module system, though one is planned), and was different from the command-line interface available to users.
- Failure to distinguish the implementation language from the tactic language would frustrate any future attempts to re-implement in another language.

This paper describes the tactic language which is implemented in Ergo 5. We have a number of design aims:

- The tactic language should be simple but powerful.
- The semantics should be clean, so that we can reason about tactics.
- There should be a single language that is usable as the command-line interface, for writing tactics, and for presenting proofs.
- The tactic language should be clearly distinguished from the implementation language.

We have implemented an instantiation of *Angel* (see Section 2) for Ergo 5. This implementation raised some interesting issues about the handling of sub-proofs that are not directly addressed in the Angel core. These issues, and their resolution in the Ergo 5 implementation of Angel, are discussed in Sections 3 and 4. Section 5 demonstrates how the Gumtree tactic constructs can be used to define some simple general-purpose proof procedures. In Section 6 we describe the implementation of the tactic language, and Section 7 shows how the tactics are used in an example proof. A final section discusses the context of the work, emphasising the advances over earlier work, and points to future research directions.

2   Angel

Since the Edinburgh LCF project first described *tactics* as programs for directing proof tools (Gordon, Milner, and Wadsworth, 1979), the notion has become very widespread. Many systems implement some form of tactic language for directing proofs. Sometimes, following the LCF style, tactics construct possible proofs which are later validated. In other systems the tactics form an extension of the set of primitive inference rules. The soundness of each tactic application is assured by ensuring that its only interaction with the proof under construction is by application of the primitive inference rules.

Angel (Martin, Gardiner, and Woodcock, 1996) is a generic tactic language. Initially it was intended to support proofs in the second of these styles, that is, providing a framework for the composition of primitive inference rules in the construction of backwards (goal-directed) proofs, but it turns out to be more general than this. Term rewriting, for example, which in Cambridge LCF (Paulson, 1987) is directed by a separate set of operators from those which describe tactics, could also be described using Angel. Program refinement is an instance of term rewriting (Nickson and Hayes, 1996), and we expect Ergo 5's implementation of Angel to support refinement tactics as well as proof tactics.

The concept of a *goal* is central to Angel's semantics, and to its genericity. A goal is an object of the underlying system that can be manipulated by an inference rule. An execution of a basic Angel tactic maps a single goal to another. In a term-rewriting system, a goal is a term of the logic, and a tactic application yields a new goal (the transformed term). In Ergo 5, a goal is a sequence of of open proof nodes; an application of a tactic yields a new sequence of open nodes. Section 3 describes the new *parallel composition* operator of the Gumtree language, which can analyse goals (node sequences) into subgoals (subsequences or individual nodes).

The paper of Martin et al. (1996) describes Angel in an abstract way, independent of any particular implementation. It gives denotational semantics and a (more accessible) axiomatic semantics, in the form of a large collection of laws which allow one tactic to be transformed into another. Also presented are some commonly used derived tactics—together with derived transformation rules. Further derived tactics are discussed in Martin's thesis (1994).

Because Angel is a small language, its semantics is quite clean and easy to reason about. Nevertheless it is able to describe a large class of useful algorithms. The language is named Angel because the account in (Martin et al., 1996) makes tactics angelically nondeterministic. That is, a tactic is a relation between goals, rather than a single function. Thus we may arrange that a compound tactic will fail only if there is no possible path from input to output: an implementation must avoid dead-end paths, which is typically achieved by a backtracking search. The angelic style removes the need for the tactic programmer to program backtracking explicitly.

The principal constructs of Angel are:

Rules The atomic tactic '**rule** R' applies an inference rule. If applied in the domain of the rule, it will map a goal to a single new goal. If a rule is applied outside its domain, it fails.

Special atomic tactics The atomic tactic **skip** always succeeds exactly once, leaving its goal unchanged. The atomic tactic **fail** always fails.

Sequential composition A sequential composition t_1 ; t_2 first applies t_1 and then applies t_2 to the resulting goal. If t_1 fails, the whole composition fails; if t_2 fails, other alternatives in t_1 may be explored.

Alternation An alternation $t_1 \mid t_2$ applies either t_1 or t_2, succeeding if either one succeeds.

Cut the cut operator $!\, t$ prunes away all but the first alternative from t. That is, if some subsequent tactic fails, no alternatives within t will be explored.

We adopt the following order of precedence for operator binding: function application (such as application of **rule** to its argument) binds most tightly, followed by cut, then sequential composition. The parallel combinator $\|$ (to be defined in Section 3) is next, and alternation binds most loosely.

These semantics may be expressed by viewing a tactic as a function mapping a single input goal to a sequence of alternative output goals. Recursive tactics are given meaning using fixpoints. A forthcoming paper generalises the semantic description of Angel tactics, discussing which of the following laws hold in different circumstances (for instance, if a notion of a single-threaded state is added to the denotation of the tactics).

A collection of nineteen laws illustrates the interactions of the basic Angel tactic constructs. Some are straightforward: **skip** ; $t = t$, and **fail** ; $t = $ **fail**, etc. The right-distributive law is perhaps the most indicative of the angelic semantics: we have $(t_1 \mid t_2)\, ;\, t_3 = (t_1\, ;\, t_3) \mid (t_2\, ;\, t_3)$. This is not the case in (say) LCF or HOL.

In addition to these basic constructs, Angel also provides for *structural combinators* which permit tactics to be applied to sub-expressions of the goal. The set of structural combinators provided by an Angel implementation will generally depend on the nature of the inference system that underlies Angel. For example, the logic that underlies a refinement tool will have inference rules that allow refinement of the components of a compound program fragment; an instantiation of Angel for a refinement tool might include structural combinators corresponding to program constructors, as in the tactic language of Red (Vickers, 1990).

In a theorem prover supporting backwards inference in the style of natural deduction, such as Ergo 5, (Utting, 1996), the input of a typical fundamental inference step is a goal to be proved (the *conclusion* of the inference rule), and the result is a set or sequence of *premisses*. The account of Angel in (Martin et al., 1996) describes a *parallel* tactic combinator for dealing with proof trees which branch in this way. A different approach has been taken in Ergo 5; this is described in Section 3, below.

3 Gumtree for Tactic Programming

The tactic-based interface for Ergo 5 is called *Gumtree*. We have decided to use a single language for the interactive prover interface, for tactic programming and for the representation of proofs. This is a language based on Angel, with various adaptations. A major development is in the treatment of parallel subgoals, described in the following sections. Other changes make the language easier to use in an interactive context. We believe that the majority of the laws of Angel will also apply to Gumtree—see (Martin, Nickson, and Utting, 1997). We hope that these laws will permit tactic optimisation, as well as enhancing users' understanding of the semantics of the tactics they write.

Gumtree implements all of the tactic combinators described above. Ergo 5 is a generic prover with no inbuilt core logic. The genericity means that it is not possible to provide a fixed set of Angel structural combinators corresponding to object-level connectives. Instead, we introduce a single structural combinator that corresponds to the tree-structure arising from the meta-level conjunction in the premisses of inference rules. We write this operator $\|$ (pronounced 'parallel').

Open Subgoals Because in a goal-directed proof the application of a rule or tactic to a single goal may give rise to several subgoals, and it may be appropriate to deal with several at once, Gumtree's goals are *sequence* of subgoals (which are leaves of the proof tree) to record the user's current position.

For example, with an initial goal of $\vdash A \wedge B \wedge C$, two applications of the rule *and−intro* will give rise to a new sequence of subgoals: $\vdash A \quad \& \quad \vdash B \quad \& \quad \vdash C$.

Gumtree's Parallel In the paper of Martin et al., the parallel operator and meta-conjunction used in this way were strictly non-associative binary operators. This has proven to be sometimes hard to use. For example, following the application of a large tactic which produces three subgoals, it is necessary to know how they are grouped, before further tactics or rules can be applied. In the example above, the structure of the proof tree (and any tactics which are to be subsequently applied) would depend on the bracketing used in the term $A \wedge B \wedge C$, which is unfortunate. A completed proof branch is denoted by a special distinguished subgoal (). Special processing is needed to arrange that for any subgoal g, we have that $g\&()$ is transformed to g.

In Gumtree we allow $\|$ and $\&$ to be *associative*, and to remove empty subgoals seamlessly. Thus, in effect, we treat the set of current subgoals as a sequence, and any collection of tactics which is to operate on them as a sequence, too. Added subtlety comes from the fact that in a tactic such as $(t_1 \| t_2)$, either (or both) of t_1 and t_2 may in turn contain a parallel composition—and by the associative property, therefore, this tactic may apply to a sequence of (say) five subgoals. Further discussion of this design decision, and the resulting tactic laws, can be found in the technical report (Martin et al., 1997).

Tactic Arity The manner in which these subgoals are associated with the tactics is clearly critical. When the tactics exhibit nondeterministic behaviour, the complication is multiplied. To explain how parallel tactics are composed, we define a notion of tactic *arity*. The arity of a deterministic tactic is a pair $m \rightarrow n$; a tactic with this arity will transform a goal with m subgoals into one with n subgoals. A system of arity-checking assists in pre-execution correctness checks for such tactics, and allows certain optimisations (see Section 6).

Behaviour of Parallel Informally, parallel behaves as follows: when the tactic $(t_1 \| \ldots \| t_n)$ is applied to the goal $(g_1 \& \ldots \& g_m)$, tactic t_1 (of arity $j \to k$) operates on some initial subsequence of the g_i—say $g_1 \& \ldots \& g_j$—returning some sequence $h_1 \& \ldots \& h_k$, and leaving t_2 to operate on some initial sequence of $g_{j+1} \& \ldots \& g_m$, etc. Eventually all the h_i are concatenated to form a new goal, provided the last tactic is of a suitable arity to operate on all the remaining subgoals.

When one or more tactics in the parallel composition is nondeterministic, each of the possible outcomes must be tried. The outcomes may have different arities, so that arities of nondeterministic tactics may not be fixed.

Fixed-arity tactics The tactic **skip** is fully general; it will apply to any sequence of subgoals, and succeed, leaving them unchanged. Sometimes it is useful to have less general versions of **skip** which match specific numbers of subgoals. Thus we define **zero** which succeeds when the current list of subgoals is empty, and fails otherwise, and **one** which succeeds (leaving the goal unchanged) when there is exactly one current subgoal. Then we may define

$$two = \textbf{one} \| \textbf{one}$$

and more generally

$$\begin{aligned} \textit{fix } \, 0 &= \textbf{zero} \\ \textit{fix } \, (n+1) &= \textit{fix } \, n \| \textbf{one} \end{aligned}$$

These tactics obey a number of interesting laws. Details may be found in the technical report of Martin et al. (1997).

Examining terms Angel includes a powerful construct (π) that can be used to write tactics whose behaviour depends on the detailed structure of the goals to which they are applied. For Gumtree, we have adopted a simpler construct. The tactic **term**(T) succeeds if the current goal has a single subgoal, and the term associated with that subgoal is unified with T. Success or failure of **term** can be used to write tactics that conditionally branch based on the terms to which they are applied; the variable bindings resulting from unification can provide parameters for rule applications.

4   Gumtree for User Interface

As well as being used for programming tactics, Gumtree is the language used as Ergo's command line interface. This imposes some additional requirements on the language, and leads to some further departures from Angel.

The command line interface displays a prompt showing the current set of open nodes. Initially this is just the node which will be the root of the proof tree, whose sequent is the sequent to be proved. The user enters a Gumtree command; this may be a Gumtree tactic (such as a rule application, or a sequential or parallel composition, or a user-defined tactic), or a special interface command.

Executing tactics If the user command is a tactic, the tactic is executed, with the goal sequence shown in the prompt as input. If the tactic fails, a message is printed and the new prompt is the same as the old. If the tactic succeeds, the new prompt shows the goals that resulted from the tactic application. If there may be alternative solutions to the tactic, just the first solution is shown. The alternatives may be explored on backtracking.

Special commands The system maintains a history of all commands that have contributed to the proof. The **history** command displays this history as a numbered list. The **retry** command triggers backtracking to explore alternative solutions for previous commands; if there are no alternatives for the most recent command, that command fails and disappears from the history. Subsequent **retry** commands will explore alternatives for the next most recent command. With a numeric argument corresponding to one of the numbered history entries, **retry**(N) starts the process of finding alternatives from the specified command; those after it fail immediately. The **commit** command discards any alternative solutions for the last command in the history, as if the command were surrounded by a cut operator; with a numeric argument it forms the sequential composition of all commands beyond that one in the history, and commits the resulting composition. The command **undo** is equivalent to **commit** followed by **retry**; with an argument N, it is equivalent to **commit**(N) and **retry**. These special interface commands cannot be used in tactics.

Global commands Other global Ergo commands can be used from the Gumtree prompt. The user can define named tactics with parameters, examine the state of the proof in various formats, alter the information displayed in the prompt, and save and load portions of proof trees in files. These global commands are not directly available in tactics (though they can be executed by escaping to Prolog: see Section 6).

Navigating the proof tree Because interactive users find it is sometimes convenient to focus attention on only part of the proof tree, two further commands are added to the Gumtree tactic language to support the user interface. The **defer** command temporarily removes the open proof nodes to which it is applied from the goal sequence, as if they had been proved. The **select**(*Nodes*) command makes the specified sequence of nodes current, deferring the sequence that was previously current and awakening any previously deferred nodes that are selected. Node selection and tactic execution may be combined, as *Nodes* ::: *tactic*. To facilitate proof tree navigation using these commands, nodes may be named. The command **name**(*Name*) associates the name with the proof node to which it is applied; **names**([*Name*$_1$,...,*Name*$_n$]) names many nodes simultaneously. The name will thenceforth be used in place of the number to identify the node.

These navigation commands are intended for use in interactive proof scripts. Their use in general-purpose tactics is discouraged as it makes analysis (and hence checking and optimisation) of tactics difficult.

5   Derived Tactics

Using the tactics we have defined, some derived tactics may be constructed. The following tactics arise in most tactic-based systems, though the names are not well standardised. The equations below may be viewed as definitions of the derived tactics, though the tactics may also be implemented directly (for example, for efficiency's sake).

try t attempts to apply t, but succeeds whether t is successful or not. *exhaust t* applies t repeatedly; as many times as possible.

$$try\ t = t \mid \mathbf{skip}$$
$$exhaust\ t = t\ ; exhaust\ t \mid \mathbf{skip}$$

Variations on these, such as exhaustive application subject to a minimum or maximum number of applications, are also possible.

We also define tactics which apply their arguments to the current parallel goals: *every t* will apply t to every available subgoal; *tryevery* is a robust version of *every* which applies t wherever possible, and leaves unchanged any goal to which t is not applicable:

$$every\ t = (t \parallel every\ t) \mid \mathbf{zero}$$
$$tryevery\ t = every(t \mid \mathbf{one})\ .$$

If we wish to try to apply t not just to each of the current goals, but also to any resulting subgoals, we can employ a form of breadth-first search. First, we define *any t*, which applies t exactly once to some of the current goals, and *some t*, which behaves like *tryevery t*, but fails if *all* of the parallel applications of t fail.

$$any\ t = \mathbf{skip} \parallel t \parallel \mathbf{skip}$$
$$some\ t = every(any\ t)$$

Now, the breadth-first application of t is simply the exhaustive application of *some t*: try to apply t to each of the current subgoals. If it does not apply to any of them, then stop. Otherwise, we have a new list of subgoals to which the procedure can be applied recursively.

$$bfs\ t = exhaust(some\ t)$$

We may also define a depth-first application of t, and expect that for a deterministic t the outcome should be the same as that for the breadth-first application. The relative efficiencies will depend on the nature of t.

$$dfs\ t = every(t\ ; dfs\ t \mid \mathbf{one})$$

6 Implementation

Ergo is implemented in Qu-Prolog (Robinson and Hagen, 1997).[1] Qu-Prolog provides *implicit parameters* (Cheng, Robinson, and Staples, 1991), which behave like global variables, except that updates are undone on failure (as with normal Prolog variables), so that the Prolog backtracking search can be used. Implicit parameters are efficiently implemented in Qu-Prolog using destructive update (with trailing of values), but their semantics can be cleanly described by supposing that every Prolog procedure that refers to an implicit parameter has an extra argument providing the value of that parameter, and every procedure that updates the parameter has an extra argument for returning the new value.

A recent Qu-Prolog development implements *indexed implicit parameters*, which are associative arrays with the same properties as implicit parameters. Individual elements of indexed implicit parameters can be accessed and updated in near-constant time.

The Ergo proof tree is represented by an indexed implicit parameter (Utting, 1996). Each node has a unique identity, which is its index into the indexed implicit parameter. The value stored there is a term of the form:

$$node(\textit{Term}, \textit{Context}, \textit{Rule}, \textit{SubNodes})$$

The sequent represented by this node is $\textit{Context} \vdash \textit{Term}$. If the proof node has been justified, *Rule* is the name of the inference rule used, and *SubNodes* is a list (possibly empty) of the nodes representing the premisses of the rule. If the node is open, *Rule* and *SubNodes* are undefined.

The interface to the proof tree abstract data type is via the procedures

proof_start(*Term*, *Context*, *Id*)

(which creates the open root node *Id* of a new proof tree) and

proof_step(*Rule*, *Id*, *SubNodes*)

(which applies *Rule* to justify the node *Id* and returns the list of new nodes *SubNodes*). Gumtree tactics use these procedures to modify the proof structure. Additionally, the tactic interpreter must maintain the sequence of current goals, partitioning it appropriately among the branches of parallel compositions.

Essentially, tactics are viewed as nondeterministic functions between current goal sequences, that modify proof trees by 'side effect' (work is in progress to adapt the abstract model of Angel to incorporate side-effects, using monads (Moggi, 1989)). Nondeterministic functions are easily represented in Prolog as procedures with two arguments: one corresponding to the input and one to the output. The proof tree is modified (apparently by 'side effect') using the procedures for updating implicit parameter arrays.

The tactic **skip** is the identity operation; it leaves both the proof tree and current goal sequence unchanged. Its implementation in Prolog can be described by the clause:

---

[1] This section assumes some familiarity with Prolog, but not with Qu-Prolog.

```
tactic(skip, Goals, Goals).
```

The tactic **one** maps a singleton goal sequence to itself, without modifying the proof tree:

```
tactic(one, [Goal], [Goal]).
```

This will fail unless the input goal sequence (passed as the second argument) unifies with [Goal], i.e., it must be a sequence of length one. The output goal sequence (returned as the third argument) is the same as the input. This generalises to *fix n*:

```
tactic(fix(N), Goals, Goals) :-
  length(Goals, N).
```

The tactic that applies a rule accepts a single input, to which the rule is applied, and returns the premisses of the rule:

```
tactic(rule(R), [InGoal], OutGoals) :-
  proof_step(R, InGoal, OutGoals).
```

Sequential composition simply 'threads' the goal sequence through the composed tactics, while alternation is represented directly by Prolog nondeterminacy, and ! by Prolog cut.

```
tactic((T1;T2), InGoals, OutGoals) :-
  tactic(T1, InGoals, MidGoals),
  tactic(T2, MidGoals, OutGoals).
tactic((T1|T2), InGoals, OutGoals) :-
  tactic(T1, InGoals, OutGoals).
tactic((T1|T2), InGoals, OutGoals) :-
  tactic(T2, InGoals, OutGoals).
tactic(!(T), InGoals, OutGoals) :-
  tactic(T, InGoals, OutGoals),
  !.
```

Note that the two tactics in an alternative composition may have different arities, so by analysing arities (perhaps during compilation) it may be possible to decide that one clause is inapplicable, reducing the need for search.

6.1 Parallel Composition

A parallel composition of two tactics must partition the input goal sequence between the composed tactics, and concatenate the results:

```
tactic((T1||T2), InGoals, OutGoals) :-
  append(InGoals1, InGoals2, InGoals),
  tactic(T1, InGoals1, OutGoals1),
  tactic(T2, InGoals2, OutGoals2),
  append(OutGoals1, OutGoals2, OutGoals).
```

This implementation would be hopelessly inefficient: for each possible partitioning of `InGoals`, it tries `T1` and (perhaps) `T2`. Instead, we augment `tactic/3` with a further argument, and allow tactics to operate on any prefix of the input goal sequence, returning the unused input as well as the output. Now, sequential composition passes the result of $T1$ as input to $T2$, while parallel composition passes the unused goals.

```prolog
tactic(fix(N), InGoals, UnusedGoals, OutGoals) :-
  %% OutGoals = the initial segment of InGoals, of length N;
  %% UnusedGoals = the remainder.
  length(OutGoals, N),
  append(OutGoals, UnusedGoals, InGoals).
tactic(rule(R), [InGoal|UnusedGoals], UnusedGoals, OutGoals) :-
  %% The rule is applied to the first InGoal, leaving the
  %% remainder unused.  The OutGoals are the premisses of the
  %% rule application.
  proof_step(Rule, InGoal, OutGoals).
tactic(term(T), [InGoal|UnusedGoals], UnusedGoals, [InGoal]) :-
  %% T is unified with the term associated with  the first InGoal,
  %% leaving the unused.  The single output goal is the first InGoal.
  proof_node_term(InGoal, T).
tactic((T1;T2), InGoals, UnusedGoals, OutGoals) :-
  %% The unused goals are those left unused by T1;
  %% the output goals of T1 are passed to T2, and it must use
  %% all of them.
  tactic(T1, InGoals, UnusedGoals, MidGoals),
  tactic(T2, MidGoals, [], OutGoals).
tactic((T1|T2), InGoals, UnusedGoals, OutGoals) :-
  %% Simple nondeterministic choice.
  tactic(T1, InGoals, UnusedGoals, OutGoals).
tactic((T1|T2), InGoals, UnusedGoals, OutGoals) :-
  tactic(T2, InGoals, UnusedGoals, OutGoals).
tactic(!(T), InGoals, UnusedGoals, OutGoals) :-
  tactic(T, InGoals, UnusedGoals, OutGoals),
  !.
tactic((T1||T2), InGoals, UnusedGoals, OutGoals) :-
  %% The unused goals of T1 are passed to T2; any that T2
  %% does not use are unused by the composition.
  %% are merged.
  tactic(T1, InGoals, Unused1, Out1),
  tactic(T2, Unused1, UnusedGoals, Out2),
  append(Out1, Out2, OutGoals).
```

The remaining occurrences of **append** (which leads to non-linear time behaviour) can be eliminated using standard Prolog transformations (accumulator pairs and difference lists).

To apply a tactic, the user interface simply calls `tactic/4`, passing the current goal sequence as *InGoals*, ensuring that no goals are left unused, and obtaining the new goal sequence from *OutGoals*:

```prolog
do_command(T, InGoals, OutGoals) :-
```

```
tactic(T, InGoals, [], OutGoals).
```

6.2   Compilation of Tactics

The above is a simple tactic interpreter. Using the meta-programming facilities (**term_expansion** and definite clause grammars) available in most Prolog systems, including Qu-Prolog, we can modify the definitions to produce a tactic compiler. Instead of passing a tactic to **tactic/4** for execution, the compiler uses the clauses of **tactic/4** (suitably annotated) as a database of translation rules. For example, the tactic *fix N* is translated directly to

```
length(OutGoals, N), append(OutGoals, UnusedGoals, InGoals).
```

Recursive calls on **tactic/4** can almost always be handled at compile time (i.e., as recursive calls on the translator); for example, the tactic

rule $R1$; (**one** $\|$ **rule** $R2 \|$ (**rule** $R3$; *fix* 2)); *fix* 4

is translated to:

```
%% rule(R1)
  InGoals = [InR1|UnusedR1],
  proof_step(R1, InR1, OutR1),
%% ;(one||
  length(OutOne, 1),
  append(OutOne, UnusedOne, OutR1),
%% rule(R2)||
  UnusedOne = [InR2|UnusedR2],
  proof_step(R2, InR2, OutR2),
%% rule(R3)
  UnusedR2 = [InR3|UnusedR3],
  proof_step(R3, InR3, OutR3),
%% ;fix(2)
  length(OutFix2, 2),
  append(OutFix2, UnusedFix2, OutR3),
  UnusedFix2 = [],
%% )
UnusedR3 = [],
append(OutOne, OutR2, OutMidA),
append(OutMidA, OutFix2, OutMid),
%% ;fix(4)
  length(OutFix4, 4),
  append(OutFix4, UnusedFix4, OutMid),
  UnusedFix4 = [],
UnusedGoals=UnusedR1,
OutGoals = OutFix4.
```

The technique is similar to a Hidden Accumulator Grammar (Tarau, Dahl, and Fall, 1995).

If this tactic is to be applied anywhere but as a component of a parallel composition (e.g. as a 'top-level' tactic), the compiler can determine that

UnusedGoals must be empty. By carefully analysing arities (easy since this example is fully deterministic) the above can be simplified to:

```
InGoals = [InR1],
proof_step(R1, InR1, [OutGoal1,InR2,InR3]),
proof_step(R2, InR2, [OutGoal2]),
proof_step(R3, InR3, [OutGoal3,OutGoal4]),
OutGoals = [OutGoal1,OutGoal2,OutGoal3,OutGoal4].
```

The arity of the whole tactic has been determined $(1 \to 4)$, as have the arities of the rules $(R1 : 1 \to 3; \ R2 : 1 \to 1; \ R3 : 1 \to 2)$. If any of these arities disagrees with information already available to the compiler, it can issue a warning that the tactic must fail. Use of *fix*, as in this example, often allows dramatic optimisations.

6.3   Proof navigation

To handle the extra Gumtree tactics for proof navigation (Section 4) requires further minor changes to the representation of tactics.

Deferring and selecting nodes To deal with deferred nodes, the tactic system needs an additional pair of parameters, representing the initial and final lists of deferred nodes. The list is initially empty, and all constructs apart from **defer** and **select** thread the list unchanged. To defer a node, it is moved from the input list to the deferred list; to select nodes, they are moved into the current goal list:

```
tactic(defer, In, Unused, Out, InDef, OutDef) :-
  %% Defer the first input node; no output.
  In = [Node|Unused],
  Out = [],
  append(InDef, [Node], OutDef).
tactic(select(Nodes), In, Unused, Out, InDef, OutDef) :-
  %% separate available nodes (input or deferred) into
  %% outputs (selected nodes) and new list of deferred
  %% nodes (the remainder).
  append(In, InDef, Available),
  difference(Available, Nodes, OutDef),
  Out = Nodes,
  Unused = [].
```

Naming nodes The names for nodes are Prolog (meta-)variables. The command **name**(*Name*) simply unifies *Name* with the number of the input node. Unlike other Prolog systems, Qu-Prolog retains the names of metavariables throughout a session, so the user can refer to the node by the name of its metavariable. Ergo simply arranges to substitute the name of the metavariable for the node number whenever it appears in output.

6.4 Escape to Prolog

Within the simple, clean Gumtree language, it is possible to execute certain Prolog code safely. The code can directly affect neither the proof (the forthcoming module system for Qu-Prolog will enforce this) nor the goal sequences (since these parameters are hidden from Gumtree users). It *can* influence the flow of control in the tactic (by failing or by succeeding on backtracking), and can bind proof variables that are used in later proof steps. The syntax is

```
{G}
```

where G is a Prolog goal; the idea and syntax are borrowed from Definite Clause Grammars (Pereira and Warren, 1980), and it is implemented in the same way.

7 Example

This section illustrates the use of the Gumtree interface for Ergo 5. We will discuss proofs of the following theorem of classical propositional logic:

$$(A \vee (B \Leftrightarrow C)) \Leftrightarrow ((A \vee B) \Leftrightarrow (A \vee C)) \ .$$

We will use some classical sequent calculus rules as our primitive inference rules (Gentzen, 1969), with minor modifications because Ergo 5 uses sequents with a single consequent. Thus, the implication-left and not-left rules will be

$$\frac{\vdash p \vee r \qquad q \vdash r}{p \Rightarrow q \vdash r} \ implies-left \qquad\qquad \frac{\vdash q \vee p}{\neg\, p \vdash q} \ not-left \ .$$

Clearly, this is not a particularly taxing activity for a proof tool, and decision procedures for propositional calculus are known which are far more efficient than those presented here. We use this logic merely as a readily-accessible example of the application of Gumtree.

The following sections will be expressed using Ergo 5's concrete syntax. The logical connectives are clear enough. The tactic combinators are mostly as they have been presented previously. The language's sequents appear as `hyps ---> consequent`. Hypothesis lists are presented separately from the sequents themselves; see below. When discussing interaction with the tool, output is in `typewriter` font and user inputs are presented in sans serif.

7.1 Fully Interactive Proof

Having initiated an Ergo 5 proof with the proposition above as our goal, the system prints a list of open goals and prompts for a command to apply at the single open node (number 1):

```
    1::: A or (B <=> C) <=> (A or B <=> A or C)
[1]:::
```

Clearly, our first step must be to apply the rule of equivalence introduction:

```
[1]::: rule(iff_intro).
```

and the system responds with

```
    2::: (A or (B <=> C) => (A or B <=> A or C)) and
((A or B <=> A or C) => A or (B <=> C))
[2]:::
```

The next major connective is the conjunction, so we apply and-introduction, but we notice that this will give two subgoals, each of which is an implication, so we may save time by following the and-introduction with an implication-introduction in each of the subgoals:

```
[2]::: rule(and_intro);every(rule(implies_intro)).
    hyp 1:::  A or (B <=> C)
    hyp 2:::  A or B <=> A or C
    7::: hyp=[1] ---> A or B <=> A or C
    8::: hyp=[2] ---> A or (B <=> C)
[7, 8]:::
```

Now the response is considerably more complicated. Firstly, we have a listing of hypotheses which appear in some of the following sequents, then a list of open subgoals. Finally, the prompt now reflects the fact that there are two 'current' subgoals.

    The two goals are quite different in form, so we will deal with them separately. Goal 7 is similar to the goal we started with, and, recognising that the same pattern of reasoning may be useful repeatedly, we define a tactic:

```
[7, 8]::: tactic local === rule(iff_intro);rule(and_intro);every(rule(implies_intro)).
```

The goal state and the prompt remain unchanged. Now we may apply the tactic to the goals we have chosen to work on (goal 7):

```
[7, 8]::: [7] ::: local.
    hyp 1:::  A or (B <=> C)
    hyp 2:::  A or B <=> A or C
    hyp 3:::  A or B
    hyp 4:::  A or C
    12::: hyp=[1, 3] ---> A or C
    13::: hyp=[1, 4] ---> A or B
    8::: hyp=[2] ---> A or (B <=> C)
[12, 13]:::
```

Notice that goal 8 remains open, but not part of the current context. The hypotheses may be simplified by application of the or-left rule,

```
[12, 13]::: every(rule(or_left)).
    hyp 1:::  A or (B <=> C)
    hyp 2:::  A or B <=> A or C
    hyp 5:::  A
    hyp 6:::  B
```

```
    hyp 7::::  A
    hyp 8::::  C
    14::: hyp=[1, 5] ---> A or C
    15::: hyp=[1, 6] ---> A or C
    16::: hyp=[1, 7] ---> A or B
    17::: hyp=[1, 8] ---> A or B
    8::: hyp=[2] ---> A or (B <=> C)
[14, 15, 16, 17]:::
```

and some of the resulting subgoals discharged by the rule or-intro-L followed by
the rule of assumption.

```
[14, 15, 16, 17]::: some(rule(or_intro_L);rule(assump)).
    hyp 1:::  A or (B <=> C)
    hyp 2:::  A or B <=> A or C
    hyp 6:::  B
    hyp 8:::  C
    15::: hyp=[1, 6] ---> A or C
    17::: hyp=[1, 8] ---> A or B
    8::: hyp=[2] ---> A or (B <=> C)
[15, 17]:::
```

7.2 Increasing Automation

Clearly, this proof will take some time with this level of user intervention. We
have used some of the general purpose tactics, but a little more programming
can considerably reduce the effort required.

Restarting the proof, we may observe from the above that the tactic **local**
may be applied several times. In fact, it would be a useful general-purpose replacement for the rule **iff_intro**. The proof may begin by applying **local** ex-
haustively.

```
[1]::: !(bfs(local))
```

The result is the same as the result of the sequence of steps taken above, up to
the formation of the subgoals which were numbered 8, 12, and 13 above.

Concentrating on the subgoals 12 and 13, if we define a tactic which applies
any of the left-hand-side (antecedent) inference rules, as appropriate,

```
tactic lefts === rule(or_left) | rule(iff_left) |
                 rule(implies_left) | rule(and_left) |
                 rule(not_left).
```

we may simplify all the antecedents at once, using **bfs(lefts)**. This produces
twenty subgoals, all similar to:

```
    hyp 5:::   A
    hyp 9:::   A
    hyp 29:::  C
    hyp 25:::  B
    hyp 30:::  C
```

```
hyp 11:::  A
16::: hyp=[5, 9] ---> A or C
36::: hyp=[5] ---> B or (C or (A or C))
37::: hyp=[5, 29] ---> C or (A or C)
38::: hyp=[5, 25] ---> B or (A or C)
39::: hyp=[5, 25, 30] ---> A or C
```

And each of these can be solved by some number of instances of **or_intro** followed by **assump**.

```
[16, 36, 37, 38, 39, 18, 40, 41, 42, 43, 20, 44, 45, 46, 47, 22, 48,
49, 50, 51]::: every(exhaust(rule(or_intro_L) | rule(or_intro_R));rule(assump))
```

The other half of the proof tree may be approached similarly.

The result is quite a complex formal proof; not one that we would wish to produce without a significant measure of automation.[2] Here it is in diagrammatic form, each arc denoting the application of one primitive rule. This graph is produced via Ergo's daVinci interface (Fröhlich, 1996).

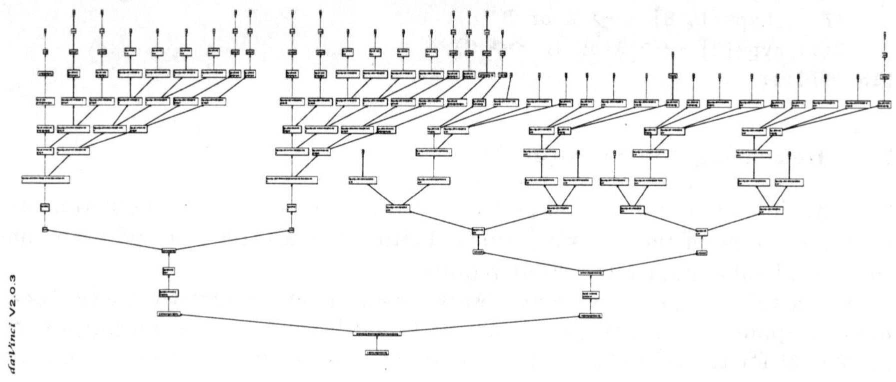

7.3   Fully Automatic Propositional Proofs

The tactics above point the way towards a fully-general tactic for proving tautologies. In the most general Gentzen-style system, this is entirely straightforward: one applies all the introduction rules, and all the left-hand-side (antecedent) rules, and then applies assumption. With the single-conclusion sequents of Ergo 5, slightly more ingenuity is needed. Nothing is claimed about the efficiency of the following solution, but it demonstrates the style of tactics which we are able to write using Gumtree.

Difficulty arises because the rule for or-introduction must swap one of the disjuncts into the antecedent.[3]

---

[2] An alternative (or complement) to automation is a well-developed set of lemmas. For decidable theories like propositional calculus, full automation is a better solution.

[3] The rules which simply discard one of the disjuncts (as used previously) are not sufficiently general, as they cannot be used, for example in a proof of $\vdash p \vee \neg\, p$. Therefore, we use the rule $constructive - or \mathrel{\widehat=} \mathbf{from}(\neg\, p \vdash q)\ \mathbf{infer}\ (\vdash p \vee q)$.

Conversely, the rule for *not* in the antecedent must re-introduce the negated term as a disjunct in the consequent. These two are, then, virtually mutual inverses, and must not be applied indiscriminately if cycles are to be avoided. Therefore, we create a tactic which will apply any one of the left- or right-introduction rules, *but* will apply *not−left* only if the negated term is not atomic—that is, if one of the introduction rules may subsequently be applied to it.

$$
\begin{aligned}
tactic \;\; props === \; &!(\textbf{rule}(and\_intro) \\
&\mid \textbf{rule}(not\_intro) \\
&\mid \textbf{rule}(constructive\_or) \\
&\mid \textbf{rule}(iff\_intro) \\
&\mid \textbf{rule}(implies\_intro) \\
&\mid \textbf{rule}(true) \\
&\mid (\textbf{rule}(not\_left) \; ; \textbf{rule}(constructive\_or); \\
&\qquad !(\textbf{rule}(not\_intro) \mid \textbf{rule}(implies\_intro) \mid \\
&\qquad \textbf{rule}(iff\_intro) \mid \textbf{rule}(and\_intro) \mid \\
&\qquad \textbf{rule}(constructive\_or))) \\
&\mid \textbf{rule}(and\_left) \\
&\mid \textbf{rule}(or\_left) \\
&\mid \textbf{rule}(implies\_left) \\
&\mid \textbf{rule}(iff\_left) \\
&\mid \textbf{rule}(false\_left)).
\end{aligned}
$$

Following application of *bfs(props)* no propositional connectives will be left in any of the subgoals, except that some antecedents will be negated. It suffices, then, to apply the rule of assumption and a rule of contradiction to determine which subgoals may be discharged. The final tactic, then, is

$$!(bfs(props)) \; ; some(\textbf{rule}(contrad) \mid \textbf{rule}(assump)) \; .$$

This tactic also has the benefit that if it is applied to a goal which is not a tautology it will simplify it into subgoals containing only atomic (non-propositional) terms.

7.4   Recording Proof Information

When the partial proof at Section 7.1 is saved, Ergo 5 uses the tactic language to record the proof. The proof record is a tactic whose structure exactly reflects the structure of the proof tree: the proof tree nodes are applications of **rule**; sequential composition connects the outputs of one rule to the input of the next rules; and parallel compositions represent the branches that occur when a rule has multiple premisses. Interspersed with rule applications are `annotate` commands, which are automatically generated by the user interface to annotate discharged nodes with the user-level command that discharged the node.

The annotations in the proof representation can be used to automatically generate a proof script, which is a cleaned-up version of the commands typed at the user interface:

```
  prove(A or (B <=> C) <=> (A or B <=> A or C)).
  rule(iff_intro).
  rule(and_intro);every(rule(implies_intro)).
  names([N_2,N_3]).
N_2:::  local.
  names([N_4,N_5]).
 [N_4, N_5]:::
  every(rule(or_left)).
  names([N_6, N_7, N_8, N_9]).
 [N_6, N_7, N_8, N_9]:::
  some(rule(or_intro_L);rule(assump)).
```

Since we did not explicitly name nodes in this proof, the proof script generator
has invented names (N_i) wherever the proof branched, to make it easier to
trace the flow of proof terms through the script.

Where the user invoked high-level tactics, and where generality (and alternation) were present in the user commands, these are seen in the proof script
too. By contrast, another presentation of the proof eliminates the annotations
and displays just the lowest-level rule applications:

```
rule(iff_intro);
rule(and_intro);
(
  rule(implies_intro);
  rule(iff_intro);
  rule(and_intro);
  (
    rule(implies_intro);
    rule(or_left);
    (
      rule(or_intro_L); rule(assump)
    ||
      one
    )
  ||
    rule(implies_intro);
    rule(or_left);
    (
      rule(or_intro_L); rule(assump)
    ||
      one
    )
  )
||
  rule(implies_intro);
  one
)
```

This presentation yields a tactic that is closely fitted to the goal at hand, and
will not normally be more widely applicable.

Both presentations support reuse of proofs. The proof script can be used to reapply the proof steps, one command at a time; it may freely be edited, for example to adapt it to a related but different proof. The lower level presentation is more suited to fully automatic reapplication, for example when rebuilding a theory (perhaps after trivial changes in its axiomatic basis). It may also be suitable as input to an external proof checking tool, in case additional certification of proofs is needed. Tactics are free to add their own annotations to proof nodes (distinct from the command annotations inserted by the user interface), and it is foreseen that these may be used to provide other presentations of proofs, perhaps to support more powerful or flexible facilities for reuse.

8   Discussion and Conclusion

This paper has described the Gumtree interface for Ergo 5. We have been careful to make a single language be the basis of the tactic programming mechanism, the interface, and the tool's proof storage and presentation. Certain primitives are more useful in one of these situations, and less so in the others. By presenting the user with a single language, we simplify the process of learning to use the tool and maximise the opportunities for reuse. Storing both the proof tree and the commands used to create it in a single structure gives us the greatest possible opportunity to reconstruct the proof—either in the presence of changes to the definitions of the tactics used to construct the proof, or of changes to the primitive (or derived) rules used in the proof.

We have been careful to preserve *Angel*'s simple axiomatic semantics for the tactics in Gumtree. We find this to be an aid to understanding which will in turn be an aid in constructing correct[4] tactics. Our early experiences confirm that this is a useful paradigm. Gumtree adapts and extends the notion of *parallel* composition of tactics, and we expect to be able to give an axiomatic semantics for this. We believe that this is the first full implementation of a theorem-proving system with a tactic language which has an independent semantics; that is, one whose semantics is not given operationally.

Related Work Many systems implement tactics with broadly the same semantics as Angel. In particular, much of the language described here may be regarded as a subset of Isabelle's tactic language (Paulson, 1989). Isabelle does not take the general approach to subgoals which we have presented, but instead, designating subgoals by positive integers, it provides tacticals for combining values of type `int->tactic`. These tacticals allow effects similar to our *every*, *some*, etc. Support is provided for treating subgoals as a stack; our approach again appears to generalise this.

Closer to the system described here is the work of Felty (1993) on tactics in logic programming languages. She describes the common tactic combinators in much the same way as we have, but again adopts a different approach to parallel

---

[4] Observe that correct tactics are those which perform as expected; tactics cannot, by their construction, be *unsound*.

goals (branching trees). She has the tactic `maptac`, which applies a tactic over a structured (collection of) goal(s), permitting a greater variety of structure than the simple list goals we permit here. `maptac` applies only a single tactic, however, and does not offer the structuring permitted by our parallel operator.

Future Work A separate report will consider how the laws of Angel are preserved and extended in the Gumtree implementation. Further generalisation of the denotational semantics of Angel using monads (Moggi, 1989) will assist in this. As we develop application theories for Ergo 5, we plan to write Gumtree versions of relevant proof procedures, expecting the semantics of Angel to assist in validating these procedures. We are also developing Gumtree tactics for window inference (Robinson and Staples, 1993), so that Ergo 5 can be used in a style similar to that of previous Ergo versions, which formed the basis of the program refinement tool (Carrington, Hayes, Nickson, Watson, and Welsh, 1996).

Acknowledgements

We are grateful to several reviewers for their comments on an earlier draft of this paper.

Andrew Martin notes that the tactic *props* presented above was developed jointly with Stephen Brien. It is of interest because we wrote it for an earlier partial implementation of Angel. It is, therefore, the first truly portable Angel tactic.

References

Carrington, D., Hayes, I., Nickson, R., Watson, G., and Welsh, J. (1996). A tool for developing correct programs by refinement, *in* He Jifeng (ed.), *Proc. BCS 7th Refinement Workshop, Bath, UK*, Electronic Workshops in Computing, Springer, pp. 1–17. Also available as Technical Report UQ-SVRC-95-49, Software Verification Research Centre, University of Queensland.
    URL: *http://www.springer.co.uk/eWiC/Workshops/7RW.html*

Cheng, A. S. K., Robinson, P. J., and Staples, J. (1991). Higher level meta programming in Qu-Prolog 3.0, *Proceedings of the 1991 International Conference for Logic Programming*, Paris.

Felty, A. (1993). Implementing tactics and tacticals in a higher-order logic programming language, *Journal of Automated Reasoning* 11: 43–81.

Fröhlich, M. (1996). Real world applications of daVinci: Theorem prover Ergo.
    URL: *http://www.informatik.uni-bremen.de/~davinci/applications/ergo.html*

Gentzen, G. (1969). Investigations into logical deduction, *in* M. E. Szabo (ed.), *The Collected Papers of Gerhard Gentzen*, North-Holland, pp. 68–131.

Gordon, M. J. C., Milner, R., and Wadsworth, C. P. (1979). *Edinburgh LCF: A Mechanised Logic of Computation*, Vol. 78 of *LNCS*, Springer-Verlag.

Martin, A. (1994). *Machine-Assisted Theorem-Proving for Software Engineering*, D.Phil. thesis, University of Oxford. Also available as Technical Monograph PRG-121, ISBN 0-902928-95-3, Oxford University Computing Laboratory Wolfson Building, Parks Road, Oxford, OX1 3QD, UK.

Martin, A., Nickson, R., and Utting, M. (1997). Improving Angel's parallel operator: Gumtree's approach, *Technical Report 97-15*, Software Verification Research Centre, The University of Queensland, QLD 4072, Australia. To appear.

Martin, A. P., Gardiner, P. H. B., and Woodcock, J. C. P. (1996). A tactic calculus, *Formal Aspects of Computing* **8**(4): 479–489. An abridged version appears in the journal; the full version is available at the Formal Aspects of Computing FTP site, ftp://ftp.cs.man.ac.uk/pub/fac.

Moggi, E. (1989). Computational lambda-calculus and monads, *Proceedings Fourth Annual Symposium on Logic in Computer Science*, IEEE Computer Society Press, Washington, D.C.

Nickson, R. and Hayes, I. (1996). Supporting contexts in program refinement, *Technical Report 96-29*, Software Verification Research Centre, Department of Computer Science, The University of Queensland. Accepted for publication in *Science of Computer Programming*.

Paulson, L. C. (1987). *Logic and Computation—Interactive Proof with Cambridge LCF*, Cambridge University Press.

Paulson, L. C. (1989). The foundation of a generic theorem prover, *Journal of Automated Reasoning* **5**: 363–397. Also University of Cambridge Computer Laboratory Technical Report No. 130.

Pereira, F. C. N. and Warren, D. H. D. (1980). Definite clause grammars for language analysis – a survey of the formalism and a comparison with augmented transition networks, *Artificial Intelligence* **13**: 231–278.

Robinson, P. and Hagen, R. (1997). Qu-Prolog 4.2 reference manual, *Technical Report 97-11*, Software Verification Research Centre, The University of Queensland.

Robinson, P. and Staples, J. (1993). Formalizing a hierarchical structure of practical mathematical reasoning, *Journal of Logic and Computation* **3**(1): 47–61.

Tarau, P., Dahl, V., and Fall, A. (1995). Backtrackable state with linear assumptions, continuations and hidden accumulator grammars, *Technical Report 95-2*, Departement d'Informatique, Université de Moncton.
URL: *http://clement.info.umoncton.ca/html/state.html*

Utting, M. (1996). An architecture for a unified refinement/proof tool, *Fifth Australasian Refinement Workshop*, Software Verification Research Centre, The University of Queensland.

Vickers, T. (1990). An overview of a refinement editor, *Proceedings of the Fifth Australian Software Engineering Conference*, pp. 39–44.

Whitwell, K. (1992). A tactical environment for an interactive theorem prover, *Technical Report 92-7*, Software Verification Research Centre, The University of Queensland.

Supporting Data Refinement in a Program Refinement Tool

Jamie Shield, Ray Nickson, and David Carrington

Software Verification Research Centre
School of Information Technology
The University of Queensland
Australia 4072

{jims,nickson,davec} @ it.uq.edu.au

Abstract. This paper addresses the practical aspects of supporting data refinement in the refinement calculus. We analyse the difficulties associated with defining and applying data refinement rules with a computer-based tool. Our analysis is illustrated by discussion of choices made when extending an existing refinement tool to support data refinement.

1 Introduction

Ensuring that programs satisfy their specifications is a major objective for the software engineering community. One approach for achieving this objective is refinement.

> "Refinement is the process of moving from abstract specifications to less abstract specifications, via some data or operation transformation which allows the behaviour of the abstract system to be simulated by the more refined system" [Lan96, p47].

There are many approaches to refinement including B [Lan96], VDM [Jon90] and the refinement calculus. The refinement calculus, which has a calculational style, was developed independently by Back [Bac80], Morgan [Mor88b] and Morris [Mor87] via the extension of Dijkstra's [Dij76] weakest precondition calculus to include abstract specification statements.

Some tools supporting the refinement calculus exist but none of them has achieved widespread use. The objective of this paper is to address issues associated with tool support for data refinement within the refinement calculus, something that has received little attention from tool developers.

This paper is organized in the following manner. Section 2 provides the necessary background on the refinement calculus with a focus on tools and data refinement. Section 3 reviews requirements for tools supporting data refinement from a user's perspective. Section 4 focuses on requirements at the implementation level and discusses the decisions we made when extending a program refinement tool. Currently this support is limited to data refinement of blocks. Our approach can be extended for data refinement of modules.

2　Refinement Calculus

This section introduces procedural and data refinement and the corresponding tool support that exists for them. There are three components to the refinement calculus:

Specification Constructs: A traditional imperative programming language is extended with specification constructs, most importantly an abstract specification statement. This extension produces a wide-spectrum language that allows programs to contain a mix of specification and implementation detail. Only programs without any specification constructs are executable; such programs are referred to as code.

Refinement Relation: The refinement relation captures formally the notion that one program is substitutable for another in any context. Because the refinement relation is transitive, program development can proceed by producing a sequence of programs beginning with the initial specification and ending with an executable program.

Refinement Rules: The rules of the refinement calculus are derived from the semantics of the wide spectrum language and the refinement relation. Applying these rules allows more refined programs to be generated without having to prove from first principles that the refinement relation holds. Because most rules are schemas with generic parameters, there is often an applicability condition that needs to be established to ensure that the particular rule application is valid.

2.1　Procedural Refinement

Procedural refinement involves transforming a system so that the algorithm used by the program becomes more explicit, more implementation oriented, or more deterministic. Procedural refinement rules typically transform a single specification statement within a partially refined program to a programming language construct. Such constructs include primitives like assignment statements and compositions like loop structures. For compositions, the form and content of the sub-components are determined by the original specification statement and the refinement rule. In the case of a loop, the loop body is a new specification statement.

Carrington et al. [CHN+96] discuss some tools that support the procedural refinement calculus including the Program Refinement Tool (PRT) [CHN+95]. PRT is constructed on top of the customisable and interactive theorem prover, Ergo [NTU96]. PRT uses an innovative logic developed to manage manipulation of expressions involving both logical and programming language variables [NH95].

2.2　Data Refinement

Data refinement is typically used to replace abstract or non-supported data structures (such as a heap) with more concrete data structures (such as an array representation of a heap). This change of representation is performed by replacing some of the variables in a program. Even within the refinement calculus, there are multiple approaches to data refinement including those by Morris [Mor89,Mor90], Back [BvW90,Bac88] and Morgan [MG90,Mor88a]. Because PRT is based on Morgan's formulation of the refinement calculus as presented in [Mor94], we concentrate primarily on that approach. A brief comparison with Back and von Wright's approach appears in Sect 3.1.

Morgan's refinement calculus supports data refinement of modules. Module declarations in [Mor94] encapsulate local variable declarations, procedure declarations and initialisations. Their purpose is to provide data abstractions. Typically these modules also have associated initial conditions and possibly invariant conditions. An example module is shown in Fig. 1. This module determines the parity of the initial segment of 1-bits in a stream of bits. This example is not realistic but was chosen to illustrate the concepts of data refinement.

$$
\begin{aligned}
&\textbf{m}\textit{odule Odd} \\
&\quad \textbf{var } l : \mathbb{N}; \\
&\quad \textbf{procedure } \textit{SetBits}(\textbf{value } b : \textbf{seq } 0..1) \;\; \widehat{=} \\
&\qquad \{\exists\, j : \textit{dom } b \bullet b[j] = 0\} \\
&\qquad \big[\; \textbf{var } \; i : \mathbb{N} \bullet \\
&\qquad\quad l, i := 0, 0; \\
&\qquad\quad \textbf{do } b[i] = 1 \rightarrow \\
&\qquad\qquad i, l := i + 1, l + 1 \\
&\qquad\quad \textbf{od} \\
&\qquad \big] \\
&\quad \textbf{procedure } \textit{Parity}(\textbf{result } p : \mathbb{B}) \;\; \widehat{=} \\
&\qquad p := \textit{odd}(l) \\
&\quad \textbf{end}
\end{aligned}
$$

Fig. 1. Module example

A module is refined either by refinement of one or more of its exported (externally visible) procedures or by state transformation of the local variables declared inside the module. The latter technique is known as data refinement. Morgan defines data refinement as a two stage process first involving the augmentation of all constructs within the module to introduce the concrete variables and the coupling invariant. The coupling invariant shows the relationship between the abstract and the concrete variables. This is followed by an auxiliarisation and diminution stage in which the abstract variables are made auxiliary and subsequently removed.

We data refine our example module to use a Boolean rather than a natural number for the parity checking process.

abstract variable	$l : \mathbb{N}$	
concrete variable	$s : \mathbb{B}$	
coupling invariant	$s = odd(l)$	

Augmentation: A declaration is added to the module for the variable s, and the module's initialisation and each procedure body are augmented. Our module has no explicit initialisation clause; this is equivalent to **initially** *true*. Augmenting the initialisation requires conjoining the coupling invariant (Law 17.5)[1].

Each assignment statement is augmented. Law 17.8 states that

$$w := E \textbf{ becomes } w, c := E, F$$

provided

$$CI \Rrightarrow CI[w, c \backslash E, F]$$

where c is the concrete variable (or variables) and CI is the coupling invariant.

Thus in SetBits, $l, i := 0, 0$ **becomes** $l, i, s := 0, 0, false$ and $l, i := l+1, i+1$ **becomes** $l, i, s := l + 1, i + 1, \neg s$. Morgan [Mor94, p.166 and p.172] makes it clear that these rules for augmentation and diminution are not refinement laws.

For the assignment in Parity, we convert the original assignment to a specification statement (Law 1.7) and then augment it (Law 17.6):

$$p, s : [true, s = odd(l), p = odd(l)]$$

The result of the augmentation is shown in Fig. 2.

Diminution: The declaration of l is removed from the module. The augmented initialisation is diminished by existentially quantifying the abstract variable yielding

$$\exists l : \mathbb{N} \bullet s = odd(l)$$

which is satisfied for every value in the type of s, hence is equivalent to *true*.

Each assignment $w, a := E, F$ can be diminished to $w := E$ provided E contains no references to a. This is the case for both assignments in SetBits. The specification statement in Parity can be refined using the strengthen postcondition, reduce frame and weaken precondition rules to

$$p : [p = s]$$

which refines to the assignment $p := s$ with no need for further diminution. The guard requires no transformation since it contains no auxiliary variables. The result of the diminution is shown in Fig. 3.

---

[1] All laws cited here are taken from [Mor94].

$$\textbf{module } Odd$$
$$\textbf{var } s : \mathbb{B};\ l : \mathbb{N};$$
$$\textbf{procedure } SetBits(\textbf{value } b : \textbf{seq } 0..1)\ \ \widehat{=}$$
$$\{\exists j : dom\ b \bullet b[j] = 0\}$$
$$[\![\ \textbf{var } i : \mathbb{N} \bullet$$
$$l, i, s := 0, 0, false;$$
$$\textbf{do } b[i] = 1 \rightarrow$$
$$i, l, s := i + 1, l + 1, \neg\, s$$
$$\textbf{od}$$
$$]\!]$$
$$\textbf{procedure } Parity(\textbf{result } p : \mathbb{B})\ \ \widehat{=}$$
$$p, s : [true, s = odd(l), p = odd(l)]$$
$$\textbf{end}$$

Fig. 2. Result of augmenting Fig. 1

The rules for augmentation and diminution are not algorithm refinement laws; Morgan describes them as "transformations for which we use the word *'becomes'* ". Using such rules requires the user to apply them systematically throughout the scope of the abstract variables. As a consequence, there exist intermediate stages (when the augmentation or diminution has been applied to some but not all components of the module) that cannot be formally related to either the initial or final versions of the module. This potential for discontinuity can be a problem when considering tool support. An alternative approach is to use a one step rule that combines both augmentation and diminution, such as Morgan's data refinement rule [MG90] for an arbitrary specification statement.

2.3   Data Refinement of Blocks

A module declaration within some block is equivalent to that block with the declarations within the module placed with similar declarations within the block. The module initialisation is converted to a command that is placed at the beginning of the block body. For example, we can convert the module of Fig. 1 into the block shown in Fig. 4 (where we have also conjoined the procedure bodies as the block body). The procedure parameters are considered as part of the external context[2].

We have used this equivalence to prototype tool support for data refinement. The equivalence means that the tool does not need to support modules since data refinement can be applied to block declarations. The transformation of

---

[2] Promoting the procedure parameters to the external context requires care to avoid variable capture.

```
module Odd
    var s : 𝔹;
    procedure SetBits(value b : seq 0..1)  ≙
        {∃ j : dom b • b[j] = 0}
        ⟦ var i : ℕ •
            i, s := 0, false;
            do b[i] = 1 →
                i, s := i + 1, ¬ s
            od
        ⟧
    procedure Parity(result p : 𝔹) ≙
        p := s
end
```

Fig. 3. Result of diminishing Fig. 2

procedures is performed to avoid any dependence on tool support for procedures and parameters.

Existing tool support for data refinement with the refinement calculus is limited. We are aware only of the tactic-driven refinement tool of Groves et al. [GNU92] that provides a single rule for block data refinement.

3   Supporting Data Refinement — User View

This section covers requirements pertinent to the users of data refinement tools and demonstrates the prototype implementation.

3.1   Requirements

Data refinement tools have many requirements in common with procedural refinement tools. Carrington et al. [CHN$^+$96] discuss the requirements of both users and implementors of procedural refinement tools.

One of the major characteristics of a data refinement tool is its method. The choice of method affects the simplicity of the data refinement process, the guidance and support that is provided and also the flexibility of the data refinement. For example, while Morgan's auxiliary variables approach (as exemplified in Sect. 2.2) is relatively stringent with respect to the process that must be followed, it does provide a high level of guidance for the user. In contrast, Back and von Wright's approach [BvW90] to data refinement is finer-grained with the encoding and decoding commands allowing a "bottom-up" approach.

Another important characteristic of data refinement tools is their refinement rules. For the tool to be effective, the rules should be:

$$
\begin{aligned}
&\lVert\ \mathbf{var}\ l : \mathbb{N}\ \bullet \\
&\qquad \{\exists j : dom\ b \bullet b[j] = 0\} \\
&\qquad \lVert\ \mathbf{var}\ i : \mathbb{N}\ \bullet \\
&\qquad\qquad l, i := 0, 0; \\
&\qquad\qquad \mathbf{do}\ b[i] = 1 \rightarrow \\
&\qquad\qquad\qquad i, l := i + 1, l + 1 \\
&\qquad\qquad \mathbf{od} \\
&\qquad \rVert; \\
&\qquad p := odd(l) \\
&\rVert
\end{aligned}
$$

Fig. 4. Block form of Fig. 1

Simple: To ease the task of data refinement for the user, it is beneficial to provide simple rules. Complex rules tend to be restricted in their applicability.

Correct: For the user to be confident in the validity of the results of the refinement, the tool must only use correct rules within a sound logic.

Orthogonal: It is beneficial to the user if each rule is designated a specific task and that these roles do not overlap. If significant overlap is present, many rules may be redundant. Additionally too many rules may cause the user confusion.

Comprehensive: There should always exist at least one rule that may be applied to a certain construct in order to data refine it. The tool is incomplete and potentially unusable if a circumstance without possible refinement arises.

The ability for the tool to automate the process, where possible, is also seen to be advantageous. This automation has a direct correlation to both the aforementioned orthogonality and comprehensiveness aspects of the rules. That is, rule selection is trivial if there exists one and only one rule to apply to data refine a specific construct. It is also significantly easier to automate the application of simple rules with few associated obligations rather than cumbersome complex rules.

3.2   Preliminary Results

In this section the data refinement of the module in Fig. 1 is repeated on the block representation (Fig. 4) using the prototype tool. This should help reinforce the ideas presented in Sect. 3.1. Figure 5 shows the command used to introduce the block and its context into the tool.

After the program is introduced, the following command initiates data refinement of the block.

```
data_refine([l],[s:bools],s=odd(l)).
```

```
refinement parity ===
  [[ var l:ints @
    [[ var i:ints @
          [l,i]:=[0,0];
          (do (b at i) = 1 then
               [i,l]:=[i+1,l+1]
             od)
    ]];
    [p]:=[odd(l)]
  ]]
  in_context
        [var b:seq(0 up_to 1), pre ex j (j in domain(b) => (b at j)=0)].
```

Fig. 5. Initialisation command

This command automatically applies all the necessary augmentation and diminution steps. Figure 6 results.

With the data refinement complete, the task is to simplify the program. The first specification statement (i) can be refined by an assignment to i and s. Similarly the specification statements (ii) and (iii) are replaced with the assignments $i, s := i + 1, \neg s$ and $p := s$ respectively. The logical constant constructs are subsequently removed. With these refinements and the simplification of the guard, the final data refined program results:

```
[[ var s:bools @
        [[ var i:ints @
               [i,s] := [0,false];
               (do (b at i) = 1 then
                        [i,s]:=[i+1,not s]
               od)
        ]];
        [p]:=[s]
]]
```

Data refinement within this tool consists of identifying the abstract variables, concrete variables and coupling invariant, proving any obligations that arise and subsequently procedurally refining the resulting data refined program until the desired program is achieved. This straightforward approach has been achieved both through the choice of data refinement rules and the automation provided by the **data_refine** command. The decision was made to make the data refinement step as automatic as possible to ensure that the augmentation and diminution steps were applied consistently over all program components within the affected scope. A consequence is that the user has additional procedural refinement to perform after the data refinement but this is considered an acceptable tradeoff. For example, during the process of data refinement all assignment statements in the original program are converted to equivalent specification statements that the user has to subsequently refine back to assignments.

$$\big[\!\big[\ \mathbf{var}\ s : \mathbb{B}\ \bullet$$

$$\big[\!\big[\ \mathbf{var}\ i : \mathbb{N}\ \bullet$$

$$\big[\!\big[\ \mathbf{con}\ c1, c2\ \bullet$$

$$i, s : \begin{bmatrix} \exists\, x_1 \bullet s = odd(x_1) & \exists\, x_2 \bullet s = odd(x_2) \\ c1 = 0 & x_2 = c1 \\ c2 = 0 & i = c2 \end{bmatrix} \qquad (i)$$

$$\big]\!\big];$$

$$\mathbf{do}\ \forall\, x_0 \bullet s = odd(x_0) \Rightarrow b[i] = 1 \rightarrow$$

$$\big[\!\big[\ \mathbf{con}\ c1, c2\ \bullet$$

$$i, s : \begin{bmatrix} \exists\, x_3 \bullet s = odd(x_3) & \exists\, x_4 \bullet s = odd(x_4) \\ c2 = x_3 + 1 & x_4 = c2 \\ c1 = i + 1 & i = c1 \end{bmatrix} \qquad (ii)$$

$$\big]\!\big]$$

$$\mathbf{od}$$

$$\big]\!\big];$$

$$\big[\!\big[\ \mathbf{con}\ c\ \bullet$$

$$p, s : \begin{bmatrix} \exists\, x_5 \bullet s = odd(x_5) & \exists\, x_6 \bullet s = odd(x_6) \\ c = odd(x_5) & p = c \end{bmatrix} \qquad (iii)$$

$$\big]\!\big]$$

$$\big]\!\big]$$

Fig. 6. Result of data refining Fig. 4 using the tool

4   Supporting Data Refinement — Implementation View

This section covers the requirements and the tool influences on the PRT implementation decisions.

4.1   Requirements

There are three specific requirements the implementor should consider when providing support for data refinement within a program refinement tool.

New Relation: The first of these requirements is the need for a data refinement relation that is distinct from the algorithmic refinement relation. That is, the data refinement operator $\preceq$ is distinct from the algorithmic operator $\sqsubseteq$. The data refinement relation must be defined and supported by axioms and theorems.

Coupling Invariant: The coupling invariant must be represented, and available for all data refinement steps. This can be achieved by encompassing the block being data refined with an invariant construct specifying the coupling invariant. The program window inference proof paradigm [NH95] used by PRT allows context to be represented as a list of predicates. This mechanism can be used to represent the coupling invariant, which can consequently be used for proof and/or transformation at any stage during the data refinement. However, if the tool did not support such context handling (or invariant introduction constructs), the tool developer may need to, for example, introduce an internal tool variable that temporarily holds the coupling invariant throughout the data refinement process.

Rules: The choice of rules has a large influence upon the success of the data refinement tool and Sect. 3.1 discussed some of the users' requirements for rules. Morgan's rules as used in the example in Sect. 2.2 are an obvious initial choice. As mentioned previously, when a module (or block) that contains more than one program construct is data refined using these rules, all constructs must be data refined consistently. This can be achieved by an appropriate choice of rules, by fully automating the process of applying rules, or by allowing flexible application of rules but having the tool check the consistency.

4.2 Rule Set

We now review the rules that has been used in PRT and state their origins.

Assignments: For the data refinement of all assignments to be consistent, we adopted an approach from [Nic93]. Nickson suggests procedurally refining the assignment to a specification statement and subsequently performing data refinement on the specification. One disadvantage of this rule is that the user is left with the task of re-introducing (if desired) an assignment statement. However, the consistency of the approach and the flexibility with which the user may deal with the resulting program suggest that this technique is advantageous.

Specifications: For coupling invariant CI, abstract variable a and concrete variable c, the rule for data refining specifications is taken from [MG90]. The following data refinement is always valid:

$$a, x : [pre, post] \preceq c, x : [(\exists\, a \bullet CI \wedge pre), (\exists\, a \bullet CI \wedge post)]$$

Guards: Morgan [MG90] is the inspiration for the rule for data refining guards: Given a coupling invariant CI, abstract guards G_i, and abstract statements S_i, let the concrete guards G_i' and concrete statements S_i' be such that

$G_i' = (\forall\, a \bullet CI \Rightarrow G_i)$ and
$S_i \preceq S_i'$

Then provided $(\exists\, a \bullet CI \wedge (\bigvee i \bullet G_i)) \Rightarrow (\bigvee i \bullet G_i')$, the following data refinement is valid:

$$\textbf{if }([]\,i \bullet G_i \rightarrow S_i)\textbf{ fi } \preceq \textbf{ if}([]\,i \bullet G_i' \rightarrow S_i')\textbf{ fi}$$

Other language constructs are data refined by recursively data refining all their subcomponents. For instance,

$$\frac{S \preceq S',\, T \preceq T'}{S;\ T \preceq S';\ T'}$$

All subcomponents must be data refined consistently using the same sets of abstract and concrete variables and the same coupling invariant.

4.3   Tool Considerations

This section discusses some of the advantages and some constraints that the tool placed on the design and implementation of the data refinement extension. We were constrained to Morgan's version of the refinement calculus since the tool was to be based on PRT. It would be interesting to pursue other avenues, such as Back's approach.

The programming window inference proof paradigm [NH95] of PRT is restricted to a sequential application of refinement rules. Consequently the parallel application of Morgan's data refinement rules [Mor94] to the constructs within blocks (or modules), while necessary, is unsupported. Until such parallel rule application becomes possible, the tool must ensure that all constructs are data refined and not allow only some of the constructs to be data refined. The tool enforces this property by allowing rules to be applied only via the `data_refine` command.

The implementation of the rules was hampered by the lack of support for arbitrary length variable vectors. Rules such as the simple specification rule used for data refinement of multiple assignment had to be broken down into many different rules. In the final implementation, there existed a rule for data refining a single assignment, a two variable assignment and so on.

$$x := e \sqsubseteq \big[\!\big[\ \textbf{con } c \bullet x\!:\,[\,c = e\ ,\ x = c\,]\,\big]\!\big]$$

$$x1, x2 := e1, e2$$

$$\sqsubseteq\ \big[\!\big[\ \textbf{con } c1, c2 \bullet$$

$$x1, x2\!:\begin{bmatrix} c1 = e1 & x1 = c1 \\ c2 = e2 \ , & x2 = c2 \end{bmatrix}$$

$$\big]\!\big]$$

These rule families could be reunited by writing tactics that automatically applied the appropriate rule; however, this is cumbersome.

One of the implementation goals was to restrict the changes to the original PRT as much as possible. By introducing rule families, it was possible to avoid expanding PRT to handle arbitrary length variable vectors. Despite the powerful mechanisms of the original implementation that allowed high level refinement abstractions, the program window inference paradigm [NH95] was modified [Nic96] to deal with the data refinement relations.

PRT's refinement rules are written in a simple pattern matching language, and this hampers generalisation of the rules. This, however, was not entirely disadvantageous as it forced the re-design of the rule set to include simpler rules. Also the mere fact that a language for refinement rules existed shows that PRT's support for the extension was relatively high. This support is also exemplified by PRT's significant ability to handle contextual information. This feature was used for both representing the coupling invariant and for proving applicability conditions.

5 Conclusions and Future Work

Although some limited support for data refinement now exists, it would be useful to include support for the refinement of modules. Such a structuring mechanism would lift the current restriction on the size of refinable systems within the prototype too. However, refinement of modules requires that module structures are supported by the tool. A stepping stone to this goal is adding procedure parameters to the tool as discussed in [Shi96].

Another desirable feature is for the rule set to be enhanced by the provision of more specialised rules. For instance, the inclusion of rules for functional data refinement [Mor94, Sect. 17.7] would allow this technique to be used where applicable. Functional data refinement is a special case of data refinement where an abstraction function from the concrete to the abstract variables has been identified. These rules would simplify the obligations to be discharged albeit in a restricted set of circumstances. Also, for example, Nickson [Nic93] describes laws for data refining special cases of specification statements that could be included. These rules have separated the 'Data Refine Specification Statement' into three cases depending upon whether the abstract variables are in the frame of the specification statement being data refined. While such rules have the advantage of easing the discharging of obligations, they unfortunately are also harder to prove, implement, automate and their applicability is limited. Care is also required not to introduce so many rules that the rule base becomes overwhelming and consequently difficult to use.

Another possibility is to extend the tool's logic so that n-step correct relations are supported as well as 1-step correct relations [Nic96]. This would simplify using the augmentation and diminution laws in [Mor94] as explained in Sect. 2.2.

The rule set encompassed by the prototype tool implementation can also be consolidated by replacing the rule families by generalised rules. Such generalised rules may suffer the same proof difficulties as the specialised rules mentioned above.

Although data refinement is typically applied to many program constructs, i.e., an entire module, and is inherently not a sequential task, it would be interesting to experiment with the use of such sequential methods as Back and von Wright's encoding and decoding commands [BvW90]. The advantage of their approach is that it works in a bottom-up fashion. The user applies a rule to a single construct (i.e., an assignment) and consequently data refines it. Incrementally, the data refinement is 'worked out' to the edges of the variable introduction block where the abstract variables are replaced with concrete variables. As with any bottom-up method, a disadvantage with this approach is the lack of guidance offered to the user.

Morgan's rules for the two-stage auxiliary variables approach are not trivial to implement in a tool because of the need to ensure overall consistency. This, combined with the desire to produce simple, correct, comprehensive and orthogonal rules led us to a different set of rules. While the rules used by the prototype may not be ideal, they do have the advantage that they allow a relatively high degree of automation as exemplified by Sect. 2.2.

It is interesting to note that the refinement of the Odd module did not require data refinement of the guard yet the tool's version of the refinement did perform data refinement of the guard. More sophisticated rule application commands would easily overcome this problem and hence remove the need for simplifying the guard after data refinement.

The prototype rules are relatively successful at achieving the user's requirements as listed in Sect. 3.1. For instance, orthogonality and comprehensiveness of the rules[3] is evident by the fact that each program construct is refined by exactly one rule. However, the rules in the rule set are at present unproved; and the rules allowing sequential application of the refinements of components, when the theory suggests they should be done in parallel, are reliant upon the tool ensuring consistency of the entire data refinement. Further work is required to refine the rule set and complete the associated proofs of the rules. The ability to automate the rule applications in Sect. 2.2 adds support to the claim that the rules are relatively simple, as desired.

Even if an ideal rule set is chosen, often constraints of the tool lead to the actual implementation being less than desired. The requirement for rule families and various other issues require the developer to balance carefully the need to extend the tool and the mere implementation of data refinement. The careful selection of this balance is essential to the success of data refinement support.

We have shown that the (apparently simple) process of performing data refinement by hand does not correspond directly to comparably simple tool support. The choice of data refinement method, rules and the context in which they are applied are vital to the success of the tool.

---

[3] Excluding program constructs that aren't supported by the tool, e.g. modules.

Acknowledgements

We thank Ian Hayes for providing the motivation for the work and his very useful comments on early drafts of the paper. Geoff Watson provided assistance with PRT. Finally, we thank the reviewers for their helpful suggestions for improving the paper.

References

[Bac80] R.J.R. Back. *Correctness Preserving Program Refinements: Proof Theory and Applications.* Tract 131, Mathematisch Centrum, Amsterdam, 1980.

[Bac88] R. J. R. Back. Data refinement in the refinement calculus. Series A 68, Inst. för Informationsbehandling, Åbo Akademi, Turku, Finland, 1988.

[BvW90] R. J. R. Back and J. von Wright. Command lattices, variable environments and data refinement. Series A 102, Åbo Akademi – Departments of Computer Science and Mathematics, 1990.

[CHN+95] D. Carrington, I. Hayes, R. Nickson, G. Watson, and J. Welsh. The PRT User Manual. version 1.03. Technical Report 95-56, Software Verification Research Centre, The University of Queensland, 1995.

[CHN+96] D. Carrington, I. Hayes, R. Nickson, G. Watson, and J. Welsh. A tool for developing correct programs by refinement. In He Jifeng, editor, *Seventh Refinement Workshop*, Workshops in Computing. BCS FACS, Springer-Verlag, 1996. Also available as TR-95-49, Software Verification Research Centre, The University of Queensland.

[Dij76] E. W. Dijkstra. *A Discipline of Programming.* Academic Press, 1976.

[GNU92] L. Groves, R. Nickson, and M. Utting. A tactic driven refinement tool. In Cliff B. Jones, Roger C. Shaw, and Tim Denvir, editors, *Fifth Refinement Workshop*, Workshops in Computing, pages 272–297. BCS FACS, Springer-Verlag, 1992.

[Jon90] C.B. Jones. *Systematic Software Construction using VDM.* Prentice Hall, 1990.

[Lan96] K. Lano. *The B Language and Method: A Guide to Practical Formal Development.* Springer Verlag, 1996.

[MG90] C. Morgan and P. Gardiner. Data refinement by calculation. *Acta Informatica*, 27:481–503, 1990. Reprinted in [MV94].

[Mor87] J. M. Morris. A Theoretical Basis for Stepwise Refinement and the Programming Calculus. *Science of Computer Programming*, 9:287–306, 1987.

[Mor88a] C. Morgan. Auxiliary variables in data refinement. Technical report, Programming Research Group, Oxford University, 1988. In [MV94].

[Mor88b] C.C. Morgan. The Specification Statement. *ACM Transactions on Programming Languages and Systems. Reprinted in [MV94]*, 10(3), July 1988.

[Mor89] J. M. Morris. Laws of data refinement. *Acta Informatica*, 26:287–308, 1989.

[Mor90] J. M. Morris. Piecewise data refinement. In E. W. Dijkstra, editor, *Formal Development of Programs and Proofs*, chapter 10, pages 117–137. Addison-Wesley, 1990.

[Mor94] C. Morgan. *Programming from Specifications.* Prentice Hall, second edition, 1994.

[MV94] C. Morgan and T. Vickers, editors. *On the Refinement Calculus.* Springer-Verlag, 1994.

[NH95]   R. Nickson and I. Hayes. Program Window Inference. Technical Report 95-29, Software Verification Research Centre, The University of Queensland, 1995.

[Nic93]   R. Nickson. *Tool Support for the Refinement Calculus*. PhD thesis, Victoria University of Wellington, 1993.

[Nic96]   Ray Nickson. Window inference for data refinement. In *Fifth Australasian Refinement Workshop*. Software Verification Research Centre, The University of Queensland, 1996.

[NTU96]   R. Nickson, O. Traynor, and M. Utting. Cogito Ergo Sum: Providing structured theorem prover support for specification formalisms. In Kotagiri Ramamohanarao, editor, *Cogito Ergo Sum: Providing Structured Theorem Prover Support for Specification Formalisms*, volume 18(1) of *Australian Computer Science Communications*, pages 149–158, 1996.

[Shi96]   J. Shield. Addition of Parameterised Procedures to PRT, CS465 Project Report. Interactive verification honours project report, 1996.

The Mechanical Verification of Solid State Interlocking Geographic Data

Andrew Simpson, Jim Woodcock and Jim Davies

Programming Research Group
Oxford University Computing Laboratory
Wolfson Building
Parks Road
Oxford OX1 3QD
United Kingdom

Abstract. We describe how the formal notation Communicating Sequential Processes, together with the refinement checker Failures Divergences Refinement, has been applied to the analysis of safety invariants for geographic databases associated with Solid State Interlocking railway signalling systems. We present a means for modelling geographic databases and their safety invariants in terms of sequences of events. We also present a refinement-based method for reducing the complexity of the verification of geographic data, which helps to make the automation of the task at hand tractable.

1   Introduction

Solid State Interlocking (SSI) railway signalling systems [3, 4] control a large percentage of the railway networks of many countries, including Great Britain, Australia, Belgium and South Africa. SSI is a computerised control system used on railways which replaces the traditional relay-based electromechanical systems. One of its functions is to grant routes to trains. This involves the locking of points and sub-routes, and the setting of signals. As is the case for all signalling systems, the fundamental requirement of the system is that it should reduce the possibility of hazards, such as collisions—which are possible if two trains can occupy the same section of track—and derailments—which are possible if points can move under a train [6].

At their simplest, SSIs can be viewed as consisting of generic programs—which are concerned with the transfer of data between interlockings and the trackside—and geographic databases—which are concerned with local data, such as signals and points.

Because of the state explosion involved—the internal state typically has 2^n possible configurations [7][1]—the mechanical verification of safety invariants for geographic data has generally been considered to be of too great a complexity

---

[1] Where n is the number of components, e.g., points, signals, etc. with which the system is concerned.

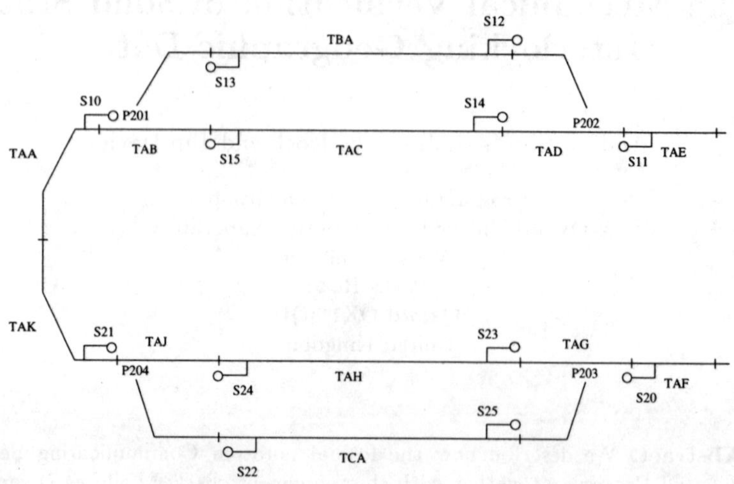

Fig. 1. The Open ALVEY

for the current generation of automated proof tools. Indeed, it is precisely this difficulty in decomposing the problem at hand that has limited the success of other approaches (e.g., CCS [7]).

In this paper, we outline how geographic data for such systems may be described in terms of Communicating Sequential Processes (CSP) [5], and suggest an approach for separating safety invariants into a series of smaller, more manageable proofs. This involves identifying the components with which we are concerned, and, via the refinement checker Failures Divergences Refinement (FDR) [8], proving that the relevant properties hold of these components.

2   Geographic Data

British Rail's Geographic Data Language (GDL) [1] provides a means of describing signalling rules for railway junctions in terms of its track circuits, points, signals, routes and sub-routes. These concepts are introduced in this section, and are illustrated by the Open ALVEY interlocking of Figure 1.

Track circuits, e.g., TAA, divide the track into sections, with train detection devices informing the interlocking if a section is *occupied* or *clear*, represented by 'o' and 'c' respectively.

Points, e.g., P201, steer trains across junctions, and can be in one of two positions at any time—*controlled normal* ('cn') or *controlled reverse* ('cr'). Taking Figure 1 as an example, if a train is travelling over track circuit TAA towards track circuit TAB and points P201 are in controlled normal position, then the train will continue along TAB towards track circuit TAC. On the other hand, if points P201 are in controlled reverse position, then the train will turn off towards track circuit TBA. In addition, Boolean checks may be performed on sets of points. These checks determine whether or not points are able to move into controlled normal or controlled reverse position, and are termed *free to go normal* and *free to go reverse* respectively. A set of points is said to be *controlled free to go normal* ('cfn') (respectively, *controlled free to go reverse* ('cfr')) either if it is free to go normal (respectively, free to go reverse), or if it is controlled normal (respectively, controlled reverse).

Signals, e.g., S10, can allow or refuse the entry of a train onto particular sections of track and are situated in advance of the section which they control. For example, signal S20 controls entry onto the section made up of track circuits TAG, TAH and TCA. We do not consider the behaviour of signals in the following; as such, we do not concern ourselves with their state space.

Routes are sections of track between two signals, which proceed from an entry signal to an exit signal. For example, in Figure 1, route R13 is the section of track between signal S13 (the entry signal) and signal S21 (the exit signal). This route spans three track circuits: TAB, TAA and TAK, and contains points P201. A route may be in one of two states at any time: *set* ('s') or *unset* ('xs'). Note that there are two routes which start at signal S10. One is associated with track circuits TAB and TBA, and the other is associated with track circuits TAB and TAC. These routes are labelled R10A and R10B respectively.

Sub-routes are subsections of routes which are associated with specific track circuits. The naming convention for sub-routes is as follows. Given any track circuit, there may be several sub-routes over that circuit, with that number being dependent upon the possible entries and exits for that track circuit. As an example, track circuit TAB has three entry and exit points: at TAA, TAC and TBA. These entry and exit points are labelled clockwise from a 12.00 position. Thus, entry (or exit) from (or to) circuit TBA is labelled A, entry (exit) from (to) TAC is labelled B, and C is associated with entry (exit) from (to) TAA. It follows that sub-route UAB-AC is associated with track circuit TAB, with entry from track circuit TBA and exit at track circuit TAA. In addition, sub-route UAB-CB has entry from TAA and exit at TAC, and sub-route UAB-BC runs from TAC to TAA. Finally, the entry and exit points for UAB-CA are TAA and TBA respectively. As with other system components, sub-routes may be in one of two states at any time: *locked* ('l') or *free* ('f').

3 Conditional Checks

The setting of routes, freeing of sub-routes, and movement of points are all dependent upon conditional checks performed upon geographic data.

The first conditional checks which we describe are those associated with the setting of routes. As an example, route R14 runs from signal S14 over track circuits TAD and TAE and points P202. The condition concerned with setting route R14 is written

$$\text{Q14 } \textbf{if} \quad \text{P202 cfn UAE-AB f UAD-AB f}$$
$$\textbf{then } \text{R14 s P202 cn UAD-BA l UAE-BA l}$$

and is interpreted in the following way. Before route R14 can be set, points
P202 must be controlled free to go normal, and sub-routes UAE-AB and UAD-
AB must both be free. If any of these conditions are not met, then the route
cannot be granted. Alternatively, if each of these Boolean checks evaluates to
true, then route R14 can be set. This involves moving points P202 to controlled
normal position (if they are not already in that position), and locking sub-routes
UAD-BA and UAE-BA.

We assume that the locking of routes is atomic, i.e., two routes cannot be in
the process of being locked simultaneously. Also, for the purposes of modelling,
we apply no conditions upon routes becoming unset, and assume that any route
can enter that state unconditionally and at any time.

A sub-route can become free when a check performed upon conditions associated with it evaluates to true. In the above example, when route R14 is set,
sub-routes UAD-BA and UAE-BA are both locked. The latter can become free
when the following condition is met.

$$\text{UAE-BA f if TAE c UAD-BA f UAD-CA f}$$

That is, UAE-BA can become free when track circuit TAE is clear, and sub-
routes UAD-BA and UAD-CA are both free.

The following rule states that sub-route UAD-BA can become free when
track circuit TAD is clear and route R14 is unset.

$$\text{UAD-BA f if TAD c R14 xs}$$

A number of conditions must be met before points can move. Such conditions
are referred to as points free to move data. Referring to Figure 1, the *berth track
circuit* of points P201—track circuit TAB—must be clear before P201 can move
in either direction. If it were not, a derailment might occur. In addition, before
these points can move into controlled normal position, sub-routes UAB-AC and
UAB-CA must both be free. Furthermore, both UAB-BC and UAB-CB must
both be free before points P201 are free to go reverse.

Such a condition is written thus:

$$\text{P201N TAB c UAB-AC f UAB-CA f}$$
$$\text{P201R TAB c UAB-BC f UAB-CB f}$$

Here, points P201 are free to go normal if and only if track circuit TAB is clear
and sub-routes UAB-AC and UAB-CA are both free.

4   Safety Invariants

There are a number of safety invariants which must hold of all SSI geographic
databases. Examples of such properties are given below.

1. For every track circuit, no more than one of the sub-routes passing over it
   should be locked for a route at any time.

2. If a sub-route over a track section containing points is locked, then the points are correctly aligned with that sub-route.
3. If a route is set, then all of its sub-routes are locked.
4. If a track circuit containing points is occupied, then the points are locked.
5. If a sub-route is locked for a route, then all sub-routes ahead of it on that route are also locked.

If any invariants are not satisfied for a particular interlocking, then there is an error in the geographic data. In the next section, we describe how conditional checks on geographic data can be described in terms of CSP processes.

5   A CSP Approach

5.1   Basic Types and Functions

Referring to Figure 1, we introduce the set of track circuits, $\mathcal{T}$, the set of points, $\mathcal{P}$, the set of signals, $\mathcal{S}$, the set of routes, $\mathcal{R}$, and the set of sub-routes, $\mathcal{U}$,—*for this specific interlocking*—thus:

$$\mathcal{T} = \{\,\text{TAA, TAB, TAC, TAD, TAE, TAF,}$$
$$\text{TAG, TAH, TAJ, TAK, TBA, TCA}\}$$

$$\mathcal{P} = \{\,\text{P201, P202, P203, P204}\}$$

$$\mathcal{S} = \{\,\text{S10, S11, S12, S13, S14, S15,}$$
$$\text{S20, S21, S22, S23, S24, S25}\}$$

$$\mathcal{R} = \{\,\text{R10A, R10B, R11A, R11B, R12, R13, R14, R15,}$$
$$\text{R20A, R20B, R21A, R21B, R22, R23, R24, R25}\}$$

$$\mathcal{U} = \{\,\text{UAA-AB, UAA-BA, UAB-AC, UAB-BC, UAB-CA,}$$
$$\text{UAB-CB, UAC-AB, UAC-BA, UAD-AB, UAD-AC,}$$
$$\text{UAD-BA, UAD-CA, UAE-BA, UAF-BA, UAG-AB,}$$
$$\text{UAG-AC, UAG-BA, UAG-CA, UAH-AB, UAH-BA,}$$
$$\text{UAJ-AC, UAJ-BC, UAJ-CA, UAJ-CB, UAK-AB,}$$
$$\text{UAK-BA, UBA-AB, UBA-BA, UCA-AB, UCA-BA}\}$$

We represent an interlocking, $\mathcal{I}$, in terms of a quintuple $\mathcal{I} = (\mathcal{T}, \mathcal{P}, \mathcal{S}, \mathcal{R}, \mathcal{U})$.

We also define a number of functions which define the relationships between these components. The first such function,

$$locks : \mathcal{R} \rightarrow \mathbb{P}\mathcal{P}$$

relates routes and points. Given some route $r \in \mathcal{R}$, $locks(r)$ identifies the set of points which are locked when r is set. For example, $locks(\text{R10A}) = \{\text{P201}\}$.

The injection sr,

$$sr : \mathcal{T} \rightarrowtail \mathbb{P}\mathcal{U}$$

relates a track circuit $t \in \mathcal{T}$ to the set of sub-routes associated with that circuit, while the function

$$berth : \mathcal{P} \to \mathbb{P}\,\mathcal{T}$$

returns, for all $p \in \mathcal{P}$, the berth track circuits of points p. Although such a case does not arise in this example, it should be noted that it is possible for a point to be associated with more than one track circuit.

The surjection *circuit* returns the berth track circuit of a sub-route.

$$circuit : \mathcal{U} \twoheadrightarrow \mathcal{T}$$

As an example, $circuit(\text{UAB-AC}) = \text{TAB}$.

The injection *subroutes*,

$$subroutes : \mathcal{R} \rightarrowtail \mathrm{seq}\,\mathcal{U}$$

relates a route $r \in \mathcal{R}$ to its sub-routes in the order in which they are occupied. For example, $subroutes(\text{R13}) = \langle \text{UAB-AC}, \text{UAA-AB}, \text{UAK-AB} \rangle$.

Finally, the injections *norm* and *rev*

$$norm : \mathcal{P} \rightarrowtail \mathbb{P}\,\mathcal{U}$$
$$rev : \mathcal{P} \rightarrowtail \mathbb{P}\,\mathcal{U}$$

relate points to the sub-routes associated with them when they are in controlled normal or controlled reverse position respectively.

5.2   CSP and FDR

We assume familiarity with the notation of CSP; the uninitiated are referred to [5] for an introduction.

In this paper, we concentrate on the model of CSP which is often referred to as its 'standard' model—the failures-divergences model [2]. The fundamentals of this model are as follows.

Given a CSP process P,

- a *trace* of P is a finite sequence of events which P can perform;
- a *failure* of P is a pair (tr, ref) such that ref is a set of events which P may refuse to perform following an observation of the trace tr; and
- a *divergence* of P is a trace tr, following which it is possible for P to participate in an infinite, uninterrupted sequence of internal events.

We denote the sets of traces, failures and divergences of a process P by $traces(P)$, $failures(P)$ and $divergences(P)$ respectively.

A process Q is said to be a failures-divergences refinement of another process P if Q cannot behave in a manner forbidden by P. We define this relationship between processes in the following way.

Definition 1. A process Q is a *failures-divergences refinement* of another process P, denoted $P \sqsubseteq Q$, if and only if

$$\textit{failures}\,[\![Q]\!] \subseteq \textit{failures}\,[\![P]\!] \wedge \textit{divergences}\,[\![Q]\!] \subseteq \textit{divergences}\,[\![P]\!]$$

Failures-Divergences Refinement (FDR) [8] is a mechanical verification tool which allows one to test for refinement between two processes. It is this tool which has been employed in the verification tasks described in the following.

5.3 Channels

We start our CSP interpretation of SSI geographic data by defining a number of channels which represent changes in state of interlocking components.

A change in state of a route is represented by a communication on the channel *routeState*. For example, *routeState.r.req* indicates that some route $r \in \mathcal{R}$ has been (successfully) requested, *routeState.r.s* indicates that r has been set, and *routeState.r.xs* indicates that r has become unset. We choose not to record unsuccessful requests—if the state of the system is such that a particular route cannot be set, then the communication representing the successful request for that route is refused.

Communications on the channel *subrouteState* represent the change in state of a sub-route from free to locked, or vice versa.

Communications on the channel *pointState* indicate whether a point has become (or ceased to be) controlled free to go normal or controlled free to go reverse, while a change in point position is represented by observations on the channel *pointPosition*.

Finally, communications on the channel *circuitState* relate to track circuits becoming clear or occupied.

5.4 Initial Assumptions

We assume that all track circuits are initially clear and that all points are initially controlled normal (and, therefore, controlled free to go normal). We also assume that, initially, no points are free to move in either direction, all sub-routes are initially locked, and all routes are initially unset. These initial conditions allow the system to start in a 'safe' state, i.e., a state in which no routes can be set.

Following this, if any track circuits are detected as being occupied or any points are detected as being in a reverse position, then the internal state is updated accordingly. In addition, provided that the associated track circuit is clear, the conditions are such that the final sub-route in a given route can be freed, which, in turn, will allow the previous sub-route in the route to be freed (again, provided that the associated track circuit is clear) and so on, until each sub-route of that route is free.

5.5 Modelling Conditional Checks

Recall the track components with which the route setting data for R14 was concerned: points P202, and sub-routes UAD-AB, UAD-BA, UAE-AB and UAE-BA. Initially, points P202 are controlled free to go normal, and sub-routes UAE-AB and UAD-AB are both locked.

The process R14*true* represents the route setting data for route R14 when points P202 are controlled free to go normal, and sub-routes UAD-AB and UAE-AB are both free.

$$R14\mathit{true} =$$
$$\mathit{routeState}.R14.req \rightarrow \mathit{pointPosition}.P202.cn \rightarrow$$
$$\mathit{subrouteState}.\text{UAD-BA}.l \rightarrow \mathit{subrouteState}.\text{UAE-BA}.l \rightarrow$$
$$\mathit{routeState}.R14.s \rightarrow R14\mathit{true}$$
$$\Box$$
$$\mathit{pointState}.P202.\mathit{cfn}.\mathit{false} \rightarrow R14\mathit{false}(\mathit{false}, f, f)$$
$$\Box$$
$$\mathit{subrouteState}.\text{UAE-AB}.l \rightarrow R14\mathit{false}(\mathit{true}, l, f)$$
$$\Box$$
$$\mathit{subrouteState}.\text{UAD-AB}.l \rightarrow R14\mathit{false}(\mathit{true}, f, l)$$

If there is a request for R14 in this state, then points P202 are locked in controlled normal position, and sub-routes UAD-BA and UAE-BA are both locked. Alternatively, if any of the conditions necessary for the setting of route R14 become false, then this request is blocked and the process behaves as R14*false*. Note how the above process relates to the route setting data of Section 3—*if* all of the relevant conditions are met, *then* the route may be set.

The process R14*false* is defined similarly, although a request for route R14 to be set is not allowed when the system is in this state. In this process, the value of variable x determines whether points P202 are controlled free to go normal. Similarly, the values of y and z are concerned with whether or not sub-routes UAE-AB and UAD-AB respectively are free. For example, if the value of x is *true*, then points P202 are controlled free to go normal; if the value of y (respectively z) is f, then UAE-AB (respectively UAD-AB) is free. If all of the relevant conditions are met, then the behaviour of this process becomes that of R14*true*.

$$R14\mathit{false}(x, y, z) =$$
$$\textbf{if}\quad x = \mathit{true} \wedge y = f \wedge z = f$$
$$\textbf{then}\ R14\mathit{true}$$
$$\textbf{else}\quad \mathit{pointState}.P202.\mathit{cfn}?n \rightarrow R14\mathit{false}(n, y, z)$$
$$\Box$$
$$\mathit{subrouteState}.\text{UAE-AB}?n \rightarrow R14\mathit{false}(x, n, z)$$
$$\Box$$
$$\mathit{subrouteState}.\text{UAD-AB}?n \rightarrow R14\mathit{false}(x, y, n)$$

Given the above, we define

$$R14 = R14\mathit{false}(\mathit{true}, l, l)$$

We define processes representing route setting data for other routes similarly.

Processes concerning sub-route release data are defined in a similar fashion to those associated with route setting data. As all sub-routes are initially locked and all track circuits are initially clear, we define

$$\text{UAE-BA} = \text{UAE-BA}locked(c, l, l)$$

where UAE-BA*locked* is defined thus:

$$\text{UAE-BA}locked(x, y, z) =$$
$$\textbf{if} \quad x = c \wedge y = f \wedge z = f$$
$$\textbf{then} \ subrouteState.\text{UAE-BA}.f \rightarrow \text{UAE-BA}free(c, f, f)$$
$$\textbf{else} \ \ circuitState.\text{TAE}?n \rightarrow \text{UAE-BA}locked(n, y, z)$$
$$\Box$$
$$subrouteState.\text{UAD-BA}?n \rightarrow \text{UAE-BA}locked(x, n, z)$$
$$\Box$$
$$subrouteState.\text{UAD-CA}?n \rightarrow \text{UAE-BA}locked(x, y, n)$$

Here, the values of the variables x, y and z represent the states of track circuit TAE, and sub-routes UAD-BA and UAD-CA respectively. For example, the value of x is c if TAE is clear. Sub-route UAE-BA becomes free after being locked if the value of x is c and the values of both y and z are f. If this is the case, then the process UAE-BA*free* is invoked. If this is not the case and there is a change in the state of track circuit TAE or sub-routes UAD-BA or UAD-CA, then the value of the appropriate variable is updated accordingly.

The conditional check on sub-route UAE-BA when it is free is represented by the process UAE-BA*free*. Again, the value of x represents the state of track circuit TAE, and the values of y and z represent the states of sub-routes UAD-BA and UAD-CA respectively. Note that UAE-BA can become locked at any time, at which point the process UAE-BA*locked* is invoked.

$$\text{UAE-BA}free(x, y, z) =$$
$$subrouteState.\text{UAE-BA}.l \rightarrow \text{UAE-BA}locked(x, y, z)$$
$$\Box$$
$$circuitState.\text{TAE}?n \rightarrow \text{UAE-BA}free(n, y, z)$$
$$\Box$$
$$subrouteState.\text{UAD-BA}?n \rightarrow \text{UAE-BA}free(x, n, z)$$
$$\Box$$
$$subrouteState.\text{UAD-CA}?n \rightarrow \text{UAE-BA}free(x, y, n)$$

As we have seen, it is not always the case that conditionals for sub-route release are based upon the state of other sub-routes. In the case of the first sub-route of a route, sub-route release is conditional upon the state of that route. To illustrate this point, we take sub-route UAD-BA. Here, the relevant conditions are given by the states of track circuit TAD and route R14.

$$\text{UAD-BA} = \text{UAD-BA}locked(c, xs)$$

$$
\begin{aligned}
&\text{UAD-BA}\textit{locked}(x, y) = \\
&\textbf{if} \quad x = c \wedge y = xs \\
&\textbf{then} \; \textit{subrouteState}.\text{UAD-BA}.f \rightarrow \text{UAD-BA}\textit{free}(c, xs) \\
&\textbf{else} \quad \textit{circuitState}.\text{TAD}?n \rightarrow \text{UAD-BA}\textit{locked}(n, y) \\
&\qquad\quad \square \\
&\qquad\quad \square \; n : \{\textit{req}, xs\} \bullet \textit{routeState}.\text{R14}.n \rightarrow \text{UAD-BA}\textit{locked}(x, n)
\end{aligned}
$$

Note that processes of this form synchronise only upon the *req* and *xs* values passed on channels representing changes in route state, and not on the *s* values. Recall that route setting is atomic, i.e., it cannot be the case that two routes are in the process of being set at any one time. As such, the process UAD-BA is concerned with successful requests for route R14 and when that route becomes unset, but not with when it becomes set. Therefore, we have the following.

$$
\begin{aligned}
&\{\textit{routeState}.\text{R14}.\textit{req}, \textit{routeState}.\text{R14}.\textit{xs}\} \subseteq \alpha\,\text{UAD-BA} \\
&\wedge \\
&\textit{routeState}.\text{R14}.s \notin \alpha\,\text{UAD-BA}
\end{aligned}
$$

The process UAD-BA*free* is defined thus:

$$
\begin{aligned}
&\text{UAD-BA}\textit{free}(x, y) = \\
&\qquad \textit{subrouteState}.\text{UAD-BA}.l \rightarrow \text{UAD-BA}\textit{locked}(x, y) \\
&\qquad \square \\
&\qquad \textit{circuitState}.\text{TAD}?n \rightarrow \text{UAD-BA}\textit{free}(n, y) \\
&\qquad \square \\
&\qquad \square \; n : \{\textit{req}, xs\} \bullet \textit{routeState}.\text{R14}.n \rightarrow \text{UAD-BA}\textit{free}(x, n)
\end{aligned}
$$

There are two Boolean checks concerned with every set of points. We represent points free to move data as the synchronisation of processes representing such checks. Taking points P201 as an example, if P201 are controlled normal or free to go normal, then they are controlled free to go normal. Recall that these points are free to go normal if track circuit TAB is clear, and sub-routes UAB-AC and UAB-CA are both free. In the following, *pos* represents the position of P201, while the variables x, y and z represent the states of track circuit TAB, and sub-routes UAB-AC and UAB-CA respectively.

$$
\begin{aligned}
&\text{P201}\textit{cfnTrue}(\textit{pos}, x, y, z) = \\
&\textbf{if} \quad \textit{pos} \neq cn \wedge (x = o \vee y = l \vee z = l) \\
&\textbf{then} \; \textit{pointState}.\text{P201}.\textit{cfn}.\textit{false} \rightarrow \text{P201}\textit{cfnFalse}(\textit{pos}, x, y, z) \\
&\textbf{else} \quad \textit{pointPosition}.\text{P201}?n \rightarrow \text{P201}\textit{cfnTrue}(n, x, y, z) \\
&\qquad\quad \square \\
&\qquad\quad \textit{circuitState}.\text{TAB}?n \rightarrow \text{P201}\textit{cfnTrue}(\textit{pos}, n, y, z) \\
&\qquad\quad \square \\
&\qquad\quad \textit{subrouteState}.\text{UAB-AC}?n \rightarrow \text{P201}\textit{cfnTrue}(\textit{pos}, x, n, z) \\
&\qquad\quad \square \\
&\qquad\quad \textit{subrouteState}.\text{UAB-CA}?n \rightarrow \text{P201}\textit{cfnTrue}(\textit{pos}, x, y, n)
\end{aligned}
$$

Alternatively, if this condition is contradicted, then P201*cfnFalse* is invoked.

$$P201cfnFalse(pos, x, y, z) =$$

if $pos = cn \lor (x = c \land y = f \land z = f)$
then *pointState*.P201.*cfn*.*true* $\rightarrow$ P201*cfnTrue*(*pos*, *x*, *y*, *z*)
else *pointPosition*.P201?*n* $\rightarrow$ P201*cfnFalse*(*n*, *x*, *y*, *z*)
$\square$
circuitState.TAB?*n* $\rightarrow$ P201*cfnFalse*(*pos*, *n*, *y*, *z*)
$\square$
subrouteState.UAB-AC?*n* $\rightarrow$ P201*cfnFalse*(*pos*, *x*, *n*, *z*)
$\square$
subrouteState.UAB-CA?*n* $\rightarrow$ P201*cfnFalse*(*pos*, *x*, *y*, *n*)

Only when points P201 are controlled normal, or when track circuit TAB is clear and sub-routes UAB-AC and UAB-CA are both free, will points P201 again become controlled free to go normal.

The processes concerned with the controlled free to go reverse condition for points P201 are defined similarly. Thus, we represent the points free to move data for points P201 by

$$P201 = P201cfnTrue(cn, c, l, l)$$
$$\| [\, \alpha\, pointPosition.P201 \cup \alpha\, circuitState.TAB\,] \|$$
$$P201cfrFalse(cn, c, l, l)$$

It might appear at first glance that the above involves a duplication of information and that it might be simpler to define one all-encompassing process P201 which reduces the need for parallelism and hence reduces the state space. However, it is often the case that, for a given point p, the process $p\_cfnTrue$ might be needed to verify a particular safety invariant and there is no need to consider the process $p\_cfnFalse$ (or vice versa). For this reason, it is more efficient to define two separate processes. It follows that if one or the other condition is required, we take into account that condition, while if both are required, then the parallel combination of both is invoked. This notion of requiring specific parts of a geographic database will become clearer when we explain our concept of an extraction.

Our CSP representation of an interlocking's geographic data involves the composition of processes such as those described above. In addition, our assumption regarding route setting atomicity is enforced by the process RC.

$$RC(R) = \square\, r : R \bullet routeState.r.req \rightarrow routeState.r.s \rightarrow RC(R)$$

This process ensures that, following a successful request for some route $r \in R$, no successful requests for further routes can be observed until r has been set. At present, we do not incorporate the possibility of route cancellation. However, for the purposes of modelling, we assume that routes can become unset unconditionally and at any time.

Again, note that this process does not synchronise on all values passed on the channel *routeState*:

$$\alpha\, RC(R) = \bigcup \{\, r : \mathcal{R} \mid r \in R \bullet \{\, routeState.r.req, routeState.r.s\,\}\,\}$$

We denote the composition of the conditional checks on the geographic data pertaining to an interlocking $\mathcal{I}$, together with $RC(\mathcal{R})$, by $\mathcal{I}_{\mathrm{CSP}}$.

6   Verifying SSI Safety Invariants

In this section, we illustrate how our CSP representation of SSI geographic data can be verified with respect to safety invariant 1 of Section 4.

Recall that all sub-routes are initially locked; however, they are not *locked for routes*, i.e., their locking was not brought about as part of the process of setting a route. Thus, this invariant is not invalidated at initialisation as might appear at first. However, all communications of the form *subrouteState.u.l* after initialisation do form part of the route setting process. Therefore, we must ensure that all observations of sub-routes being locked are separated by corresponding observations of the relevant sub-route being freed.

Assume a set of sub-routes $U \subseteq \mathcal{U}$. If any $u \in U$ becomes locked, then u must become free before other sub-routes can be locked.

Initially, all elements of U are locked. If at any stage the value of *Locked* is the empty set, then the process $S_1\,AllFree$ is invoked. On the other hand, if there are still sub-routes which are locked, then there is a choice between a locked sub-route becoming free and a free sub-route becoming locked.

$$
\begin{aligned}
S_1\,Init(&Locked, Free) = \\
&\mathbf{if} \quad\;\; Locked = \emptyset \\
&\mathbf{then}\;\; S_1\,AllFree(Free) \\
&\mathbf{else}\;\;\; \Box\, u : Locked \bullet subrouteState.u.f \rightarrow \\
&\qquad\qquad\qquad\qquad S_1\,Init(Locked \setminus \{u\}, Free \cup \{u\}) \\
&\qquad\;\; \Box \\
&\qquad\;\; \Box\, u : Free \bullet subrouteState.u.l \rightarrow \\
&\qquad\qquad\qquad\qquad S_1'(Locked, Free \setminus \{u\}, u)
\end{aligned}
$$

If a sub-route is locked for a route, then it must be freed before another can be locked in this manner.

$$
\begin{aligned}
S_1'(&Locked, Free, u) = \\
&subrouteState.u.f \rightarrow S_1\,Init(Locked, Free \cup \{u\}) \\
&\Box \\
&\Box\, v : Locked \bullet subrouteState.v.f \rightarrow S_1'(Locked \setminus \{v\}, Free \cup \{v\}, u)
\end{aligned}
$$

The process $S_1\,AllFree$ describes the required behaviour of the sub-routes contained in U after all have been released from their initial state.

$$
\begin{aligned}
S_1\,AllFree(U) = \Box\, u : U \bullet\; &subrouteState.u.l \rightarrow \\
&subrouteState.u.f \rightarrow S_1\,AllFree(U)
\end{aligned}
$$

In the case of TAK, we have $sr(\text{TAK}) = \{\text{UAK-AB}, \text{UAK-BA}\}$. Recalling the convention for naming sub-routes, UAK-AB runs from track circuit TAA towards track circuit TAJ, while UAK-BA runs in the opposite direction. Thus, safety invariant 1 for track circuit TAK states that UAK-AB and UAK-BA should not be locked for routes simultaneously.

By composing the processes $S_1(U)$ and $CHAOS_\Sigma$, where

$$CHAOS_\Sigma = (\sqcap a : \Sigma \bullet a \to CHAOS_\Sigma) \sqcap Skip$$

we obtain a process which states that no two sub-routes from the set U can be simultaneously locked, while imposing no restrictions upon other events.

$$S_1(U) = S_1 \, Init(U, \emptyset) \, |[\bigcup \{u : U \bullet \alpha \, subrouteState.u\}]| \, CHAOS_\Sigma$$

Definition 2. Given an interlocking $\mathcal{I} = (\mathcal{T}, \mathcal{P}, \mathcal{S}, \mathcal{R}, \mathcal{U})$, safety invariant 1 holds of $\mathcal{I}$ if and only if $\forall t : \mathcal{T} \bullet S_1(sr(t)) \sqsubseteq \mathcal{I}_{\text{CSP}}$.

The Open ALVEY interlocking consists of several million possible configurations—as such, we look for ways to reduce the complexity of the problem.

Recall from Section 3 that a sub-route becomes locked when a route passing over it is set. Given a track circuit $t \in \mathcal{T}$, we need take into account only the route setting data for those routes which may lock sub-routes contained in the set $sr(t)$.

Referring to Figure 1, four routes travel over track circuit TAK: R13, R15, R22 and R24. The route setting data for these routes is as follows.

> Q13 **if**   P201 cfr UAA-BA f UAB-CA f
>   **then** R13 s P201 cr UAB-AC l UAA-AB l UAK-AB l

> Q15 **if**   P201 cfn UAA-BA f UAB-CB f
>   **then** R15 s P201 cn UAB-BC l UAA-AB l UAK-AB l

> Q22 **if**   P204 cfr UAJ-CB f UAK-AB f
>   **then** R22 s P204 cr UAJ-BC l UAK-BA l UAA-BA l

> Q24 **if**   P204 cfn UAJ-CA f UAK-AB f
>   **then** R24 s P204 cn UAJ-AC l UAK-BA l UAA-BA l

Referring to the above, only routes R22 and R24 can lock sub-route UAK-BA. Observing the conditions for these routes to be set, we see that before this sub-route can be locked by either route, sub-route UAK-AB must be free. Therefore, we need take no other processes into account in ensuring that UAK-AB and *then* UAK-BA cannot be locked—if sub-route UAK-AB is locked, then neither route R22 nor route R24 can be set, and, therefore, sub-route UAK-BA cannot be locked.

Examining the conditions for routes R13 and R15 to be set, we see that before either route can be set, sub-route UAA-BA must be free. The sub-route release data for this sub-route is given by

$$\text{UAA-BA f } \textbf{if} \text{ TAA c UAK-BA f}$$

Therefore, we need take into account only processes representing route setting data for routes R13 and R15, and sub-route release data for sub-route UAA-BA to ensure that UAK-BA and *then* UAK-AB cannot be locked—if sub-route UAK-BA is locked, then UAA-BA cannot be freed, and, therefore, neither route R13 nor route R15 can be set.

In this case, only five processes need to be considered to ensure that safety invariant 1 holds for track circuit TAK. The justification for this is based upon the fact that the events with which we are concerned—the behaviour of sub-routes associated with track circuit TAK—can only ever occur in $\mathcal{I}_{\text{CSP}}$ with the co-operation of processes representing route setting data for routes R13, R15, R22 and R24. Composition with further processes will only serve to reduce the set of possible behaviours for these components, while introducing further independent components and thus expanding the state space of the check to be performed.

Given our approach to constructing CSP processes corresponding to SSI conditional checks, we define the process

$$\begin{aligned}
\text{TAK}imp = \quad & (RC(\{\text{R13}, \text{R15}, \text{R22}, \text{R24}\}) \\
& \| \\
& ((\text{R13} \,|[\, \alpha \, subrouteState.\text{UAA-BA} \,]|\, \text{R15}) \\
& \| \\
& (\text{R22} \,|[\, \alpha \, subrouteState.\text{UAK-AB} \,]|\, \text{R24}))) \\
& \| \\
& \text{UAA-BA}
\end{aligned}$$

If $S_1(sr(\text{TAK})) \sqsubseteq \text{TAK}imp$, then we can claim that safety invariant 1 holds of $\mathcal{I}_{\text{CSP}}$ for track circuit TAK. This is because all relevant behaviours of the sub-routes contained in the set $sr(\text{TAK})$ are considered by the process $\text{TAK}imp$.

We processes such as UAA-BA*locked* the *offset* for TAK*imp*, and we term processes such as TAK*imp* an *extraction* for track circuits TAK.

In the following, we let $\gamma_{\mathcal{I}}(sr(t))$ denote the extraction from $\mathcal{I}_{CSP}$ of those processes associated with the setting of sub-routes over a track circuit $t \in \mathcal{T}$.[2]

The following proposition formalises our intuition that if safety invariant 1 holds of the relevant extraction, then it holds of the interlocking as a whole.

Proposition 3. *Given an interlocking* $\mathcal{I} = (\mathcal{T}, \mathcal{P}, \mathcal{S}, \mathcal{R}, \mathcal{U})$,

$$\forall t : \mathcal{T} \bullet S_1(sr(t)) \sqsubseteq \gamma_{\mathcal{I}}(sr(t)) \Rightarrow S_1(sr(t)) \sqsubseteq \mathcal{I}_{\text{CSP}}$$

---

[2] A suitable algorithm for generating such extractions is described in [9]; this is omitted due to limitations on space.

Some care is needed in determining which events that the processes involved in extractions should synchronise upon. For example, in the extraction $\gamma_{\mathcal{I}}(sr(\text{TAK}))$, the alphabets of the processes representing route setting data for routes R13 and R15 have four events in common: the event representing the release of sub-route UAA-BA, and those representing the locking of sub-routes UAA-AB, UAA-BA and UAK-AB.

These processes must communicate on *subrouteState*.UAA-BA—the state of this sub-route helps to determine whether routes R13 and R15 can be set, so it is essential that the two process's view of that sub-route's state is identical. On the other hand, they do not synchronise on the locking of UAA-AB and UAK-AB. This is because the locking of these sub-routes for route R13 is of no consequence to the route setting data for route R15; similarly the locking of these sub-routes for route R15 is irrelevant to the route setting data for R13. Therefore, there is no need for the processes representing route setting data for these routes to synchronise on the communications *subrouteState*.UAA-AB.*l* and *subrouteState*.UAK-AB.*l*.

As a further example, processes R22 and R24 have the events representing the release of sub-route UAK-AB, and the locking of UAK-AB, UAK-BA and UAA-BA in common. For similar reasons to those outlined above, these processes synchronise on the channel *subrouteState*.UAK-AB only.

Synchronisation between the two pairs of routes is slightly more complicated. Routes R13 and R15 are both concerned with the state of sub-route UAA-BA. As routes R22 and R24 both lock UAA-BA, these pairs of processes must synchronise on *subrouteState*.UAA-BA.*l*. Similarly, routes R22 and R24 are both concerned with the state of sub-route UAK-AB. As routes R13 and R15 both lock sub-route UAK-AB, the pair of processes representing route setting data for these routes must synchronise on *subrouteState*.UAK-AB.*l*. These are the only events upon which the two pairs of routes synchronise.

We also compose these processes with RC and UAA-BA to give us the process against which we verify safety invariant 1 for track circuit TAK.

Via FDR such checks on individual track circuits can be carried out quickly and efficiently, with the complexity of checks for our simple interlocking ranging in size from only 232 states (for track circuits TAC, TAH, TBA and TCA) to 14592 states (for track circuits TAA and TAK). The complexity of such verification is illustrated in Table 1—these checks can be accomplished in a matter of seconds on a standard workstation.

Note that verifying this safety invariant for track circuit TAK involves a check of the same size as that required to verify its 'mirror image' circuit, TAA.

The verification of this invariant for other track circuits follows similar patterns. Points P201 (respectively P203 and P204) provides the offset for track circuit TAB (respectively TAG and TAJ), and the size of checks for these are of a similar order as those for TAD, which has the offset points P202.

Checks for track circuits TAC, TAH, TBA and TCA require no offset at all, and, as such, are of a relatively small size. As an example, only two routes lock sub-routes over track circuit TCA: R10B and R11A. The route setting data for

Track circuit	States
TAA	14592
TAB	3532
TAC	232
TAD	3312
TAG	3312
TAH	232
TAJ	3532
TAK	14592
TBA	232
TCA	232

Table 1. The complexity of verifying invariant 1 for the Open ALVEY

these routes is as follows.

$$\text{Q10B} \ \textbf{if} \quad \text{P201 cfn UAB-BC f UAC-AB f}$$
$$\textbf{then} \ \text{R10B s P201 cn UAB-CB l UAC-BA l}$$

$$\text{Q11A} \ \textbf{if} \quad \text{P202 cfn UAD-BA f UAC-BA f}$$
$$\textbf{then} \ \text{R11A s P202 cn UAD-AB l UAC-AB l}$$

Given this route setting data, we have

$$\gamma_{\mathcal{I}}(sr(\text{TAC})) = RC(\{\text{R10B}, \text{R11A}\}) \parallel \text{R10B} \parallel \text{R11A}$$

We can further reduce the complexity of verifying safety invariant 1 for track circuits in the following way. Consider some track circuit $t \in \mathcal{T}$ for an interlocking $\mathcal{I} = (\mathcal{T}, \mathcal{P}, \mathcal{S}, \mathcal{R}, \mathcal{U})$. If we can establish that, for every pair of sub-routes $u, v \in sr(t)$, u and v cannot be locked simultaneously, then it follows that no more than one of $sr(t)$ can be locked simultaneously.

Proposition 4. *Given an interlocking* $\mathcal{I} = (\mathcal{T}, \mathcal{P}, \mathcal{S}, \mathcal{R}, \mathcal{U})$,

$$\forall t : \mathcal{T} \bullet$$
$$(\forall u, v : sr(t) \mid u \neq v \bullet S_1(\{u, v\}) \sqsubseteq \gamma_{\mathcal{I}}(\{u, v\}))$$
$$\Rightarrow S_1(sr(t)) \sqsubseteq \gamma_{\mathcal{I}}(sr(t))$$

Proof. Assume

$$\forall u, v : sr(t) \mid u \neq v \bullet S_1(\{u, v\}) \sqsubseteq \gamma_{\mathcal{I}}(\{u, v\})$$

Given any sub-route $u \in sr(t)$, our assumption states that no other sub-route $v \in sr(t)$ such that $v \neq u$ can be locked when u is locked. As such,

$$S_1(sr(t)) \sqsubseteq \gamma_{\mathcal{I}}(sr(t))$$

Sub-routes	States
(UAD-AB, UAD-AC)	128
(UAD-AB, UAD-BA)	16
(UAD-AB, UAD-CA)	128
(UAD-AC, UAD-BA)	128
(UAD-AC, UAD-CA)	16
(UAD-BA, UAD-CA)	64

Table 2. The complexity of verifying invariant 1 for track circuit TAD

Thus, the verification of safety invariant 1 for a track circuit reduces to analysing pairs of sub-routes over that track circuit. The complexity of verifying track circuit TAD in this fashion is illustrated in Table 2.

We can further reduce the complexity of verifying safety invariant 1 by reducing our checks from a mutual exclusion property on pairs of sub-routes to a mutual exclusion property on pairs of routes which are associated with these sub-routes.

To accomplish this, we need a new extraction algorithm to identify the components we need to take into account in order to verify that two routes running over the same track circuit cannot be set simultaneously.[3] We let $\delta_{\mathcal{I}}(t, r, s)$ denote the extraction associated with a pair of routes $r, s \in \mathcal{R}$, both of which pass over track circuit $t \in \mathcal{T}$.

Proposition 5. *Given an interlocking* $\mathcal{I} = (\mathcal{T}, \mathcal{P}, \mathcal{S}, \mathcal{R}, \mathcal{U})$, *a track circuit* $t \in \mathcal{T}$, *and sub-routes* $u, v \in sr(t)$ *such that* $u \neq v$, *define the sets*

$$R_u = \{r : \mathcal{R} \mid u \in \mathrm{ran}(subroutes(r))\}$$
$$R_v = \{r : \mathcal{R} \mid v \in \mathrm{ran}(subroutes(r))\}$$

which are the sets of routes which lock u *and* v *respectively. It follows that*

$$(\forall\, r_u : R_u;\ r_v : R_v \bullet S_1(\{u, v\}) \sqsubseteq \delta_{\mathcal{I}}(t, r_u, r_v))$$
$$\Rightarrow S_1(\{u, v\}) \sqsubseteq \gamma_{\mathcal{I}}(\{u, v\})$$

Proof. Assume

$$\forall\, r_u : R_u;\ r_v : R_v \bullet S_1(\{u, v\}) \sqsubseteq \delta_{\mathcal{I}}(t, r_u, r_v)$$

Take any route $r_u \in R_u$. Our assumption states that no route $r_v \in R_v$ can be set if r_u is. Therefore, v cannot be locked when u is. Similarly, whenever any $r_v \in R_v$ is set, our assumption states that u is free. Therefore,

$$S_1(\{u, v\}) \sqsubseteq \gamma_{\mathcal{I}}(\{u, v\})$$

---

[3] Such an algorithm is described in [9]; again, this is omitted due to limitations on space.

Route 1	Route 2	States
R13	R22	720
R13	R24	720
R15	R22	720
R15	R24	720

Table 3. The complexity of verifying mutual exclusion for track circuit TAK

Finally, safety invariant 1 holds of an interlocking $\mathcal{I} = (\mathcal{T}, \mathcal{P}, \mathcal{S}, \mathcal{R}, \mathcal{U})$ if and only if, for every track circuit $t \in \mathcal{T}$, no two routes can simultaneously lock sub-routes over t. That is, that the setting of such routes is mutually exclusive.

Proposition 6. *Given an interlocking* $\mathcal{I} = (\mathcal{T}, \mathcal{P}, \mathcal{S}, \mathcal{R}, \mathcal{U})$, *if*

$$\forall t : \mathcal{T}; \ u, v : \mathcal{U}; \ r, s : \mathcal{R} \mid$$
$$u \in sr(t) \wedge v \in sr(t) \wedge u \neq v \wedge$$
$$r \in \{x : \mathcal{R} \mid u \in \mathrm{ran}(subroutes(x))\} \wedge$$
$$s \in \{x : \mathcal{R} \mid v \in \mathrm{ran}(subroutes(x))\} \bullet S_1(\{u, v\}) \sqsubseteq \delta_{\mathcal{I}}(t, r, s)$$

then safety invariant 1 holds of $\mathcal{I}$.

Proof. Assume

$$\forall t : \mathcal{T}; \ u, v : \mathcal{U}; \ r, s : \mathcal{R} \mid$$
$$u \in sr(t) \wedge v \in sr(t) \wedge u \neq v \wedge$$
$$r \in \{x : \mathcal{R} \mid u \in \mathrm{ran}(subroutes(x))\} \wedge$$
$$s \in \{x : \mathcal{R} \mid v \in \mathrm{ran}(subroutes(x))\} \bullet S_1(\{u, v\}) \sqsubseteq \delta_{\mathcal{I}}(t, r, s)$$

By Proposition 5,

$$\forall t : \mathcal{T}; \ u, v : \mathcal{U} \mid u \in sr(t) \wedge v \in sr(t) \wedge u \neq v \bullet$$
$$S_1(\{u, v\}) \sqsubseteq \gamma_{\mathcal{I}}(\{u, v\})$$

By Proposition 4,

$$\forall t : \mathcal{T} \bullet S_1(sr(t)) \sqsubseteq \gamma_{\mathcal{I}}(sr(t))$$

By Proposition 3,

$$\forall t : \mathcal{T} \bullet S_1(sr(t)) \sqsubseteq \mathcal{I}_{\mathrm{CSP}}$$

Finally, by Definition 2, safety invariant 1 holds of $\mathcal{I}$.

The complexity of such checks for track circuit TAK is illustrated in Table 3.

7   Discussion

We have described how, via CSP representations of SSI geographic data, a specific safety invariant can be verified of such data by a series of small mechanical checks. Four further safety invariants were mentioned in Section 4; these invariants can also be described in terms of CSP processes and verified via FDR. Limitations on space do not allow us to illustrate the analysis of these properties here, but this will be the subject of future papers. However, it is shown in [9] that the proof of safety invariant 1 for an interlocking helps in reducing the complexity of verifying these other invariants.

The methods described in this paper have been applied to data associated with a real interlocking. For reasons of commercial sensitivity, we have been unable to present such data in this paper. However, results have been encouraging, and the verification of the invariants described appear to be feasible for such data.

The keys to our method are the approach to compositionality possessed by CSP and the fact that failures-divergences refinement is preserved within context. Together, these factors allow us to extract and concentrate solely upon the specific SSI components with which we are concerned. In doing this, mechanical verification of such data which would otherwise be intractable, reduces to a series of small checks which can be carried out quickly and efficiently on a standard workstation.

We do not claim that our verification is by any means exhaustive—for example, the safety invariants which we have described make no mention of signals. Current work involves the inclusion of such factors. We are also investigating the verification of further safety invariants not mentioned in this paper.

Possible areas for future work include the automation of the translation from geographic data to CSP (algorithms for this task have already been developed), and the automation of the establishment of proof obligations. It is hoped that by mechanising these tasks, the development of a CSP-based fully automated approach to this problem might be possible. This will be the subject of future papers.

Acknowledgements

The authors wish to thank Smith System Engineering Limited, the Engineering and Physical Sciences Research Council and the Defence Research Agency (Malvern) for financial support. Thanks also to Clive Adams and Glyn Carter of Smith System Engineering Limited; Don Hayward of London Underground Limited; Mike Ingleby, David Mee, Richard Potts and Tony Terry of British Rail Research (Derby); and Janet Barnes, Anthony Hall and Trevor King of Praxis plc for their many valuable comments. Finally, thanks to the anonymous reviewers for several helpful suggestions.

References

1. SSI data preparation guide. British Railways Board Issue SSI8003, February 1990.
2. S.D. Brookes and A.W. Roscoe. An improved failures model for communicating processes. In S.D. Brookes, A.W. Roscoe, and G. Winskel, editors, *Proceedings of the NSF-SERC Seminar on Concurrency*, pages 281–305. Springer-Verlag Lecture Notes in Computer Science, volume 197, 1985.
3. A.H. Cribbens. Solid State Interlocking (SSI): An integrated electronic signalling system for mainline railways. *IEE Proceedings*, 134(3):148–158, 1987.
4. A.H. Cribbens and I.H. Mitchell. The application of advanced computing techniques to the generation and checking of SSI data. In *Railway Engineers' Forum Meeting, London, 1991*, pages 54–66. IRSE, 1992.
5. C.A.R. Hoare. *Communicating Sequential Processes*. Prentice-Hall International, 1985.
6. M. Ingleby and D.J. Mee. A calculus of hazard for railway signalling. In *Proceedings of the IEEE Workshop on Industrial-Strength Formal Specification Techniques*, 1995.
7. M.J. Morley. Modelling British Rail's interlocking logic: Geographic data correctness. Technical Report ECS-LFCS-91-186, Department of Computer Science, University of Edinburgh, 1991.
8. A.W. Roscoe. Model checking CSP. In A.W. Roscoe, editor, *A Classical Mind: Essays in honour of C.A.R. Hoare*, pages 353–378. Prentice-Hall International, 1994.
9. A.C. Simpson. Safety through security. DPhil thesis, Programming Research Group, Oxford University Computing Laboratory, 1996.

Refinement of Infeasible Real-Time Programs

Mark Utting[1] and Colin Fidge[2]

[1] Department of Computer Science,
School of Computing and Mathematical Sciences,
The University of Waikato, Hamilton, New Zealand.
E-mail: marku@cs.waikato.ac.nz
[2] Software Verification Research Centre,
School of Information Technology,
The University of Queensland, Queensland, Australia.
E-mail: cjf@it.uq.edu.au

Abstract. Embedded real-time programs can be succinctly specified using timed traces. Each sequentially executed statement acts to define a distinct trace segment. An elegant way of defining the effect of such statements is as trace 'coercers' that impose constraints on existing, but underspecified, traces. Unfortunately this model fails the usual refinement calculus feasibility test. Here we overcome this by proving that the coercive model is equivalent to a trace 'extending' model that does pass the test. The proof is itself interesting because it adopts non-standard data refinement techniques.

1   Introduction

The overall behaviour of an embedded real-time program is most succinctly specified using *timed traces* which record the value of each system variable at each moment in time. Ideally, each such trace requirement should then be refinable to a sequence of actions on consecutive trace segments that collectively achieve the total desired trace.

A number of ways of sequentially constructing traces can be envisaged. For instance, an initially empty trace can be successively *extended* by adding new segments to the end. This constructive model is straightforward, but its definition proves to be awkward because the domains of the traces change with time. A more elegant alternative is not to change the length of the trace at all. Instead we start with a complete, but underdefined, trace and each sequential action adds, through *coercion*, a new constraint on each trace segment. This model involves fewer specification variables, has simpler definitions, and enjoys some powerful refinement properties.

Our overall goal is to develop a real-time refinement calculus that is as close to the well-known 'untimed' refinement calculus as possible. However, with our coercive semantics, most of the statements of our executable target language fail the feasibility test of the untimed refinement calculus. That is, they do not satisfy the *Law of the Excluded Miracle* [4, p. 18]. Although infeasible (or *nonstrict*) program fragments often appear in the untimed refinement calculus, the conventional wisdom is that they cannot appear in executable programs, but must be removed by various program transformations before executable code is reached [20].

In contrast, we want our infeasible statements to be part of our target executable language! When we first designed the coercive semantics, we found the infeasibility of the target language statements worrying. Example programs suggested that when the infeasible statements are combined into a complete process, the resulting process is usually feasible, but we could not see any obvious program transformations that could be used to transform the infeasible statements into feasible ones, nor did we have a formal proof of the feasibility of a complete process.

In this paper, we solve this problem by proving that a complete process, containing infeasible statements, is equivalent to another process that contains only feasible statements. Since the resulting process contains only feasible statements, it is easy to prove that it is feasible. This proves that our real-time target language is indeed implementable.

In practice, developers using our real-time refinement calculus work entirely within the coercive semantics, because of its simplicity and convenient properties. The final data refinement step does not affect the structure of the program and can be done automatically once a process has been refined into the target language.

Finally, our equivalence proof is interesting in itself. It uses ideas from data refinement, but the equivalence that we need cannot be achieved using the standard data refinement proof techniques of upward or downward simulation. Instead, we use a different technique that involves induction over sequences of our target language statements.

2   Background: A Timed Specification Language

Embedded real-time systems must interact repeatedly with their environment, with each interaction taking place at a predetermined time. Experience has shown that the most convenient way of modelling such behaviour is via 'timed traces'. A number of specification and refinement models use this approach, including the *timed refinement calculus* [17, 19, 15], the *Temporal Agent Model* (TAM) [27, 25, 26], the *duration calculus* [32, 33] and *SIGNAL* [14]. Here we use the timed refinement calculus because it is closest in spirit to the familiar (untimed) refinement calculus [21, 24].

A *timed variable*

$$v : \mathbb{T} \to T$$

in the timed refinement calculus represents the history of a system variable, of type T, as a function from the absolute time domain $\mathbb{T}$ to the value of the variable at each time [16].

The timed refinement calculus uses predicate-transformer semantics. A *specification statement* consists of a *frame*, listing timed variables *constructed* by this statement, an *assumption* (cf., precondition), describing timed variables constructed by the environment, and an *effect* (cf., postcondition), defining the timed variable traces constructed by this statement [17, pp. 6–7].

In the untimed calculus, variables in the frame may be *modified* by a specification statement, and the postcondition may refer to both their initial and final values [21, ch. 8]. Timed refinement calculus specification statements *construct* entire timed traces, so there is only one 'copy' of each timed variable [17, pp. 6–7]. A constructed variable may appear in the effect predicate, but cannot appear at all in the assumption.

The timed refinement calculus offers a small number of specification statement operators, most notably parallel composition [17, p. 14], a feature not found in the untimed calculus. Conversely, however, the notion of sequential composition is missing from the timed refinement calculus. It has therefore been used to date only for developing high-level system designs consisting of a number of parallel processes [19, 18, 16, 6].

We have recently extended the timed refinement calculus with a notion of sequential composition, so that individual process specifications can be refined to sequential statements, thus supporting a smooth transition between 'parallel' and 'sequential' real-time refinements [29, 10, 11]. In doing so it was desirable to preserve the elegant timed calculus notion of there being a single trace for each system variable (i.e., no distinct 'initial' and 'final' trace values), yet provide rules as close as possible to those of the untimed calculus [21].

To this end we have defined a notion of sequentially-composed timed-trace specification statements, distinguished by the decoration '$\star$', for use when refining a process specification created using the timed refinement calculus [29]. Let $\mathbf{p}$ be the list of timed variables constructed by a process specification, and $\mathbf{e}$ be variables from the environment of the process. Our *sequential timed specification statement* is then denoted

$$\star\,\mathbf{v}\colon \big[A\,,E\big]\ .$$

Here $\mathbf{v}$, a sublist of $\mathbf{p}$, lists timed variables that may be modified by this statement. Assumption predicate A and effect predicate E may both refer to variables from $\mathbf{e}$ and $\mathbf{p}$. (For ease of expression below, we often treat a list of variables $\mathbf{w}$ as a single element, on the understanding that declarations and substitutions containing $\mathbf{w}$ stand for concurrent occurrences, and predicates involving $\mathbf{w}$ denote the conjunction of the predicate for each element of $\mathbf{w}$.)

Most importantly, this specification statement implicitly introduces a distinguished specification variable τ, denoting the current time or 'now'. This allows us to explicitly refer to the absolute time at which the statement begins and ends. Unlike the timed-trace variables, τ is destructively updated and obeys the usual zero-subscript convention [21, ch. 8]. Assumption A can use τ to refer to the time at which the statement begins. Effect E can use τ_0 to refer to the time at which the statement begins, and τ to refer to, or constrain, the time at which the statement ends. Thus, for some timed variable v, $v(\tau_0)$ and $v(\tau)$ in the effect predicate refer to v's initial and final values, respectively. (An elegant syntactic convention, that makes our calculus look even closer to the untimed one, is to allow 'v_0' as a shorthand for '$v(\tau_0)$' and 'v' for '$v(\tau)$', but for clarity in this paper we do not employ this notation.)

To complete the analogy with the untimed refinement calculus, we also introduce timed assertions [21, p. 11],

$$\star\{A\}\;,$$

and coercions [21, p. 165],

$$\star[E]\;,$$

whose predicates may refer to or constrain τ. As explained below, timed assertions do not consume time [11, §2.3], whereas timed coercions may [11, §2.2].

$$v := X \stackrel{\mathrm{def}}{=} \star\, v\colon \big[\mathrm{true}\,,\, v(\tau) = X(\tau_0)\big]$$

$$\mathbf{skip} \stackrel{\mathrm{def}}{=} \star\, \{\mathrm{true}\}$$

$$\mathbf{delay\text{-}until}\, X \stackrel{\mathrm{def}}{=} \star\, \big[\mathrm{true}\,,\, X(\tau_0) \leqslant \tau\big]$$

$$\mathbf{get\text{-}time}(v) \stackrel{\mathrm{def}}{=} \star\, v\colon \big[\mathrm{true}\,,\, v(\tau) \in \tau_0 \mathinner{\ldotp\ldotp} \tau\big]$$

$$\mathbf{read}(v, x) \stackrel{\mathrm{def}}{=} \star\, v\colon \big[\mathrm{dom}\, x = \mathbb{T}\,,\, v(\tau) \in x(\!| \,\tau_0 \mathinner{\ldotp\ldotp} \tau\, |\!)\big]$$

$$\mathbf{deadline}\, X \stackrel{\mathrm{def}}{=} \star\, \big[\tau_0 = \tau \wedge \tau \leqslant X(\tau_0)\big]$$

Fig. 1. Atomic target language constructs.

2.1   A timed target language

Given such a specification statement, a 'timed' target language can then be defined as a well-understood specification subset corresponding to common programming language constructs [26, 25, 29]. Figure 1 defines a number of simple 'atomic' target language constructs (using Z mathematical notation [31] in predicates). Let v be a variable from **p**, i.e., one that is constructed by the current process, and x a variable from **e**, i.e., one from the process's environment. Expression X may refer to variables from **p** only, thus ensuring that its value does not change while it is being evaluated. Furthermore, X must be an *untimed expression*, which does not dereference the timed variables [19, §1.1], because programming language expressions leave the dimension of time implicit. Let an untimed expression X be instantiated at absolute time t as $X(t)$. For instance, if X is '$v + 1$', then $X(7)$ denotes '$v(7) + 1$'.

The definition of assignment in Fig. 1 thus tells us that the value of v when the statement ends must equal the value of (stable) expression X, evaluated at the moment that the statement began [25, p. 11]. The **skip** construct is defined trivially as a statement that does nothing and takes no time. More interesting is the time-specific **delay-until** construct which prevents changes to any **p** variables until after the absolute time defined by time-valued expression X. It tells us that the final time τ must be at least as great as expression X. The **get-time** construct reads the value of a (perfect!) real-time clock, placing the result in variable v. The actual value saved is merely *somewhere* between absolute times τ_0 and τ, so the quicker the statement executes, the more accurate the result! The **read** construct allows us to read the value of a variable x from the environment into variable v. The effect tells us that the final value of v is *some* value that x had between times τ_0 and τ [25, p. 9]. The assumption ensures that there is always a defined value of x to read.

None of the target language constructs considered so far has stated how far apart τ_0 and τ may be, i.e., how much time the statement may consume. This seems like a severe underspecification. The **deadline** construct [10, 9] is used to solve this. As shown in Fig. 1 it is a coercion that constrains the current time to be less than X, but does not itself consume time [11]. It can be placed after a sequence of statements to restrict their overall finishing time. Also, it can be used in conjunction with the **delay-until** construct to 'bracket' a code segment with a precise duration.

Note that although all the target language constructs considered above restricted their effects to times between τ_0 and τ, higher-level specifications often wish to refer to past variable values, before time τ_0, or even future values, after time τ. However, such references must be eliminated during refinement to conform with the above target language constructs.

$$S_1 \; ; S_2 \stackrel{\text{def}}{=} S_1 \circ S_2$$

$$\textbf{if} \, (\llbracket i \bullet G_i \to S_i) \, \textbf{fi} \stackrel{\text{def}}{=} \star \big\{ (\vee \, i \bullet G_i(\tau)) \big\} \, ; \big(\llbracket i \bullet \star \big[G_i(\tau) \big] \, ; S_i \, ; delay \big)$$

$$\textbf{do} \, G \to S \, \textbf{od} \stackrel{\text{def}}{=} (\llbracket n : \mathbb{N} \bullet (\star \big[G(\tau) \big] \, ; S)^n) \, ; \star \big[\neg \, G(\tau) \big]$$

Fig. 2. Compositional target language constructs.

Figure 2 shows definitions for some familiar compositional target language constructs. Let S be a (specification) statement, and G an untimed boolean expression on variables in **p**. $S_1 \, \llbracket \, S_2$ denotes demonic choice between predicate transformers S_1 and S_2 in the usual way [22, p. 15]. The

$$delay \stackrel{\text{def}}{=} \star [\text{true}]$$

coercion is used in these definitions to introduce a computation time overhead. It changes no **p** variables, but allows time to pass [11, §2.5]. It is inserted at points where additional computational overheads are expected in a practical implementation of the constructs.

Sequential composition is defined in the usual way as composition of predicate transformers. Here 'o' is backward composition of the predicate-transformer functions denoted by S_1 and S_2 [7, p. 115][1, p. 9].

For the alternative construct **if**...**fi** we have adopted a 'timed' version of the usual definition [2, p. 9]. It first asserts that at least one of the guards G_i must be true when the construct begins. It then demonically chooses an alternative with true guard G_i and performs action S_i. The coercion in the chosen alternative may consume some (unspecified) amount of time to account for the time taken by an implementation to evaluate the guard and reach S_i. Also we have introduced a final delay to allow for the time taken to exit the construct once the chosen S_i has finished.

Our definition of the iterative construct **do**...**od** is also unconventional. Rather than the usual recursive fixed-point definition [4] we adopt a **wlp** definition [5, p. 185], so that it is easy to unfold loops into a demonic choice. A weakness of this approach is that it is equivalent to the usual weakest precondition semantics only for loops that always terminate, but since we are interested in terminating processes only in this paper, this is acceptable. Let S^n denote n sequential instances of statement S. Then Fig. 2 shows that a loop can be defined as the demonic choice of n repetitions, provided that the total behaviour leaves guard G false, and each iteration begins with G true. Note that the coercion before the loop body S may consume time, representing the time required to evaluate G at each successful iteration. The final coercion may also consume time, representing the computational overhead of evaluating the false guard and exiting the construct.

For the definition of iteration in Fig. 2 to be valid we must have some guarantee that the loop will terminate. This is not explicitly enforced above. However, the usual refinement calculus loop-introduction rule [21, ch. 5] requires the programmer to show termination via a variant function. For the purposes of this paper, we implicitly assume that the loop was developed using such a rule. Although unconventional, the unfolding definitions of alternatives and iteration in Fig. 2 are useful in the proof below.

Finally, since our sequential specification statement operates only on existing traces, we need some way of defining traces initially when specifying a complete process, in order to complete the link back to trace-constructing timed refinement calculus specification statements [17]. We therefore introduce a *process* construct to our target language:

$$\textbf{proc p} : \textbf{T} \bullet S \textbf{ corp} .$$

This creates a new process, consisting of a number of timed variables **p**, of type **T**, sets the initial value of τ to zero, then executes the statement S within the new process. After S terminates (recall that we limit ourselves to terminating processes in this paper), the process construct passes out the final value of τ and the completed traces to the environment of the process. Semantic definitions for this process construct are given in the following sections.

3   Semantics

Our challenge now is to give a semantics to $\star\mathbf{v}\colon[A\,,E]$ statements that allow them to define timed-trace segments between times τ_0 and τ. There are several possible approaches.

1. A *coercive* model begins with a full-length, but underspecified trace, and each statement constrains the segment between times τ_0 and τ.
2. A trace *extending* model begins with an empty trace and each statement then adds a new segment, of length $\tau - \tau_0$, to the end.
3. A trace *overriding* model begins with a full-length, but underspecified trace, and each statement updates the trace by overwriting the segment between τ_0 and τ.
4. A *chop* semantics assumes that each statement is a (single) predicate defining the behaviour of a (full-length) trace between times τ_0 and τ, and that sequential composition denotes logical conjunction of these constraints [32, 26].
5. A trace *concatenation* model assumes each statement defines its own disjoint trace segment, of length $\tau - \tau_0$. Sequential composition then concatenates these segments to create the total trace.

We reject approaches 4 and 5 as incompatible with predicate-transformer semantics. They require us to give an unusual semantics to the sequential composition operator, whereas we wish to preserve the usual (and desirable) view of S_1 ; S_2 denoting functional composition [7, p. 115][1, p. 9]. As already noted, approach 2 is a straightforward, but awkward model, and approach 1 is our preferred model. In practice, approach 3 is similar to 2. The remainder of this section describes the first two approaches in depth.

3.1   A coercive semantics

In the *coercive* semantics, specification statements and other target language constructs introduce constraints on existing timed traces. The only variable that is updated in the usual way [21] is the special current-time variable τ. This approach results in an elegant calculus, since there is only one set of timed variables (rather than distinct initial and final values for each sequential statement), is close in spirit to the 'parallel' timed refinement calculus [17], and enjoys some useful refinement rules, such as the ability to freely carry assertions forward through sequential composition [29].

We define a function $\mathcal{C}$ that maps the syntactic constructs of our target language into predicate transformers. It is sufficient to give the semantics of our specification statement, since all atomic (Fig. 1) and compositional (Fig. 2) target language constructs are defined using it.

For timed variables $\mathbf{w}$ and a (typically continuous) set of absolute times T, the predicate

$$stable(\mathbf{w}, T) == (\forall\, t_1, t_2 : T \bullet \mathbf{w}(t_1) = \mathbf{w}(t_2))$$

states that $\mathbf{w}$ is unchanging during these times. Then the coercive semantics for our sequential timed specification statement, expressed as a standard specification statement [22], is

$$\mathcal{C}(\star\,\mathbf{v}\colon [A\,,E]) \stackrel{\mathrm{def}}{=} \tau\colon [A\,,E \wedge \tau_0 \leqslant \tau \wedge \mathit{stable}(\mathbf{p}\setminus\mathbf{v}, \tau_0 \mathinner{\ldotp\ldotp} \tau)]\ .$$

The statement on the right makes three properties of the timed sequential specification statement explicit.

- The special variable τ appears explicitly in the frame. Moreover, timed variables $\mathbf{v}$ do *not* appear. Thus τ is the *only* variable that can change!
- A conjunct is added to ensure that time cannot go backwards, i.e., τ cannot be less than τ_0.
- A conjunct is added to ensure that all timed process variables from $\mathbf{p}$, other than those in $\mathbf{v}$, have the same value from times τ_0 to τ.

Definitions of timed assertions,

$$\star\,\{A\} \stackrel{\mathrm{def}}{=} \star\,[A\,,\tau_0 = \tau]\ ,$$

and coercions,

$$\star\,[E] \stackrel{\mathrm{def}}{=} \star\,[\mathrm{true}\,,E]\ ,$$

are similar to the ones for the standard calculus [21, pp. 11, 165], but explicitly consider the passage of time [11]. Predicates A and E may both refer to τ, and E to τ_0. However a timed assertion does not consume time [11, §2.3], whereas a timed coercion may [11, §2.2].

In the coercive semantics, the **proc p : T** $\bullet$ S **corp** construct creates undetermined traces for process variables $\mathbf{p}$, sets the starting time to zero, and then 'executes' statement S. We can regard the construct as sequential composition of three predicate transformers, **proc p : T**, S and **corp**.

Here we give a semantics to the **proc p : T** and **corp** components. Since this involves creating new variables we must refer to their underlying predicate transformers. Let $\mathbf{PRED_w}$ be the set of predicates over variables $\mathbf{w}$ [2, §2].

As before, let $\mathbf{e}$ be the variables in the environment of the process of interest. Then the predicate-transformer semantics of the process creation component is

$$\mathcal{C}(\mathbf{proc\ p : T}) \stackrel{\mathrm{def}}{=} \lambda\,P : \mathbf{PRED_{e,p,\tau}} \bullet (\forall\,\mathbf{p} : \mathbb{T} \to \mathbf{T};\ \tau : \mathbb{T} \bullet \tau = 0 \Rightarrow P)\ .$$

This creates full-length traces for the process variables $\mathbf{p}$, but the traces contain demonically-chosen arbitrary values. The starting time τ is set to zero.

In the coercive semantics the process termination component does not need to do anything to complete the trace, i.e.,

$$\mathcal{C}(\mathbf{corp}) \stackrel{\mathrm{def}}{=} \mathbf{skip}\ .$$

(Current time τ is left equal to the time at which S terminated.)

3.2 A trace extending semantics

The trace *extending* semantics is closer to the standard refinement calculus [21] in that there are initial and final values for all variables. However the definition involves more conjuncts, to account for the changing domain of the traces, and the notion of 'updating' trace variables that represented 'behaviour over all time' in the parallel timed refinement calculus [17] seems unnatural.

Let function $\mathcal{E}$ define the trace extending semantics. For our specification statement this is

$$\mathcal{E}(\star \mathbf{v}\colon [A\,,\,E]) \stackrel{\mathrm{def}}{=} \tau, \mathbf{p}\colon \left[A \atop \mathrm{dom}\,\mathbf{p} = 0\mathinner{\ldotp\ldotp}\tau \middle| \begin{matrix} E \\ \tau_0 \leqslant \tau \\ \mathrm{dom}\,\mathbf{p} = 0\mathinner{\ldotp\ldotp}\tau \\ \mathbf{p}_0 = (0\mathinner{\ldotp\ldotp}\tau_0) \vartriangleleft \mathbf{p} \\ stable(\mathbf{p}\setminus\mathbf{v}, \tau_0\mathinner{\ldotp\ldotp}\tau) \end{matrix} \right].$$

The standard specification statement on the right has the following features.

- Both τ and all process variables $\mathbf{p}$ appear in the frame and are thus updated by the statement.
- A conjunct is added to the assumption requiring that all trace variables in $\mathbf{p}$ initially have the same length, being defined from 0 until starting time τ. (It would be nonsensical to have different length 'histories' at the same moment in time.)
- A conjunct is added to the effect predicate to state that time cannot go backwards.
- The second new conjunct added to the effect predicate states that the final values of all trace variables $\mathbf{p}$ have the same length, ranging up until time τ. (Again it would be nonsensical for the statement to terminate with different length histories.) Even variables in $\mathbf{p}$ that are *not* in $\mathbf{v}$ have their traces extended.
- The third new effect conjunct ensures that the initial segment of each trace $\mathbf{p}$, between times 0 and τ_0, is the same as in $\mathbf{p}_0$. In other words, we cannot rewrite the past!
- The final extra effect conjunct requires process variables $\mathbf{p}$, other than those mentioned in $\mathbf{v}$, to be unchanged, i.e., to exhibit the same value throughout execution of this statement.

Definitions of timed assertions and coercions in the extending model follow the approach shown in Section 3.1 above. However note that predicates A and E cannot refer to 'future' values, after time τ, in extending specifications because this would involve references to timed trace variables outside their domain. This highlights a major practical distinction between the two models, and explains why the coercive model is more expressive.

As in Section 3.1, we divide the process construct into sequential components. The trace extending semantics of the process creation component is

$$\mathcal{E}(\mathbf{proc}\ \mathbf{p} : \mathbf{T})$$
$$\stackrel{\mathrm{def}}{=} \lambda P : \mathbf{PRED}_{\mathbf{e},\mathbf{p},\tau} \bullet (\forall \mathbf{p} : \mathbb{T} \nrightarrow \mathbf{T};\ \tau : \mathbb{T} \bullet \tau = 0 \wedge \mathrm{dom}\,\mathbf{p} = \{0\} \Rightarrow P).$$

This initialises all process variables $\mathbf{p}$ to traces that contain just a single demon-ically-chosen initial value, defined at time zero.

The process termination component takes each finite trace in $\mathbf{p}$ and extends it to a total trace (to conform with 'parallel' timed-trace specifications [17]) by adding an arbitrary tail:

$$\mathcal{E}(\mathbf{corp}) \stackrel{\text{def}}{=} \mathbf{p} \colon \left[\operatorname{dom} \mathbf{p} = 0 \mathbin{.\,.} \tau , \; \begin{array}{l} \operatorname{dom} \mathbf{p} = \mathbb{T} \\ \mathbf{p}_0 = (0 \mathbin{.\,.} \tau) \triangleleft \mathbf{p} \end{array} \right] .$$

As usual for extending specifications, the assumption requires all $\mathbf{p}$ variables to have the same length history. The effect predicate makes each timed-trace function total, with the initial segment up until time τ_0 unchanged, and the remainder unspecified. (Again, we leave τ equal to the time the process terminated.)

4   Infeasibility

A surprising feature of the coercive semantics is that most specifications, includ-ing our target language constructs, are *infeasible*. The standard feasibility test (the *Law of the Excluded Miracle* [4, p. 18]) for a program P is

$$P\,false = false \ .$$

Choosing P to be the coercive semantics of our specification statement (Sec-tion 3.1), this test is equivalent to [21, p. 58]

$$(\tau = \tau_0) \wedge A \Rrightarrow (\exists\,\tau : \mathbb{T} \bullet E \wedge \tau_0 \leqslant \tau \wedge stable(\mathbf{p} \setminus \mathbf{v}, \tau_0 \mathbin{.\,.} \tau)) \ .$$

This is not true if E places any constraints on $\mathbf{v}$, other than those already in A. Similarly, the assumption A does not necessarily entail the stability requirement on other $\mathbf{p}$ variables.

This problem occurs in the coercive model because each specification state-ment *on its own* is not free to choose values for entire trace variables. Indeed, the expansion to a traditional specification statement given in Section 3.1 showed that $\mathbf{v}$ was not even in the frame! It is the combination of the effect predicates of all sequentially-composed statements that determines overall trace behaviour.

Failure to pass the feasibility test does not prevent us from applying refine-ment rules, but in the standard refinement calculus it means we can never reach implementable code [21, p. 10]. Nevertheless, in our calculus, we *can* implement some sequences of these 'infeasible' statements, by relying on restrictions im-posed on the target language (Section 2.1). Every executable process has the form $\mathbf{proc}\ \mathbf{p} : \mathbf{T} \bullet S\ \mathbf{corp}$, and every target language statement S_i within S constrains trace values only between times τ_0 (exclusive) and τ (inclusive). This means that each statement S_i operates on trace segments disjoint from those segments constrained by any other S_j. It is this disjointness that makes such statements implementable. Operationally, it allows the choice of each time-value pair $(t \mapsto x)$ in a trace to be delayed until execution reaches the statement that has $\tau_0 < t \leqslant \tau$.

For example, an assignment statement $v := X$ sets the value of v at end time τ equal to the value of expression X evaluated at start time τ_0 (as long as expression X is unchanging while the assignment executes). Assuming such a statement consumes exactly two time units, its weakest-precondition semantics with respect to some predicate P would then be

$$(v := X)\,P = (\forall\,\tau : \mathbb{T} \bullet \tau = \tau_0 + 2 \wedge v(\tau) = X(\tau_0) \Rightarrow P)\left[\tfrac{\tau}{\tau_0}\right]$$
$$= v(\tau + 2) = X(\tau) \Rightarrow P\left[\tfrac{\tau+2}{\tau}\right]\,.$$

Note that an isolated assignment, like $v := 3$, fails the usual feasibility test:

$$(v := 3)\ \text{false} = v(\tau + 2) \neq 3$$
$$\neq \text{false}\,.$$

However, when a sequence of assignments, such as '$x := 1\,;\,x := 2\,;\,x := x + 3$', is embedded in a process, we have:

$$(\mathbf{proc}\ x : \mathbb{N} \bullet x := 1\,;\,x := 2\,;\,x := x + 3\ \mathbf{corp})\,P$$
$$= (\forall\,x : \mathbb{T} \to \mathbb{N};\ \tau : \mathbb{T} \bullet$$
$$\tau = 0 \Rightarrow$$
$$x(2) = 1 \Rightarrow$$
$$x(4) = 2 \Rightarrow$$
$$x(6) = x(4) + 3 \Rightarrow P\left[\tfrac{\tau+6}{\tau}\right])$$
$$= (\forall\,x : \mathbb{T} \to \mathbb{N} \bullet x(2) = 1 \wedge x(4) = 2 \wedge x(6) = 5 \Rightarrow P\left[\tfrac{6}{\tau}\right])\,.$$

Note that the semantics of this program allows us to deduce its finishing time ($\tau = 6$), and the value of x at various times up until then, but we cannot deduce anything about *future* values of x. This is exactly as one would expect. Furthermore, it is easy to see that the program is feasible:

$$(\forall\,x : \mathbb{T} \to \mathbb{N} \bullet x(2) = 1 \wedge x(4) = 2 \wedge x(6) = 5 \Rightarrow \text{false}\left[\tfrac{6}{\tau}\right])$$
$$= \neg\,(\exists\,x : \mathbb{T} \to \mathbb{N} \bullet x(2) = 1 \wedge x(4) = 2 \wedge x(6) = 5)$$
$$= \text{false}\,.$$

This illustrates how, for some process specification $\mathbf{proc}\ p :\!\bullet\! \mathbf{T} \bullet S\ \mathbf{corp}$, all the statements in S cooperate to constrain the choice of traces. The requirement that τ cannot be decreased is used to guarantee that each statement constrains a unique segment of each trace.

5 Correspondence Between Semantics

In this section, we prove that for any process, constructed using our timed target language constructs, its coercive semantics is equivalent to its trace extending semantics. This justifies our use of the coercive semantics during refinement; the coercive semantics is simply an abstraction of the more constructive extending semantics. Our goal is thus to prove the following theorem.

Theorem 1. *For any process*

$$\textbf{proc p} : \textbf{T} \bullet S \textbf{ corp} ,$$

such that S is constructed from target language constructs (Figs. 1 and 2) only,

$$\mathcal{C}(\textbf{proc p} : \textbf{T} \bullet S \textbf{ corp}) = \mathcal{E}(\textbf{proc p} : \textbf{T} \bullet S \textbf{ corp}) .$$

That is, the infeasible coercive semantics of the process is equivalent to a feasible trace extending semantics.

In proofs, we mainly use Morgan's specification statement, which is semantically defined as [22, Lem. 1]

$$\textbf{w} \colon [A\,, E] \stackrel{\text{def}}{=} \lambda P : \textbf{PRED}_\textbf{u} \bullet A \wedge (\forall \textbf{w} \bullet E \Rightarrow P) \left[\tfrac{\textbf{w}}{\textbf{w}_0}\right] ,$$

where this statement updates a sublist $\textbf{w}$ of the variables $\textbf{u}$ it can access.

As demonstrated in Fig. 1, all atomic timed target language constructs can be expressed as $\star\,\textbf{v} \colon [A\,, E]$ where $\textbf{v}$ is a sublist of $\textbf{p}$, A is in $\textbf{PRED}_{\tau,\textbf{p},\textbf{e}}$, and E is in $\textbf{PRED}_{\tau,\tau_0,\textbf{p},\textbf{e}}$. Moreover, the atomic constructs have the important property that their A and E predicates access timed traces between times 0 and τ only. It is this property that allows us to prove the data refinement result. We call such predicates *future-independent*.

Definition 2. A predicate $P \in \textbf{PRED}_{\textbf{e},\textbf{p},\tau,\tau_0}$ is *future-independent* when

$$\forall\,\textbf{e}, \textbf{p} : \mathbb{T} \to \textbf{T};\ \tau, \tau_0 : \mathbb{T} \bullet \tau_0 \leqslant \tau \Rightarrow \left(P \Leftrightarrow P\left[\tfrac{(0..\tau)\triangleleft\textbf{p}}{\textbf{p}}\right]\right) .$$

That is, it specifies properties of traces $\textbf{p}$ up until time τ only.

Our goal is to prove a data refinement relationship between the semantics, because they use different representations of timed variables. However, the standard techniques for proving data refinement, while preserving the structure of a program [8], do not work for this application.

- Most formulations of data refinement require the output variables of the abstract program to be the same as its input variables, whereas our **proc ... corp** construct *creates* trace variables $\textbf{p}$, and time variable τ, and leaves them visible in the final state.

  However, this is not a serious problem, since Back and von Wright have defined a more flexible definition of data refinement, which uses separate *encoding* and *decoding* programs to convert between the various state spaces [3, 30].

- The most common data refinement technique uses *downward simulation* [7] and this is known to be incomplete for the kind of data refinement we need, where a non-deterministic choice is moved from early in the execution sequence to later. For instance, it might seem that one way to achieve our desired data refinement is to use downward simulation, with retrieve relation

$$\mathbf{p}_e = (0 \ldots \tau) \lhd \mathbf{p}_c$$

linking abstract, full-length, coercive traces $\mathbf{p}_c$ with concrete, partially-completed, extending traces $\mathbf{p}_e$. Unfortunately, our attempts to do this showed that it results in specification statements that are still infeasible.

The usual solution to this problem is to use *upward simulation* [12], or the complete rule developed by Gardiner and Morgan [7]. However, these have the restriction that the abstract program must not contain unbounded non-determinism, whereas our abstract program does—it chooses an entire trace non-deterministically!

Therefore, we must resort to the underlying definition of data refinement to prove the result we want.

To do this we use Gardiner and Morgan's definition of data refinement [7]. They define a datatype as a triple (I, OP, F), where I and F are programs called the initialisation and finalisation, respectively, and OP is a set of programs. They say that (I, OP, F) is data refined by (I', OP', F') if, for every program S expressible as a composition of program-language operators,

$$I \,;\, S(OP) \,;\, F \sqsubseteq I' \,;\, S(OP') \,;\, F' \,.$$

Let *Atom* denote all 'atomic' timed specification statements, assertions and coercions, including target language constructs of the form illustrated in Fig. 1. For our application, we then choose

$$I = \mathcal{C}(\mathbf{proc}\ p : T)$$
$$OP = \{s : Atom \bullet \mathcal{C}(s)\}$$
$$F = \mathcal{C}(\mathbf{corp}) \,,$$

and

$$I' = \mathcal{E}(\mathbf{proc}\ p : T)$$
$$OP' = \{s : Atom \bullet \mathcal{E}(s)\}$$
$$F' = \mathcal{E}(\mathbf{corp}) \,.$$

Our proof of Theorem 1 then proceeds as follows. The lemma used is presented after the main proof.

Proof Firstly, we re-express $\mathbf{proc}\ \mathbf{p} : \mathbf{T} \bullet S\ \mathbf{corp}$ as

$$(\llbracket i \bullet \mathbf{proc}\ \mathbf{p} : \mathbf{T} \bullet S_{i,1} \,;\, \ldots \,;\, S_{i,n_i}\ \mathbf{corp}) \,,$$

where each $S_{i,j}$ is a specification statement, and n_i denotes the number of such specification statements in the ith alternative. To do this we unfold all **if**...**fi** and **do**...**od** constructs within S, using the definitions in Fig. 2, and unfold all atomic target language statements, using the definitions in Fig. 1. Then we unfold any *delay*, $\star[\ldots]$ and $\star\{\ldots\}$ statements (these can only have been introduced by unfolding **if**...**fi** and **do**...**od**) into their specification statement equivalents. This results in a program that is equivalent to S, but contains only demonic choice operators and *sequential timed specification statements*. Furthermore, the assumption and effect predicates in all those specification statements are future-independent (this is easily verified by inspection of all the predicates in Fig. 1 and Fig. 2).

It remains to move the choice operators to the outermost level. This is done by appealing to the following distributive properties of demonic choice over conjunctive programs [30].

$$(\bigsqcap i \bullet P_i) \,;\, Q = (\bigsqcap i \bullet P_i \,;\, Q)$$
$$Q \,;\, (\bigsqcap i \bullet P_i) = (\bigsqcap i \bullet Q \,;\, P_i)$$
$$(\bigsqcap i \bullet P_i) \,;\, (\bigsqcap j \bullet Q_j) = (\bigsqcap i, j \bullet P_i \,;\, Q_j)$$

Secondly, since basic properties of equality give us

$$(\bigwedge i \bullet P_i = P_i') \Rightarrow (\bigsqcap i \bullet P_i) = (\bigsqcap i \bullet P_i'),$$

it is sufficient to prove that, for all i,

$$\mathcal{C}(\textbf{proc } \mathbf{p} : \mathbf{T} \bullet S_{i,1} \,;\, \ldots \,;\, S_{i,n_i} \textbf{ corp})$$
$$= \mathcal{E}(\textbf{proc } \mathbf{p} : \mathbf{T} \bullet S_{i,1} \,;\, \ldots \,;\, S_{i,n_i} \textbf{ corp}).$$

We do this by induction over the sequence $S_{i,1} \ldots S_{i,n_i}$.

Base case: Assume $n_i = 0$, that is, the sequence $S_{i,1} \,;\, \ldots \,;\, S_{i,n_i}$ is empty. Let P be a member of $\textbf{PRED}_{\mathbf{e},\mathbf{p},\tau}$. We prove that the weakest precondition of the coercive program with respect to P is the same as that of the trace extending program.

$$\mathcal{C}(\textbf{proc } \mathbf{p} : \mathbf{T} \bullet \textbf{corp}) \, P$$

$$= \text{``By definition of } \mathcal{C}\text{''}$$
$$(\forall \mathbf{p} : \mathbb{T} \to \mathbf{T}; \ \tau : \mathbb{T} \bullet \tau = 0 \Rightarrow P)$$

$$= \text{``By predicate calculus''}$$
$$(\forall \mathbf{p} : \mathbb{T} \twoheadrightarrow \mathbf{T}; \ \tau : \mathbb{T} \bullet \tau = 0 \wedge \mathrm{dom}\,\mathbf{p} = \{0\}$$
$$\Rightarrow (\forall \mathbf{p} : \mathbb{T} \twoheadrightarrow \mathbf{T} \bullet \mathrm{dom}\,\mathbf{p} = \mathbb{T} \wedge \mathbf{p}_0 = \{0\} \lhd \mathbf{p} \Rightarrow P)\,[\tfrac{\mathbf{p}}{\mathbf{p}_0}])$$

$$= \text{``By definition of } \mathcal{E}\text{''}$$
$$(\mathcal{E}(\textbf{proc } \mathbf{p} : \mathbf{T}) \,;\, \mathcal{E}(\textbf{corp})) \, P$$

$$= \text{``By definition of } \mathcal{E}\text{''}$$
$$\mathcal{E}(\textbf{proc } \mathbf{p} : \mathbf{T} \bullet \textbf{corp}) \, P$$

Inductive case: We assume

$$\mathcal{C}(\textbf{proc p}: \textbf{T} \bullet S_{i,1}\,;\ldots;\,S_{i,k}\ \textbf{corp})$$
$$= \mathcal{E}(\textbf{proc p}: \textbf{T} \bullet S_{i,1}\,;\ldots;\,S_{i,k}\ \textbf{corp})\,,$$

and then seek to prove that

$$\mathcal{C}(\textbf{proc p}: \textbf{T} \bullet S_{i,1}\,;\ldots;\,S_{i,k}\,;\,S_{i,k+1}\ \textbf{corp})$$
$$= \mathcal{E}(\textbf{proc p}: \textbf{T} \bullet S_{i,1}\,;\ldots;\,S_{i,k}\,;\,S_{i,k+1}\ \textbf{corp})\,.$$

Let $S_{i,k+1} = \star\textbf{p}\colon\big[A\,,E\big]$, where A and E are future-independent predicates.

$$\mathcal{C}(\textbf{proc p}: \textbf{T} \bullet S_{i,1}\,;\ldots;\,S_{i,k+1}\ \textbf{corp})$$

$=$ "Split the process construct into sequential components"
$$\mathcal{C}(\textbf{proc p}: \textbf{T})\,;\mathcal{C}(S_{i,1})\,;\ldots;\mathcal{C}(S_{i,k})\,;\mathcal{C}(S_{i,k+1})\,;\mathcal{C}(\textbf{corp})$$

$=$ "Since $\mathcal{C}(\textbf{corp}) = \textbf{skip}$, which is the identity of ';'"
$$\mathcal{C}(\textbf{proc p}: \textbf{T})\,;\mathcal{C}(S_{i,1})\,;\ldots;\mathcal{C}(S_{i,k})\,;\mathcal{C}(\textbf{corp})\,;\mathcal{C}(S_{i,k+1})$$

$=$ "By inductive assumption"
$$\mathcal{E}(\textbf{proc p}: \textbf{T})\,;\mathcal{E}(S_{i,1})\,;\ldots;\mathcal{E}(S_{i,k})\,;\mathcal{E}(\textbf{corp})\,;\mathcal{C}(S_{i,k+1})$$

$=$ "By definitions of $\mathcal{E}(\textbf{corp})$ and $\mathcal{C}(\star p\colon\big[A\,,E\big])$"
$$\mathcal{E}(\textbf{proc p}: \textbf{T})\,;\mathcal{E}(S_{i,1})\,;\ldots;\mathcal{E}(S_{i,k})\,;$$
$$\textbf{p}\colon\big[\operatorname{dom}\textbf{p} = 0 \mathinner{\ldotp\ldotp} \tau\,,\operatorname{dom}\textbf{p} = \mathbb{T} \wedge \textbf{p}_0 = (0\mathinner{\ldotp\ldotp}\tau)\lhd\textbf{p}\big]\,;$$
$$\tau\colon\big[A\,,E \wedge \tau_0 \leqslant \tau\big]$$

$=$ "By Corollary 4 below (E does not contain $\textbf{p}_0$)"
$$\mathcal{E}(\textbf{proc p}: \textbf{T})\,;\mathcal{E}(S_{i,1})\,;\ldots;\mathcal{E}(S_{i,k})\,;$$
$$\tau,\textbf{p}\colon\left[\begin{matrix}\operatorname{dom}\textbf{p} = 0\mathinner{\ldotp\ldotp}\tau & & \operatorname{dom}\textbf{p} = \mathbb{T} \\ \left(\begin{matrix}\forall\textbf{p}:\mathbb{T}\rightarrowtail\textbf{T}\bullet \\ (\operatorname{dom}\textbf{p} = \mathbb{T}\,\wedge \\ \textbf{p}_0 = (0\mathinner{\ldotp\ldotp}\tau)\lhd\textbf{p}) \Rightarrow A\end{matrix}\right)\left[\tfrac{\textbf{p}}{\textbf{p}_0}\right] & {}'\,E & \begin{matrix}\textbf{p}_0 = (0\mathinner{\ldotp\ldotp}\tau_0)\lhd\textbf{p} \\ \\ \tau_0 \leqslant \tau\end{matrix}\end{matrix}\right]$$

$=$ "Changing the type of the bound variable."
$$\mathcal{E}(\textbf{proc p}: \textbf{T})\,;\mathcal{E}(S_{i,1})\,;\ldots;\mathcal{E}(S_{i,k})\,;$$
$$\tau,\textbf{p}\colon\left[\begin{matrix}\operatorname{dom}\textbf{p} = 0\mathinner{\ldotp\ldotp}\tau & & \operatorname{dom}\textbf{p} = \mathbb{T} \\ \left(\begin{matrix}\forall\textbf{p}:\mathbb{T}\rightarrow\textbf{T}\bullet \\ \textbf{p}_0 = (0\mathinner{\ldotp\ldotp}\tau)\lhd\textbf{p} \Rightarrow A\end{matrix}\right)\left[\tfrac{\textbf{p}}{\textbf{p}_0}\right] & {}'\,E & \begin{matrix}\textbf{p}_0 = (0\mathinner{\ldotp\ldotp}\tau_0)\lhd\textbf{p} \\ \\ \tau_0 \leqslant \tau\end{matrix}\end{matrix}\right]$$

$=$ "Since A and E are future-independent"
$$\mathcal{E}(\textbf{proc p}: \textbf{T})\,;\mathcal{E}(S_{i,1})\,;\ldots;\mathcal{E}(S_{i,k})\,;$$

$$\tau, \mathbf{p}: \left[\begin{matrix} \mathrm{dom}\,\mathbf{p} = 0 \mathbin{..} \tau \\[4pt] \left(\begin{matrix} \forall\,\mathbf{p} : \mathbb{T} \to \mathbf{T} \bullet \\ \mathbf{p}_0 = (0 \mathbin{..} \tau) \vartriangleleft \mathbf{p} \\ \Rightarrow A\left[\tfrac{\mathbf{p}_0}{\mathbf{p}}\right] \end{matrix} \right) \left[\tfrac{\mathbf{p}}{\mathbf{p}_0}\right] \end{matrix} \;,\; \begin{matrix} \mathrm{dom}\,\mathbf{p} = \mathbb{T} \\ \mathbf{p}_0 = (0 \mathbin{..} \tau_0) \vartriangleleft \mathbf{p} \\ E\left[\tfrac{(0..\tau)\vartriangleleft\mathbf{p}}{\mathbf{p}}\right] \\ \tau_0 \leqslant \tau \end{matrix} \right]$$

$=$ "Predicate calculus ($\mathbf{p}$ is not free in $A\left[\tfrac{\mathbf{p}_0}{\mathbf{p}}\right]$)"

$$\mathcal{E}(\mathbf{proc}\ \mathbf{p} : \mathbf{T})\,;\, \mathcal{E}(S_{i,1})\,;\, \ldots\,;\, \mathcal{E}(S_{i,k})\,;$$

$$\tau, \mathbf{p}: \left[\begin{matrix} \mathrm{dom}\,\mathbf{p} = 0 \mathbin{..} \tau \\[4pt] \left(\left(\begin{matrix} \exists\,\mathbf{p} : \mathbb{T} \to \mathbf{T} \bullet \\ \mathbf{p}_0 = (0 \mathbin{..} \tau) \vartriangleleft \mathbf{p} \end{matrix} \right) \left[\tfrac{\mathbf{p}}{\mathbf{p}_0}\right] \Rightarrow A \right) \end{matrix} \;,\; \begin{matrix} \mathrm{dom}\,\mathbf{p} = \mathbb{T} \\ \mathbf{p}_0 = (0 \mathbin{..} \tau_0) \vartriangleleft \mathbf{p} \\ E\left[\tfrac{(0..\tau)\vartriangleleft\mathbf{p}}{\mathbf{p}}\right] \\ \tau_0 \leqslant \tau \end{matrix} \right]$$

$=$ "Since every finite trace has an infinite extension"

$$\mathcal{E}(\mathbf{proc}\ \mathbf{p} : \mathbf{T})\,;\, \mathcal{E}(S_{i,1})\,;\, \ldots\,;\, \mathcal{E}(S_{i,k})\,;$$

$$\tau, \mathbf{p}: \left[\begin{matrix} \mathrm{dom}\,\mathbf{p} = 0 \mathbin{..} \tau \\ A \end{matrix} \;,\; \begin{matrix} \mathrm{dom}\,\mathbf{p} = \mathbb{T} \\ \mathbf{p}_0 = (0 \mathbin{..} \tau_0) \vartriangleleft \mathbf{p} \\ E\left[\tfrac{(0..\tau)\vartriangleleft\mathbf{p}}{\mathbf{p}}\right] \\ \tau_0 \leqslant \tau \end{matrix} \right]$$

$=$ "Introduce sequential composition, using Lemma 3 in reverse with $\tau' = \tau$ and $\mathbf{p}' = (0 \mathbin{..} \tau) \vartriangleleft \mathbf{p}$"

$$\mathcal{E}(\mathbf{proc}\ \mathbf{p} : \mathbf{T})\,;\, \mathcal{E}(S_{i,1})\,;\, \ldots\,;\, \mathcal{E}(S_{i,k})\,;$$

$$\tau, \mathbf{p}: \left[\begin{matrix} \mathrm{dom}\,\mathbf{p} = 0 \mathbin{..} \tau \\ A \end{matrix} \;,\; \begin{matrix} E \wedge \tau_0 \leqslant \tau \\ \mathbf{p}_0 = (0 \mathbin{..} \tau_0) \vartriangleleft \mathbf{p} \\ \mathrm{dom}\,\mathbf{p} = 0 \mathbin{..} \tau \end{matrix} \right];$$

$$\mathbf{p}: \left[\mathrm{dom}\,\mathbf{p} = 0 \mathbin{..} \tau \;,\; \begin{matrix} \mathrm{dom}\,\mathbf{p} = \mathbb{T} \\ \mathbf{p}_0 = (0 \mathbin{..} \tau) \vartriangleleft \mathbf{p} \end{matrix} \right]$$

$=$ "By definition of $\mathcal{E}(\star\mathbf{p}\colon \left[A\,,\,E\right])$ and $\mathcal{E}(\mathbf{corp})$"

$$\mathcal{E}(\mathbf{proc}\ \mathbf{p} : \mathbf{T})\,;\, \mathcal{E}(S_{i,1})\,;\, \ldots\,;\, \mathcal{E}(S_{i,k})\,;\, \mathcal{E}(\star\mathbf{p}\colon \left[A\,,\,E\right])\,;\, \mathcal{E}(\mathbf{corp})$$

$=$ "Combining the parts of the process statement"

$$\mathcal{E}(\mathbf{proc}\ \mathbf{p} : \mathbf{T} \bullet S_{i,1}\,;\, \ldots\,;\, S_{i,k}\,;\, S_{i,k+1}\ \mathbf{corp})$$

Lemma 3. *Two specification statements composed sequentially can be expressed as a single specification statement* [28]:

$$\mathbf{w}: \left[A_1\,,\,E_1\right]\,;\, \mathbf{w}: \left[A_2\,,\,E_2\right]$$
$$= \mathbf{w}: \left[A_1 \wedge (\forall\,\mathbf{w} \bullet E_1 \Rightarrow A_2)\left[\tfrac{\mathbf{w}}{\mathbf{w}_0}\right]\,,\,(\exists\,\mathbf{w}' \bullet E_1\left[\tfrac{\mathbf{w}'}{\mathbf{w}}\right] \wedge E_2\left[\tfrac{\mathbf{w}'}{\mathbf{w}_0}\right])\right].$$

This lemma can accommodate specifications with different frames by expanding the frames to some common superset and adding appropriate $x = x_0$ constraints to the postconditions. A special case of this lemma was used in the proof above for two specifications with disjoint frames.

Corollary 4. *A specification that updates* $\mathbf{p}$, *followed by one that updates* τ, *can be expressed as a single statement:*

$$\mathbf{p} \colon [A_1 , E_1] \,;\, \tau \colon [A_2 , E_2]$$
$$= \mathbf{p}, \tau \colon [A_1 \wedge (\forall \mathbf{p} \bullet E_1 \Rightarrow A_2) \left[\tfrac{\mathbf{p}}{\mathbf{p_0}}\right] , E_1 \left[\tfrac{\tau_0}{\tau}\right] \wedge E_2 \left[\tfrac{\mathbf{p}}{\mathbf{p_0}}\right])] \,.$$

Our second main theorem shows that all processes which contain only target language constructs are feasible.

Theorem 5. *A process* $\mathbf{proc}\ \mathbf{p} : \mathbf{T} \bullet S\ \mathbf{corp}$ *is feasible, using the coercive semantics, provided that* S *contains only target language constructs.*

Proof Using the trace extending semantics, it is easy to prove that all the atomic target language constructs in Fig. 1 are feasible. For example:

$$\mathcal{E}(v := X)\ \text{false}$$
$$= \text{``By defn. of } v := X \text{ and } \mathcal{E}\text{''}$$

$$\tau, \mathbf{p} \colon \left[\begin{matrix} A \\ \mathrm{dom}\,\mathbf{p} = 0 \mathrel{..} \tau \end{matrix} , \begin{matrix} E \\ \tau_0 \leqslant \tau \\ \mathrm{dom}\,\mathbf{p} = 0 \mathrel{..} \tau \\ \mathbf{p_0} = (0 \mathrel{..} \tau_0) \vartriangleleft \mathbf{p} \\ \mathit{stable}(\mathbf{p} \setminus \mathbf{v}, \tau_0 \mathrel{..} \tau) \end{matrix} \right]\ \text{false}$$

$$= (A \wedge \mathrm{dom}\,\mathbf{p} = 0 \mathrel{..} \tau) \Rightarrow \neg\, (\exists\, \tau : \mathbb{T};\ \mathbf{p} : \mathbb{T} \twoheadrightarrow T \bullet$$
$$v(\tau) = X(\tau_0) \wedge$$
$$\tau_0 \leqslant \tau \wedge$$
$$\mathrm{dom}\,\mathbf{p} = 0 \mathrel{..} \tau \wedge$$
$$\mathbf{p_0} = (0 \mathrel{..} \tau_0) \vartriangleleft \mathbf{p} \wedge$$
$$\mathit{stable}(\mathbf{p} \setminus \mathbf{v}, \tau_0 \mathrel{..} \tau)) \left[\tfrac{\tau,\mathbf{p}}{\tau_0,\mathbf{p_0}}\right]$$

$$= \text{false}$$

Since the target language compound constructs $(;, \mathbf{if} \ldots \mathbf{fi}$ and $\mathbf{do} \ldots \mathbf{od})$ all preserve infeasibility, S is feasible.

Next we prove that the process creation and termination constructs are feasible.

$$\mathcal{E}(\mathbf{proc}\ \mathbf{p} : \mathbf{T})\ \text{false}$$
$$= \text{``By defn. of } \mathcal{E}(\mathbf{proc}\ \mathbf{p} : \mathbf{T}) \text{ and the semantics of spec. stmts.''}$$
$$(\forall\, \mathbf{p} : \mathbb{T} \twoheadrightarrow \mathbf{T};\ \tau : \mathbb{T} \bullet \tau = 0 \wedge \mathrm{dom}\,\mathbf{p} = \{0\} \Rightarrow \text{false})$$
$$= \neg\, (\exists\, \mathbf{p} : \mathbb{T} \twoheadrightarrow \mathbf{T};\ \tau : \mathbb{T} \bullet \tau = 0 \wedge \mathrm{dom}\,\mathbf{p} = \{0\})$$
$$= \text{false}$$

$\mathcal{E}(\mathbf{corp})$ false

$= $ "By defn. of $\mathcal{E}(\mathbf{corp})$ and the semantics of spec. stmts."

$$(\operatorname{dom}\mathbf{p} = 0 \mathinner{.\,.} \tau) \wedge \neg\, (\exists\, \mathbf{p} : \mathbb{T} \nrightarrow \mathbf{T} \bullet \operatorname{dom}\mathbf{p} = \mathbb{T} \wedge$$
$$\mathbf{p}_0 = (0 \mathinner{.\,.} \tau) \lhd \mathbf{p})\, \left[\tfrac{\mathbf{p}}{\mathbf{p}_0}\right]$$

$= $ false

Since sequential composition preserves feasibility, we have proved the feasibility of $\mathcal{E}(\mathbf{proc\ p} : \mathbf{T} \bullet S\ \mathbf{corp})$. By Theorem 1, the $\mathcal{E}$ and $\mathcal{C}$ semantics are equivalent, so we have also proved that $\mathcal{C}(\mathbf{proc\ p} : \mathbf{T} \bullet S\ \mathbf{corp})$ is feasible.

6   Discussion

A controversial feature of our process construct is that it leaves the imaginary time variable τ visible. Although this variable could have been hidden, it was deemed potentially useful to be able to ask when a process finished. (Hooman also makes the finishing time of a terminating process observable [13, p. 804].) Parallel composition of **proc . . . corp** specifications could then make use of this. (For instance, the TAM calculus adds an arbitrary delay to the end of each parallel process process so that, when composed, a common finishing time can be found [27, p. 229].)

7   Conclusion

We have shown that our infeasible, abstract, 'coercive' real-time semantics are equivalent to a feasible, constructive, 'trace extending' one.

This is a novel use of infeasible (non-strict) programs. With our coercive semantics, atomic statements within a process are typically infeasible. Yet our proof shows that they are feasible in combination.

This is similar to Morgan's use of miraculous specifications to simplify the calculation of data refinement [20], and to express target language typing constraints [23]. However, he is able to remove the miraculous specifications by moving assertions around the program. Our result shows that an additional technique for reducing miraculous specifications to feasible specifications is to perform a data refinement.

Another interesting aspect of our proof is the use of induction over a sequence of atomic statements to show equivalence. This allowed us to prove a data refinement for which downward simulation is incomplete and upward simulation is not applicable, due to unbounded non-determinism in the abstract specification. The same technique may be applicable in other situations.

Acknowledgements We wish to thank Ray Nickson, Ian Hayes, David Carrington, and the anonymous referees for correcting errors in this work. This project was funded in part by the Information Technology Division of the Defence Science and Technology Organisation.

References

1. R.-J. R. Back. Predicate transformers and higher order logic. Technical Report Caltech-CS-TR-92-24, California Institute of Technology, 1992.

2. R.-J. R. Back. Refinement calculus, lattices and higher order logic. Technical Report Caltech-CS-TR-92-22, California Institute of Technology, 1992.

3. R.-J. R. Back and J. von Wright. Refinement calculus, part I: Sequential nondeterministic programs. In J. W. de Bakker, W. P. de Roever, and G. Rozenberg, editors, *REX Workshop for Refinement of Distributed Systems*, volume 430 of *Lecture Notes in Computer Science*, pages 42–66. Springer-Verlag, 1989.

4. E. W. Dijkstra. *A Discipline of Programming*. Prentice-Hall, 1976.

5. E. W. Dijkstra and C. S. Scholten. *Predicate Calculus and Program Semantics*. Springer-Verlag, 1990.

6. C. J. Fidge, M. Utting, P. Kearney, and I. J. Hayes. Integrating real-time scheduling theory and program refinement. In M.-C. Gaudel and J. Woodcock, editors, *FME'96: Industrial Benefit and Advances in Formal Methods*, volume 1051 of *Lecture Notes in Computer Science*, pages 327–346. Springer-Verlag, 1996.

7. P. H. B. Gardiner and C. Morgan. A single complete rule for data refinement. In C. Morgan and T. Vickers, editors, *On the Refinement Calculus*, pages 111–126. Springer-Verlag, 1994. Also in *Formal Aspects of Computing*, 5(4):367–382, 1993.

8. P. H. B. Gardiner and C. C. Morgan. Data refinement of predicate transformers. *Theoretical Computer Science*, 87:143–162, 1991.

9. S. Grundon. Timing constraint analysis for real-time programming. Honours project report, Department of Computer Science, University of Queensland, November 1996.

10. I. J. Hayes and M. Utting. Coercing real-time refinement: A transmitter. In D. J. Duke and A. S. Evans, editors, *BCS-FACS Northern Formal Methods Workshop*, Electronic Workshops in Computing. Springer-Verlag, 1997.

11. I. J. Hayes and M. Utting. Sequential real-time refinement. School of Information Technology, University of Queensland, January 1997.

12. C. A. R. Hoare, J. F. He, and J. W. Sanders. Prespecification in data refinement. *Information Processing Letters*, 25(2), May 1987.

13. J. Hooman. Extending Hoare logic to real-time. *Formal Aspects of Computing*, 6(6A):801–825, 1994.

14. A. Kountouris and C. Wolinski. A real-time HW/SW co-design approach based on the SIGNAL language and its environment. Technical Report 1053, IRISA, University of Rennes, October 1996.

15. B. Mahony. Using the refinement calculus for dataflow processes. Technical Report 94-32, Software Verification Research Centre, October 1994.

16. B. Mahony and I. J. Hayes. A case study in timed refinement: A central heater. In J. M. Morris and R. C. Shaw, editors, *Fourth Refinement Workshop*, pages 138–149. Springer-Verlag, 1991.

17. B. P. Mahony. The refinement calculus and data-flow processes. In *Proc. Second Australasian Refinement Workshop*, pages 1–28, Brisbane, September 1992.

18. B. P. Mahony and I. J. Hayes. A case-study in timed refinement: A mine pump. *IEEE Transactions on Software Engineering*, 18(9):817–826, September 1992.

19. C. Millerchip, B. Mahony, and I. J. Hayes. The generic problem competition: A whole system specification of the boiler system. Software Verification Research Centre, University of Queensland, June 1993.

20. C. Morgan. Data refinement by miracles. In C. Morgan and T. Vickers, editors, *On the Refinement Calculus*, pages 59–64. Springer-Verlag, 1994.

21. C. Morgan. *Programming from Specifications*. Prentice-Hall, second edition, 1994.

22. C. Morgan. The specification statement. In C. Morgan and T. Vickers, editors, *On the Refinement Calculus*, pages 1–21. Springer-Verlag, 1994.

23. C. Morgan and T. Vickers. Types and invariants in the refinement calculus. *Science of Computer Programming*, 14:281–304, 1990.

24. C. Morgan and T. Vickers. *On the Refinement Calculus*. Springer-Verlag, 1994.

25. D. Scholefield and H. Zedan. A standard for finite TAM. Technical Report YCS 206, Department of Computer Science, University of York, 1993.

26. D. Scholefield, H. Zedan, and He Jifeng. Real-time refinement: Semantics and application. In A. Borzyszkowski and S. Sokolowski, editors, *Mathematical Foundations of Computer Science 1993*, volume 711 of *Lecture Notes in Computer Science*, pages 693–702. Springer-Verlag, 1993.

27. D. Scholefield, H. Zedan, and He Jifeng. A specification-oriented semantics for the refinement of real-time systems. *Theoretical Computer Science*, 131:219–241, 1994.

28. M. Utting. *An Object-Oriented Refinement Calculus with Modular Reasoning*. PhD thesis, School of Electrical Engineering and Computer Science, University of New South Wales, 1992.

29. M. Utting and C. J. Fidge. A real-time refinement calculus that changes only time. In He Jifeng, editor, *BCS-FACS Seventh Refinement Workshop*. Springer-Verlag, 1996.

30. J. von Wright. *A Lattice-theoretical Basis for Program Refinement*. PhD thesis, Åbo Akademi University, Finland, September 1990.

31. J. B. Wordsworth. *Software Development with Z*. Addison-Wesley, 1992.

32. C. Zhou. Duration calculi: An overview. In D. Bjorner, M. Broy, and I. Pottosin, editors, *Formal Methods in Programming and Their Applications*, volume 735 of *Lecture Notes in Computer Science*, pages 256–266. Springer-Verlag, 1993. Extended abstract.

33. C. Zhou, C. A. R. Hoare, and A. P. Ravn. A calculus of durations. *Information Processing Letters*, 40:269–276, December 1991.

A Comparison of Modularity in B and Cogito

Geoffrey Norman Watson

Software Verification Research Centre
School of Information Technology,
The University of Queensland. Australia. 4072.

Abstract. This paper examines how some of the issues of modularity
have been tackled by two current formal development methods: B and
Cogito. This assessment of was undertaken as part of a project investigating adding modularity to the refinement calculus.

Particular attention is paid to issues relevant to the general problem of
moving from small-scale development to the larger scale, and also to the
implications of modularity for full formal development.

1 Introduction

This paper compares the modularity mechanisms of two formal development
methods: B and Cogito. This was done as preliminary work for a project investigating ways of extending the refinement calculus with a flexible modularity
mechanism.

Modularity is a key concept in modern software engineering, from structured
programming to object-orientation [Mey88]. It may be considered as having a
number of dimensions. At a basic level it is concerned with *managing complexity*,
by giving a syntactic structure to large systems and providing a mechanism for
managing the associated name space. Splitting a system into modules provides
a means for the encapsulation of data and its associated operations allowing
information hiding and data and function *abstraction*. It also facilitates a *separation of concerns* of different parts of the system. Modules are a natural unit
for *reuse* and, provided that the refinement method is compositional, they are
also independent units which can be *refined separately*. (This property is often
called *monotonicity* with respect to refinement.)

The structure of this paper is as follows: Sect. 2 gives a brief description of
the two methodologies, the modularity features are compared in Sect. 3, Sect. 4
discusses some points of interest arising from the comparison, and finally there
is an indication of future work. In making the comparison we were particularly
interested in the implications for tool support, full formal development (that
is proofs of obligations) and the implications of modularity for scaling up the
development methodologies.

2 The Methodologies

The two methodologies used in the comparison were:

- The B-Method [Abr96, B-C96]
- Cogito (a model-based program development methodology developed at the SVRC) [BKK+96, KKTW95, BKKT95],

Although we were looking primarily at modularity mechanisms, this was in the context of programming in the large, so the development methodology and in particular the integration of modularity into this methodology was an issue.

Two features of these systems which were important for our purposes were tool support and the semantics of modules. Modules introduce their own semantic complexities which have direct implications for formal development, so it was important that we had access to details of the theory on which the methodology is based. For B we have the B-Book [Abr96] and for Cogito we have access to technical reports and working papers.

It was also important that both systems have a supporting toolset, since this is essential for large-scale development. However we were interested in the general aspects of tool support for the methodology rather than user interface issues or a direct comparison of the particular tools used.

The following sections give a brief overview of the B and Cogito systems and their development methodologies. The details of syntax are not important except for the modularity constructs which are described in the comparison in Sect. 3.

2.1   B-Toolkit

An Outline of B

The B-Method is a model-based formal development methodology. The theory of B is described in detail in [Abr96], and two B toolsets are available commercially – this paper refers to the implementation in the B-Toolkit from B-Core(UK) [B-C96]. An introduction to the B-Method, from the point of view of refinement and implementation was given at the last Australasian Refinement Workshop [RŽ]. The language, semantics and the methodology have been developed together, so they form a very coherent system.

In B the module is called a *machine* and specifications and programs are written in *Abstract Machine Notation* (AMN). This is an extension to Dijkstra's Guarded Command Language. The mathematical basis of AMN is similar to Z, modelling programs in set theory, however unlike Z, AMN does not support schemas – it is organised into machines that comprise a state together with operations on that state, and composition is carried out on these machines.

As an example of a B module, Fig. 1 shows the B specification module for a simple symbol table (Fig. 2 is the Cogito version of the same specification).

The B-Toolkit is a fully integrated toolset that handles dependencies between specifications, refinements, obligations and proofs. Proof obligations are generated at the requisite points, and the B-Toolkit has a proof tool which can discharge many of these automatically. There is also an interactive mode of the prover which can be used, with guidance from the user, to discharge any obligations that were not proved automatically.

MACHINE   *SymTabAbs* (*SYM* , *VAL* , *kk*)

CONSTRAINTS

$kk \in \mathbb{N} \wedge kk < 10000$

VARIABLES

st

INVARIANT

$st \in ST \wedge$
$\text{card} (\text{dom} (st)) \leq kk$

INITIALISATION

$st := \varnothing$

OPERATIONS

update

$update (ss , vv)$ ==
    PRE $ss \in SYM \wedge vv \in VAL \wedge \text{card} (\text{dom} (st)) < kk \wedge ss \notin \text{dom} (st)$
    THEN
        $st (ss) := vv$
    END ;

lookup

$vv \longleftarrow lookup (ss)$ ==
    PRE $ss \in SYM \wedge ss \in \text{dom} (st)$
    THEN
        $vv := st (ss)$
    END

DEFINITIONS

ST == $SYM \nrightarrow VAL$

END

Fig. 1. Specification of a symbol table in B

The B proof system is not always easy to use and this has prompted the development of external provers to supplement the B-Toolkit [Pra95]. From the point of view of this survey it was unfortunate that, since it is an automatic system, the internals of the prover, such as its theory structure, are not described in the documentation.

The B Development Method

The development model used by B has three stages: specification, refinement and implementation, of which the refinement stage may be repeated, giving the sequence:

$$specification \rightarrow refinement \rightarrow \ldots \rightarrow refinement \rightarrow implementation$$

In the B-Method the overall refinement distance (specification to implementation) is usually quite small, complexity is handled by the layering of levels of implementation. In a typical B development a complex specification is refined to an implementation by one or more simpler machines, and these are then developed in turn. In this way Abstract Machines are built on top of one another incrementally to construct a layered design. The "bottom" layer of a complete B development is made up of implementation machines from a system library which can be directly translated to code.

Refinement in B is a general refinement relation, which subsumes both data refinement and algorithm refinement. The proof obligations for B refinement assure the adequacy of refined operations (algorithm refinement) under refinement of variables (data refinement).

2.2   Cogito

An Outline of Cogito

Cogito is a formal development methodology being developed at the SVRC. It is not as mature as B and does not yet have a published definition. Some details, notably of the algorithm refinement mechanism and its relationship to the rest of the methodology, are still the subject of current research. The specification language is called *Sum*, which is an enhancement of Z. The semantics of Sum is based on that of Z and the "Z Mathematical Toolkit", which are both defined by international standards. The main enhancement of Sum with respect to Z is its module mechanism, Sum modules encapsulate a state and its operations, but unlike B they are described by Z-like schemas.

Figure 2 shows the Sum specification module for the same symbol table example as in Fig. 1.

The Cogito system comprises a number of tools coordinated by the a central *Repository Manager*. The most important tools are the type-checker and the Ergo theorem prover[BBNU96]. Reasoning in Cogito is carried out in the context of an Ergo theory of the semantics of Sum within which specifications can be defined [BKU96].

$$
\begin{array}{|l}
\underline{\;SymTabAbs[SYM;\ VAL;\ kk : \mathbb{N}_1]\;}\hspace{3em} \\
\quad ST == SYM \nrightarrow VAL \\[1ex]
\quad \begin{array}{|l}
\underline{\;state\;} \\
\quad st : ST \\[1ex]
\quad \#(\mathrm{dom}\ st) \leq kk \\
\end{array} \\[2ex]
\quad \begin{array}{|l}
\underline{\;init\;} \\
\quad st' = \{\,\} \\
\end{array} \\[2ex]
\quad \begin{array}{|l}
\underline{\;op\ update\;} \\
\quad ss? : SYM \\
\quad vv? : VAL \\[1ex]
\quad \mathrm{pre}(\#\,\mathrm{dom}(st) < kk \wedge ss? \notin \mathrm{dom}\ st) \\
\quad st' = st \oplus \{ss? \mapsto vv?\} \\
\end{array} \\[2ex]
\quad \begin{array}{|l}
\underline{\;op\ lookup\;} \\
\quad ss? : SYM \\
\quad vv! : VAL \\[1ex]
\quad \mathrm{pre}(ss? \in \mathrm{dom}\ st) \\
\quad vv! = st(ss?) \\
\quad changes\_only\{\} \\
\end{array} \\
\end{array}
$$

Fig. 2. Specification of a symbol table in Sum

The Cogito Development Method

The Cogito methodology is defined by a *process model* that specifies the objects and actions that comprise a development and describes the relationships and dependencies between them. This model is defined in the Repository Manager which is responsible for handling dependencies, configurations and version control within Cogito. Repository objects have statuses, for instance a specification may have the status *not validated* or *validated*; dependent on whether the validation obligations for its well-formedness have been proved in Ergo.

The process model is a general mechanism which allows considerable flexibility. For instance, in Cogito algorithm refinement is done at the level of individual operations (schemas) and is a transformation on a single module, while data refinement is a relation between two modules. These relationships are validated by the proof obligations generated when the refinement is specified and the dependencies are handled by the Repository Manager.

In Cogito specifications, refinements and implementations are not separate kinds of object. The layered style is not so appropriate to this framework and the refinement style is a gradual evolution from initial specification to code. Development is complete when all the components can be translated to code – there is an *executable subset* of the Sum syntax for which this translation to a target implementation language can be done directly by one of the Cogito tools.

The use of process models also allows Cogito to support different development methodologies. For instance, although discharging proofs is optional in the current process model, Cogito could be used with a different model that had a stricter strategy which *forced* the user to discharge the well-formedness obligations for a module before refining it. However, currently Cogito is only used with the standard process model.

3   Comparison

3.1   Modularity Constructs

B - Syntax

There are three stages in refinement in the B-Method and the Abstract Machines for these stages are distinguished syntactically.

- MACHINES [1] are specifications.
- REFINEMENTS are refinements of MACHINES or other REFINEMENTS.
- IMPLEMENTATIONS are the final stage that can be converted directly to code, perhaps calling the implementations of other machines.

IMPLEMENTATIONS are a special type of REFINEMENT that have no state of their own. They only use an implementable subset of AMN but may call the operations of other machines, which are specified and implemented separately. The implementable subset of AMN excludes such constructs as the parallel composition operator. There are also restrictions on the AMN constructs that can be used for the initial specification (MACHINE) - for instance it cannot contain loops.

The AMN definition of a module has number of sections. The VARIABLES and INVARIANT sections define the state variables and an invariant for each module. The INITIALISATION and OPERATIONS sections define an initialisation and a set of operations on the state. The operation preconditions must be stated explicitly.

B - Structuring

B has a number of structuring mechanisms for modularity:

- In specifications one can use INCLUDES, USES and SEES.
- In refinements one can use SEES.
- In implementations one can use IMPORTS and SEES.

---

[1] In the description of syntax keywords are written in SMALL CAPITALS.

However, all these constructs can only reference MACHINES. Thus it is always the specification which is the interface to other components. The primary structuring mechanisms are INCLUDE and IMPORT and these will be discussed first.

INCLUDE This is the main mechanism for structuring specifications. A machine may INCLUDE one or more machines and its state is then the combined state of all the included machines. INCLUDED machines may INCLUDE other machines. A machine may not directly change the variables in INCLUDED machines although it can call the INCLUDED operations (all of which are accessible to the INCLUDING machine) to do so indirectly. Operations of the INCLUDED machine may be made part of the interface of the INCLUDING machine by using the PROMOTES clause.

Multiple copies of a single machine may be INCLUDED by using systematic renaming. This is done by supplying a prefix to the machine name in the INCLUDES list, this prefix is then used to reference all variables and operations from that machine. That is, the INCLUDES statement is of the form:

INCLUDE <id>.<mach>

where "<mach>" is the actual machine name, and "<id>" is an identifying tag. Components of the included machine "<mach>" are then referenced by using the prefix "<id>.". For instance, Fig. 3 shows the statements to include two copies of a symbol table in an outer specification. To reference operations on specific tables we can then use the tags, for example *TypeTab.update*, *AttrTab.lookup*.

MACHINE   *TwoTables* (*SYM* , *TYPE* , *ATTRIB* , *max_size*)

. . .

CONSTRAINTS

    max_size $\in \mathbb{N}$

. . .

INCLUDES

    TypeTab . *SymTabAbs* (*SYM* , *TYPE* , *max_size*) ,
    AttrTab . *SymTabAbs* (*SYM* , *ATTRIB* , *max_size*)

. . .

END

Fig. 3. Fragment of a B specification that imports two symbol tables

Avoiding clashes in the name space when INCLUDE is used is the responsibility of the developer. The device of systematic renaming can be used for this purpose, but is designed for the disambiguation of names when multiple copies of a single machine are being INCLUDED, and it has some disadvantages as a general renaming mechanism, for instance variables cannot be renamed selectively.

IMPORT is used for structuring *implementations*. IMPORTED machines may be developed separately. Renaming is available for IMPORTED machines via the convention that:

IMPORT <Tag>_<name>

will match the implementation machine:

Rename_<name>.imp

and the IMPORT will generate a copy of the implementation machine in which all occurrences of "**Rename**" are replaced by "**<Tag>**". For instance Fig. 4 illustrates the IMPORT of two copies of the B-Toolkit library implementation of arrays into the implementation of the symbol table. The library machine is called *Rename_Varr*, and the imported copies *Sym_Varr* and *Val_Varr* are used for storing symbols and values respectively. Operations of the imported implementations are referred to in a similar fashion, for instance, the two versions of *Rename_STO_ARR* as *Sym_STO_ARR* and *Val_STO_ARR*.

Note that INCLUDE or IMPORT statements are not allowed in *refinements*. All structuring in a refinement machine is inherited from the machine that it is refining.

Uses and SEES are similar constructs which allow access to variables in a machine defined elsewhere. Uses can only be used in specifications, while SEES can be used in specifications, refinements and implementations. In fact SEES is the only one of all of the structuring constructs that can be used in REFINEMENTS.

USES This construct allows the USING machine to refer to variables in the USED machine. The variables do not become part of the state of the USING machine and cannot be modified in any way. Thus they cannot, for example, be used on the left-hand side of an assignment. They *can*, however, appear in the invariant. A frequent use of USES is to specify common sets that are accessed by many parts of a specification. This is exemplified in Fig. 5 which shows part of the specification of the Caviar reservation system taken from [DD93]. The machine called *CommonSets* defines the given sets and it is USED by a number of other machines in the specification. (A full specification of Caviar, in Z, can be found in [Hay93].)

The operations of a USED machine are not visible to the USING machine.

SEES is similar to USES except that the variables of the SEEN machine *cannot* appear in the invariant. This restriction allows a SEEN machine to be refined independently of the machine that SEES it, whereas a USED machine may not. Modules imported or included once somewhere in a development can be SEEN elsewhere. In the final implementation the code for the SEEN machine is linked in only once. When a machine is SEEN, its query operations can be invoked, but not any operations that change its state.

```
IMPLEMENTATION     SymTabImp
REFINES

    SymTabConc

IMPORTS

    Sym_Varr ( SYM , kk ) , Val_Varr ( VAL , kk ) , High_Nvar ( kk )


OPERATIONS

    update

    update ( ss , vv )    ==
        VAR    tmp    IN
            High_INC_NVAR ;
            tmp ⟵ High_VAL_NVAR ;
            Sym_STO_ARR ( tmp , ss ) ;
            Val_STO_ARR ( tmp , vv )
        END ;


    . . .

END
```

Fig. 4. Fragment of B showing the import of two copies of the library implementation of arrays

To the casual user the proliferation of constructs in B and the rules associated with them, although logical, are somewhat daunting (the *B-Book* has ten pages of tables defining the visibility rules). Basically there are two sets of constructions: one for *specifications* using INCLUDES, USES and the "<id>." construct for renaming; and one for *implementations* using IMPORT, SEES and the automatic renaming of library machines on IMPORT. There is also a store of accumulated wisdom for their use, both in the *B-Book* and in other sources, for instance [LH] is a tutorial on B which includes the following "guidelines":

If X needs only access to the sets/constants of Y in its invariant, and to no update operations on Y, then X SEES Y.

If X needs access to the variables of Y in its invariant, and to no update operations on Y, then X USES Y.

If X needs to invoke update operations of Y in its operations, then X INCLUDES Y.

MACHINE   *CommonSets*

SETS

   DATES ; *SESSIONS* ; *TIMES* ; *VISITORS* ; *MEETINGS*

END

MACHINE   *Vsys*

USES

   CommonSets

INCLUDES

   Vpool , *HrV* , *TrV*

CONSTANTS

   dateoftime

PROPERTIES

   $dateoftime \in TIMES \rightarrow DATES$

INVARIANT

   ...

OPERATIONS

   BookHotelRoom

   BookHotelRoom (hr , vv , dset)   ==   ...

   ...

END

Fig. 5. Machine defining the given sets for Caviar, and an instance of its use.

Cogito - Sum Syntax

In Sum there is no explicit syntax to differentiate between the module definitions for specifications, refinements and implementations. "Specification" and "implementation" are seen as module attributes determined by the process model, rather than part of the syntax and semantics. There *are* two identifiable subsets of the constructs: the members of one are purely executable, and these would not normally appear in specifications; and members of the other are non-executable and so cannot appear in implementations, since they cannot be translated to code. Intermediate refinements may contain syntactic elements from both subsets. However, this distinction is only relevant to the tool that generates code. Where full formal development is required the model can enforce that all refinements are to the implementable subset before code can be generated for the module, but equally a semi-formal model could be used where parts of the refinement are left undeveloped – the code to be added by hand outside of the formal development process.

A Sum module has a distinguished *state* schema which, besides declaring the state variables, also has a predicate part which effectively supplies the invariant for the module. It also has a distinguished *initialisation* schema. Each operation is also defined by a separate schema. The schemas for operations may optionally designate an explicit precondition, otherwise there is an implicit calculated precondition as for Z.

Cogito - Structuring

Sum has broadly similar structuring mechanisms to B, but separates accessibility from what it terms *visibility*[2], that is the ability to reference by a short, unqualified name.

IMPORT This makes all entities in the IMPORTED module accessible, and it is transitive. Where IMPORT (without the AS clause) is used all the IMPORTING modules share the same copy of the IMPORTED module.

IMPORT ... AS This is used to instantiate a private copy of a module and requires the user to specify a new name for the copied module. It is obligatory if the module is parameterised and is therefore being instantiated by assigning actual values to its parameters. This mechanism is illustrated in Fig. 6 where two copies of the symbol table are imported.

To assist in controlling the name space, Sum permits renaming of individual components of IMPORTED ... AS modules.

When a module is IMPORTED, its accessible components can be referred to by qualified names, that is: `<module-name>.<name>`. This enables large collections of modules to be handled within the same name space without a problem of name clashes, or the worse problem of components becoming inaccessible due to the scoping problems. However, IMPORT is transitive so the full reference of a

---

[2] Note that the *B-Book uses* the term "visibility rules" to refer to accessibility.

```
┌─ TwoTables[SYM; TYPE; ATTRIB; max_size : ℕ1] ──────────
│  import SymTabAbs(SYM, TYPE, max_size) as TypeTab
│  import SymTabAbs(SYM, ATTRIB, max_size) as AttrTab
│  ┌─ state ─────────────────────────────────────────
│  │  TypeTab.state
│  │  AttrTab.state
│  └─────────────────────────────────────────────────
│  ┌─ init ──────────────────────────────────────────
│  │  TypeTab.init
│  │  AttrTab.init
│  └─────────────────────────────────────────────────
│  . . .
│  operations
│  . . .
└──────────────────────────────────────────────────────
```

Fig. 6. Cogito module importing two copies of the symbol table

component that is at the end of an IMPORT trail can require a number of qualifiers and become quite long. The Sum visibility mechanism overcomes this problem.

VISIBLE This makes IMPORTED items accessible via non-qualified (short) names. This can be used to reduce the need for repeated qualifiers and to make modules more readable. However, care must be taken to avoid inadvertently hiding variables. For example, if the two symbol tables in Fig. 6 were made visible, then the short names of the second would hide those of the first (although the latter could still be referenced by their full names).
VISIBLE is often used when one module is extending another. An elementary example is shown in Fig. 7 which is a Cogito module specifying a tree in terms of a directed graph. The DAG module is made visible so that its components can be referenced directly, although in this case the only component of the DAG referred to is the *neighbour* relation. This example also illustrates the renaming mechanism for components of an imported module.

Broadly the Sum IMPORT corresponds to the B IMPORT, and the Sum IMPORTED ...AS corresponds to the B INCLUDE. To reference components of IMPORTED Sum modules directly, that is without using qualified names, the IMPORT clause must be supplemented by a following VISIBLE clause.

3.2   Module Parameterisation and Genericity

B-Method In B parameters are either sets or elements of sets, operations are not allowed as parameters. The parameters can be constrained by the CON-STRAINTS clause which allows the user to specify properties of the parameters

TreeMod[Elem]

// A tree is defined to be an acyclic directed graph with a distinguished
// root node, and where each node has only one outgoing edge.
// From any node the root may be reached by following outgoing edges
// using the *neighbour* relation of the DAG.

import DAGMod(Elem){DAGstate/state, DAGinit/init} as ElemDAG

visible ElemDAG{neighbour, DAGstate, DAGinit}

state

DAGstate

$root : Elem$

$neighbour \in Elem \nrightarrow Elem$
$\forall c : \mathrm{dom}\ neighbour \bullet root \in neighbour^{+}(\!| c |\!)$

init

DAGinit

Fig. 7. Part of a Sum module for a Tree

that can be assumed in the machine. However, it is also required that the parameters be *completely independent* of each other. The reason for this is that the set parameters are among the base types of each abstract machine and the typing system requires that base types are independent. This is quite a strong requirement.

Parameterised machines are instantiated by supplying actual parameters when the machine is INCLUDED or IMPORTED. All formal parameters must be instantiated. Where a parameterised machine is USED the USES clause does not specify the parameters. In this case both the USED and USING machine must be INCLUDED in a common, enclosing machine. In this outer machine the INCLUDE clause for the USED machine will instantiate the parameters, and in this way the parameters for the USED will be provided just once by a single instantiation of this machine throughout the specification.

Parameters may appear in proof obligations, even though they are not fully defined until instantiated. Where constraint obligations include references to parameters, these obligations will sometimes not be provable in the context of the parameterised machine. In this case constraints are promoted as obligations on all instantiating machines. A typical case is where an implementation constrains an integer parameter to be less than some maximum representable integer, the obligation to establish this constraint must be met by the machine that instantiates the implementation.

Cogito In Sum parameters can be typed or untyped. They may be operations, given sets (types), or any entity defined in the Sum mathematical toolkit.

The parameters can be constrained by an arbitrary predicate given as part of the declaration of the parameter list, but there are no general restrictions such as the independence required by B. The ability to pass operations as parameters would seem to be a considerable advantage for the designer using Cogito in comparison to B.

On IMPORT parameterised modules must be fully instantiated and require the IMPORT AS clause (see 3.1). This creates an instance of the module as instantiated by the specific parameters. The theories associated with parameterised modules are promoted into obligations and postulates in the theories of modules that instantiate them.

3.3   Information Hiding

B-Method Information hiding is effected by two mechanisms. With the INCLUDES (and USES) clauses in specifications, B uses the *semi-hiding principle*. All the variables in the INCLUDED machine can be referenced in the INCLUDED machine, and the initialisation and invariants are inherited, but these variables are "read-only" and they can only be modified indirectly by using the operations of the machine in which they are defined.

With the IMPORTS (and SEES) clauses used in implementations, B uses the *full-hiding principle*. An implementation machine may have no separate state data of its own, only variables local to each operation. All the data used in the implementation is manipulated via the interfaces to IMPORTED machines. However, IMPORTED data may be referenced in the machine invariant and in loop invariants.

Cogito In Sum information cannot be fully hidden since all components of IMPORTED modules can be referenced using fully qualified names. However, direct reference, via short names, must be made explicit using the VISIBLE clause.

Cogito also has a mechanism, the CHANGES_ONLY construct, to control read/write access to data, and this is achieved within the individual operations of a module. The combination of VISIBLE and CHANGES_ONLY allows read/write control over IMPORTED data. For example, in the symbol table (Fig. 2) the *lookup* operation cannot change the data, and this is specified by the statement `changes_only{}` which changes the default to read-only. (In general `changes_only{ ... }` specifies a list of variables that are "writable".)

3.4   Module Refinement

B-Method There is a single general refinement method in B, this subsumes both algorithm and data refinement. The refining module specifies the module that it refines in the REFINES clause and proof obligations are associated with the refining module, as a justification that it is a valid refinement. Since refinements can also be refined this process generates a sequence of more refined modules (as illustrated in Sect. 2.1).

As an example, Fig. 8 is a refinement of the symbol table shown in Fig. 1 in which the symbols and values are stored in parallel arrays. Note that this module is introduced by the keyword REFINEMENT and explicitly names the module that it is refining: *SymTabAbs*. The module parameters are inherited from the refined module. Variables of *SymTabAbs*, for example the function *st*, are accessible in the invariant. In this case, the last two conjuncts of the invariant are the *coupling invariant* relating the data structures of the refinement module to those of the module it is refining.

Cogito In Cogito data refinement and algorithm refinement are separate operations.

Algorithmic refinement is done at the sub-module level, by transforming individual schemas within a module. This is done within the Ergo theory for the module, by applying transformation rules to the operation schemas, in a similar style to the refinement calculus. The result of such a transformation can be extracted from Ergo as a new Sum module which is a refinement of the original. (The tool to do this is still under development.)

For data refinement, the refined and refining modules are specified separately and the refinement relationship between them is encapsulated in a separate module which IMPORTS them both and generates the proof obligations for the refinement. The semantics of the data refinement relation in Cogito is in terms of the externally visible effects of the operations of the modules modelled as state machines [BKK$^+$96]. This is a more flexible mechanism than in the B-Method, and for example, allows that the same module can refine two different specifications by defining two separate refinement modules with the same concrete module and different abstract ones. In B the refining module must explicitly name the module that it is refining.

The predicate in the refinement module usually just specifies the abstract–concrete coupling invariant. However it can also contain a more general predicate which defines the relationship between the operations of the refined and refining modules, for instance it could define some composition of concrete operations as the refinement of a single abstract one (this is in effect a form of algorithm refinement). It is also possible in Cogito to have auxiliary operations local to a module which can be called by other operations in the module. (B does not allow an operation to call other operations from the same machine.)

Figures 9 and 10 show a data refinement of the symbol table of Fig. 2. This is the same refinement using parallel arrays as that given in the B example above. Figure 9 is the concrete version of the symbol table, and Fig, 10 shows the module that specifies the refinement relation. In Cogito the coupling invariant is specified in the data refinement module.

The data refinement operation of the process model generates the validation obligations for the data refinement, and also a new Ergo theory for the refinement module *SymTabRef* which inherits the theories of both the abstract and concrete modules. It is within this theory that the obligations are discharged.

REFINEMENT   *SymTabConc*

REFINES

   SymTabAbs

VARIABLES

   keys , vals , high

INVARIANT

   $keys \in 1 .. kk \rightarrow SYM \wedge$
   $vals \in 1 .. kk \rightarrow VAL \wedge$
   $high \in 0 .. kk \wedge$
   $\forall (ii , jj) . (ii \in 1 .. high \wedge jj \in 1 .. high \wedge ii \neq jj \Rightarrow keys (ii) \neq keys (jj))$.
   $dom (st) = ran (1 .. high \lhd keys) \wedge$
   $\forall ii . (ii \in 1 .. high \Rightarrow st (keys (ii)) = vals (ii))$

INITIALISATION

   $high := 0 \;\|$
   $keys :\in 1 .. kk \rightarrow SYM \;\|\; vals :\in 1 .. kk \rightarrow VAL$

OPERATIONS

   update

   $update (ss , vv) \quad ==$
       PRE
           $high < kk \wedge \forall ii . (ii \in 1 .. high \Rightarrow ss \neq keys (ii))$
       THEN
           $high := high + 1 \;\|\; keys (high + 1) := ss \;\|\; vals (high + 1) := vv$
       END ;

   lookup

   $vv \longleftarrow lookup (ss) \quad ==$
       PRE
           $\exists ii . (ii \in 1 .. high \wedge keys (ii) = ss)$
       THEN
           ANY   *tmp*   **WHERE**
               $tmp \in 1 .. high \wedge keys (tmp) = ss$
           THEN
               $vv := vals (tmp)$
           END
       END
   END

Fig. 8. The concrete version of the symbol table in B

$$\begin{array}{|l}
\hline
_\,SymTabConc[SYM;\ VAL;\ kk:\mathbb{N}_1]\,\rule{6cm}{0.4pt} \\[4pt]
\quad\begin{array}{|l}
\hline
_\,state\,\rule{5cm}{0.4pt} \\
keys:1\,..\,kk \rightarrow SYM \\
vals:1\,..\,kk \rightarrow VAL \\
high:0\,..\,kk \\
\hline
\forall\,ii,jj:1\,..\,high \bullet ii \neq jj \Rightarrow keys(ii) \neq keys(jj) \\
\hline
\end{array} \\[6pt]
\quad\begin{array}{|l}
\hline
_\,init\,\rule{5cm}{0.4pt} \\
high' = 0 \\
\hline
\end{array} \\[6pt]
\quad\begin{array}{|l}
\hline
_\,op\ update\,\rule{4cm}{0.4pt} \\
ss?:SYM \\
vv?:VAL \\
\hline
\mathrm{pre}(high < kk \wedge (\forall\,ii:1\,..\,high \bullet ss? \neq keys(ii))) \\
high' = high + 1 \\
keys' = keys \oplus \{high' \mapsto ss?\} \\
vals' = vals \oplus \{high' \mapsto vv?\} \\
\hline
\end{array} \\[6pt]
\quad\begin{array}{|l}
\hline
_\,op\ lookup\,\rule{4cm}{0.4pt} \\
ss?:SYM \\
vv!:VAL \\
\hline
\mathrm{pre}(\exists\,ii:1\,..\,high \bullet keys(ii) = ss?) \\
\exists\,ii:1\,..\,high \bullet (keys(ii) = ss? \wedge vv! = vals(ii)) \\
changes_only\{\} \\
\hline
\end{array} \\
\hline
\end{array}$$

Fig. 9. The concrete version of the symbol table in Sum

$$\begin{array}{|l}
\hline
_\,SymTabRef[SYM;\ VAL;\ kk:\mathbb{N}_1]\,\rule{5cm}{0.4pt} \\
import\ SymTabAbs(SYM, VAL, kk)\ as\ Abs \\
import\ SymTabConc(SYM, VAL, kk)\ as\ Conc \\[4pt]
\quad\begin{array}{|l}
\hline
_\,SymTabReln\,\rule{4cm}{0.4pt} \\
Abs.state \\
Conc.state \\
\hline
\mathrm{dom}\ st = \{ii:1\,..\,high \bullet keys(ii)\} \\
\forall\,ii:1\,..\,high \bullet (st(keys(ii)) = vals(ii)) \\
\hline
\end{array} \\
\hline
\end{array}$$

Fig. 10. The data refinement module for the symbol table in Cogito

Compositionality of Refinement. One aim of modularity is to reduce the complexity of large systems. To achieve this it is important that when a module is decomposed the refinement of each of the subsidiary modules is independent of the other modules, this property is known as *compositionality*. Without it the decomposition does not reduce the complexity of the task of refinement.

B-Method In the B-Method it can be shown that, subject to the general requirements of the method, such as that the variables in the refined and refining machines are distinct, refinement of each of the constructs of the AMN language is monotonic with respect to the others. In particular the parallel and sequential composition constructs, which partition the development are compositional.

Cogito In Cogito this property is guaranteed by Ergo by virtue of the fact that Cogito modules normally generate definitional theories which are conservative extensions of the existing Ergo theory graph. When generic modules are refined in Cogito, the theory interpretation mechanism [HNTU97] ensures the compositional property. That is that, in respect to any particular instantiation of the parameters, instantiation of the specification is refined by the instantiation of refinement.

3.5   Proof Obligations

The specification of a module incurs an obligation that the module is well formed in various ways. For instance, that where an invariant is specified the operations of the module preserve this. The refinement of a module also incurs an obligation to show that the refinement preserves the properties that the refinement relation is required to preserve.

B-Method The B-Method specifies in detail the obligations for the existence and adequacy of a module, and for module refinement [Abr96]. For the definition of a module the most important obligations are that the module invariant is established by the initialisation, and that it is preserved by each operation. Obligations are also incurred when a module is INCLUDED, for instance that the actual parameters obey any restrictions of the formal parameters in the INCLUDED machine. The obligations for refinement are more complex and are not considered in detail here. The B-Toolkit generates these obligations on request at the appropriate stage of the development process. The automatic and interactive phases of the proof support tool can then be used to discharge these obligations if required.

Cogito The Cogito proof obligations are very similar to B. They are generated as theorems to be proved in the Ergo theory associated with the module. The process model keeps track of all obligations and their proofs.

A detailed comparison of the obligations generated by the two systems was outside the scope of the survey on which this paper is based. However, one point of detail arose during the symbol table case study, the B and Sum specifications for which are shown in Figs. 1 and 2 respectively. The Cogito version generated a validation obligation that the operations could in fact be invoked for all instantiations of the parameters, and this required that the allowable maximum size of the table was non-zero. B does not generate this obligation (although in this case the library implementations of the of concrete data structures actually enforce this restriction). It may have been noticed that the parameter k is of type $\mathbb{N}$ for B, but of type $\mathbb{N}_1$ for Cogito, this is to enable this obligation to be discharged. After some discussion with the Cogito developers it was decided that this particular obligation was best made optional in Cogito, and is no longer required.

3.6 Semantics

B-Toolkit

Like the refinement calculus, AMN operations are defined in terms of predicate transformers, which are called *generalised substitutions* in B. The preconditions for operations must be stated explicitly.

Cogito

Sum has a Z-like semantics in which programs are predicates. Each operation has a calculated precondition, as for Z, however Sum also supports the explicit statement of preconditions. Asserting an explicit precondition incurs an obligation to show that it is equivalent to the calculated one.

The explicit precondition is a useful mechanism for casting the calculated precondition into a simpler form or to one that is more convenient to work with. By stating the precondition in this way the equivalence to the calculated need be proved once, and the simpler form can then be used throughout the refinement. This device is also used just for documenting the precondition rather than leaving it implicit, this makes explicit the specifier's intention as to the precondition of the operation. The obligation then tests the intention against what the specifier has actually written. Considerations such as these prompted the B Method to adopt the explicit statement of the precondition as mandatory (the proof of equivalence to the calculated one becoming one of the obligations).

4 Discussion

Syntax and Structuring

The B and Cogito languages, AMN and Sum, are *wide-spectrum languages*, including specifications and code in the same framework. They differ considerably in detail but one point of general difference is the fact that B differentiates between specifications, refinements and implementations at the syntax level, whereas Cogito considers this difference as a methodological issue.

In B, while the visibility rules are somewhat complex, they are rigid. Cogito, on the other hand, allows the designer considerable latitude in defining visibility, while retaining the ability to reference accessible components by full names. This has the disadvantage that handling the visibility constraints complicates the tools and theories. However, the motivation for the Cogito visibility rules is to allow fine control of the name space in order to support scaling to arbitrary levels of complexity. In such cases B would leave the user with the burden of avoiding clashes and managing renaming. This is part of a general concern in Cogito for building-in scalability so that the methodology can handle systems of large-scale complexity.

The two systems differ in their approach to structuring of module composition and decomposition. B has a number of different mechanisms to cover different circumstances, each with specific rules correct for that circumstance. On the other hand, Cogito has a few constructs that can be used flexibly to achieve similar results. In some respects the stance of Cogito to information hiding may be thought to be too loose, since it depends on a low-level discipline on the part of the developer that cannot be policed by the current tools. The flexibility in Sum given by the visible and changes_only constructs gives the user fine-grained control at the cost of the user having to confront this issue in every case. The Cogito philosophy is that extra effort in writing specifications is worthwhile if the end product is easier to read and understand.

Genericity

In this paper modules are considered as "compile-time" objects, and genericity is achieved by module parameterisation, with particular instances being created when a parameterised module is instantiated. The types of parameters that are allowed in B is more restricted than in Sum; the ability of Sum to parameterise modules with operations has been commented on above. However even more flexible mechanisms are possible and may be desirable depending how the the modularity mechanism is to be used. Being an extension of Z the Sum syntax does support generic definitions, as used in the Z mathematical toolkit, via an "`axiom schema`" construct.

Refinement

In respect to refinement, B has the advantage of a single general method, whereas Cogito has separate methods for algorithmic and data refinement. This is not of course unusual, since they are often treated separately in the literature. The difference between the two Cogito methods can also be described in terms of the "refinement method"–"refinement calculus" distinction [LW92]. Cogito data refinement is very much a refinement *method* – with the concrete module being developed separately from the abstract module and the refinement justified *post hoc* by the specification of the coupling invariant in the separate data refinement module. The refinement is justified by proofs of the obligations associated with the data refinement module. For algorithm refinement a more *calculational* style is used, with transformations to executable constructs being achieved by the application of rules in the Ergo theory for the module. Refinement in B, like Cogito data refinement, is an example of a refinement *method*. The calculational style has the advantage that there is a limited set of rules that are applicable at any particular point so guidance and prompting of the user is easier. On the other hand, it is often difficult to perform large steps with this style.

Practice indicates that it is common for the modular structure to differ between the specification and the implementation [Hal96], so it is pertinent to consider how easy this is to achieve in the two methods. In the B-Method the modular structure of refinements is determined by the INCLUDE structure of the original specification, and conceptually this just defines one large specification text. The modular structure of an implementation is determined by the IMPORT structure of the implementation machines. In this way the structure of specification and implementation may be quite different. The only restriction is that the operations and operation parameters of a refinement must be the same as those of the specification that it is refining.

The refinement relation between modules in Cogito is flexible enough to allow the modular structure to be modified during development even to the extent of changing the operational structure of refinements. The predicate in the linking schema in data refinement is not restricted to the coupling invariant for the data refinement but can also define the refined operations in terms of the refining ones in a fairly general way as described in Sect. 3.4.

Semantics

This investigation has not examined the semantics of either method in any detail although this issue has considerable implications for formal methods and tool support. It has been noted that B, which is a mature technology, has a well-defined semantics and theory. It has also been noted that in a number of areas B has adopted a fairly restrictive approach to issues such as information hiding, in many cases this sort of decision will have helped to simplify the semantics.

As far as Sum is concerned, the basic semantics is derived from Z. The chief extension is the mechanism used to handle the extended name space, since Z already has the machinery for schema composition which is required to support the composition of modules. A full description of the semantics has recently been completed. Its details differ somewhat from the current Z standard, since Sum is based on an earlier draft, but a simple mechanism has been developed that effects the translation of a Sum module into Z.

B and Cogito also differ in that the former adopts a predicate transformer semantics for operations while the latter has a predicate semantics. Predicate transformers are somewhat more expressive (they can, for example express *miracles* – for a discussion see [Gru93]), but this is mainly of theoretical interest and having a predicative semantics is not a limitation on systems in practice.

Methodology

In respect to development methodology, in both B and Cogito it is the containment of complexity permitted by modularisation and compositionality of refinements that enables them to be realistically proposed as methods that will scale up to large systems. Once again Cogito is more liberal than the B-Method, the Cogito methodology is defined by the process model which is in principle configurable while the B-Method has single well-defined formulation. Since the Cogito process model is one of the components still under development it is difficult to comment further.

One feature of the B-Method is its layered approach. It has been reported in the *B User Group* that this may lead to inexperienced users building a large number of machines with a complex network of relations between them. It also has been observed [BR93] that the B-Method tends to encourage the developer to introduce algorithmic details at a premature stage in the design process.

5   Future Work

As mentioned in the initial overview, the ultimate aim of this research is to extend the refinement calculus to include modularity. This would be supported by an extension of our refinement tool PRT [CHN$^+$96]. Since modular refinement presupposes the capability for data refinement, the work on extending PRT to support data refinement, which is described in [SNC97], is directly relevant to this overall project.

We have also been looking into methods of incorporating more general abstract data types (such as the opaque types discussed in [BH93]) into the modularity mechanism. Insight into this may be given by extending the comparison to include an example of an algebraic specification method, and some work has been done to include Extended ML [KST94, ST91] in the survey.

The differences between B and Cogito raise some issues which warrant further investigation. For instance at some points the B methodology adopts conventions that may seem unnecessarily restrictive, a detailed examination of the rationale for some of these choices given in the B-Book, and how the they apply in the Cogito case, and perhaps a comparative case study to examine their implications would be interesting. A distinctive feature of Cogito is the *visibility* mechanism for managing the name space, which is specifically designed to handle a perceived problem in scaling up to large-scale development. There have been a number of industrial trials of B. It would be interesting to review the literature on these in detail to ascertain whether they report difficulties with handling large name spaces in B.

References

[Abr96]   J.-R. Abrial. *The B Book - Assigning Programs to Meanings*. Cambridge University Press, 1996.

[B-C96]   B-Core(UK) Ltd. *B-Toolkit: User's Manual*, 1996.

[BBNU96]   Holger Becht, Anthony Bloesch, Ray Nickson, and Mark Utting. Ergo 4.1 reference manual. Technical Report 96-31, Software Verification Research Centre, Department of Computer Science, The University of Queensland, St. Lucia, QLD 4072, Australia, September 1996.

[BH93]   P. G. Bancroft and I. J. Hayes. Refining a module with opaque types. In Gopal Gupta, George Mohay, and Rodney Topor, editors, *Proceedings of the 16th Australian Computer Science Conference*, pages 615–624. Australian Computer Society, February 1993.

[BKK+96]   A. Bloesch, E. Kazmierczak, P. Kearney, J. Staples, O. Traynor, and M. Utting. A formal reasoning environment for Sum – a Z based specification language. *Australian Computer Science Communications*, 18(1):149–158, February 1996.

[BKKT95]   A. Bloesch, E. Kazmierczak, P. Kearney, and O. Traynor. Cogito: A methodology and system for formal software development. *International Journal of Software Engineering and Knowledge Engineering*, 5(4):599–617, December 1995.

[BKU96]   A. Bloesch, E. Kazmierczak, and M. Utting. The Sum environment in Ergo: A tutorial. Technical Report 96-22, Software Verification Research Centre, Department of Computer Science, The University of Queensland, St. Lucia, QLD 4072, Australia, September 1996.

[BR93]   J. Bicarregui and B. Ritchie. Invariants, frames and postconditions: a comparison of the VDM and B notations. In P.G. Woodcock, J.C.P.and Larsen, editor, *FME93: Industrial-Strength Formal Methods. First International Symposium of Formal Methods Europe Proceedings*, number 670 in Lecture Notes in Computer Science, pages 351–66. Springer-Verlag, Berlin, Germany, April 1993.

[CHN+96]   David Carrington, Ian Hayes, Ray Nickson, Geoffrey Watson, and Jim Welsh. A tool for developing correct programs by refinement. In *Proceedings of BCS-FACS 7th Refinement Workshop*, electronic Workshops in Computing. Springer-Verlag, July 1996.

[DD93]   R. Docherty and A. Diller. Caviar in AMN. Technical Report CSR-93-3, School of Computer Science, The University of Birmingham, May 1993.

[Gru93]   Jim Grundy. Predicative programming: A survey. In D. Bjørner, M. Broy, and I.V. Pottosin, editors, *Formal Methods in Programming and Their Applications*, volume 735 of *Lecture Notes in Computer Science*, pages 8–25. Springer Verlag, 1993.

[Hal96]   Anthony Hall. Using formal methods to develop an ATC information system. *IEEE Software*, 13(2):66–76, March 1996.

[Hay93]   I. J. Hayes, editor. *Specification Case Studies*. Prentice Hall International Series in Computer Science, 2nd edition, 1993.

[HNTU97]   Nicholas Hamilton, Ray Nickson, Owen Traynor, and Mark Utting. Interpretation and instantiation of theories for reasoning about formal specifications. In Patel M., editor, *Proceedings of the Twentieth Australasian Computer Science Conference (ACSC'97)*, pages 37–45, *Australian Computer Science Communications*, vol. 19, Feb 1997.

[KKTW95] Edmund Kazmierczak, Peter Kearney, Owen Traynor, and Li Wang. A modular extension to Z for specification, reasoning and refinement. Technical Report 95-15, Software Verification Research Centre, Department of

Computer Science, The University of Queensland, St. Lucia, QLD 4072, Australia, February 1995.

[KST94]   Stefan Kahrs, Donald Sannella, and Andrzej Tarlecki. The semantics of Extended ML: A gentle introduction. In *Proc. Intl. Workshop on Semantics of Specification Languages*, Workshops in Computing, pages 186–215. Springer, 1994.

[LH]   K. Lano and H. Haughton. Formal Development in B Abstract Machine Notation. Participant's notes for the tutorial presented at FME'96.

[LW92]   Hans J. Litteck and Peter J. L. Wallis. Refinement methods and refinement calculi. *Software Engineering Journal*, 7(3):219–229, May 1992.

[Mey88]   B. Meyer. *Object-oriented Software Construction*. Prentice Hall, 1988.

[Pra95]   C.H. Pratten. An introduction to proving AMN specifications with PVS and the AMN-PROOF tool. In Henri Habrias, editor, *Proc. Z Twenty Years on - What is its Future*, pages 149–165. IRIN-IUT de Nantes, October 1995.

[RŽ]   Ken Robinson and John Žic. Refinement and implementation in the B–Method. Presented at the *Fifth Australian Refinement Workshop*, proceedings available on the WWW at URL: http://www.cs.uq.edu.au/conferences/arw96/

[SNC97]   Jamie Shield, Ray Nickson, and David Carrington. Supporting data refinement in a program refinement tool. In L. Groves and S. Reeves, editors, *Formal Methods Pacific '97*, Discrete Mathematics and Theoretical Computer Science. Springer-Verlag, Singapore, 1997.

[ST91]   D. Sannella and A. Tarlecki. Extended ML: past, present and future. In Wusterhausen, editor, *7th Workshop on Specification of Abstract Data Types*, volume 534 of *Lecture Notes in Computer Science*, pages 297–322. Springer, 1991.

Action System Analysis of an Authentication Protocol
(Extended Abstract)

Michael Butler

Deptartment of Electronics & Computer Science, University of Southampton,
Highfield, Southampton SO17 1BJ, United Kingdom,
`M.J.Butler@ecs.soton.ac.uk`

1   Introduction

Part of the Needham-Schroeder Authentication Protocol is analysed using the
action system formalism. The aim of the Needham-Schroeder Authentication
protocol is to provide mutual authentication. The protocol uses public-key cryp-
tography. Lowe has shown that the original protocol is vulnerable to attack and
proposes a fix to prevent this attack. In this report, we use the fixed version of
the protocol in order to achieve one way authentication – that is, the recipient
of a message can be certain that the message is actually from the purported
sender.

We present an abstract specification of a simple authenticated communications service and then refine it in such a way that the communication and
authentication of messages is achieved using the fixed version of the Needham-
Schroeder Authentication Protocol. Three stages of development are presented:

1. The abstract system, which is a simple description of a service that allows
   agents to exchange authenticated messages.
2. The first refinement of the abstract system, which implements the exchange
   of each individual message using the protocol.
3. The second refinement, which introduces an explicit model of how a spy
   could fake and replay messages to attempt to violate the authentication.

2   Specification

The abstract service is described by the action system of Figure 1 (this uses
Abrial's B notation). With this service, agents are allowed to exchange text
(i.e., unencrypted) messages. An asynchronous communications channel between
two agents A and B is modelled as a bag of text $chan(A, B)$. The system has
operations for sending and receiving messages. The effect of agent A sending a
message T to agent B, by invoking $A.Send(B, T)$, is to add T to the channel
linking A to B. An agent B may invoke $B.Recv$ (i.e., may receive a message)
provided there is a message in some channel that is targetted at B; the output
of $B.Recv$, is a message for B along with its sending agent.

MACHINE   *AuthService*
VARIABLES $chan : (AGENT \times AGENT) \to bag\ TEXT$
INITIALISATION $chan := (\lambda(a,b) \mid a \in AGENT\ \wedge\ b \in AGENT \cdot \{\})$
OPERATIONS

$a.Send(b,t) \;\hat{=}$
　　SELECT $a,b \in AGENT\ \wedge\ t \in TEXT$
　　THEN $chan(a,b) := chan(a,b) + \{t\}$ END

$(a1, t1) \longleftarrow b.Recv \;\hat{=}$
　　ANY a,t WHERE $a,b \in AGENT\ \wedge\ t \in chan(a,b)$
　　THEN $chan(a,b) := chan(a,b) - \{t\}$ $\|$ $a1, t1 := a, t$ END

$Fake \;\hat{=}\; skip$

Fig. 1. Authenticated Communications Service

Messages are authenticated in the following sense: agent B can receive a message T indicated to be from agent A only if that message has previously been sent by agent A. This is because $B.Recv$ will only produce output $(A,\ T)$ provided $T \in chan(A, B)$ and $T \in chan(A, B)$ could only be true if $A.Send(B,\ T)$ had previously been invoked. This property will always be preserved by refinement.

The specification also contains an operation called *Fake* which models the attempts by a spy to send fake messages between agents. As can be seen, the specification says that the spy has no effect on the service[1].

3 Refinement

In the first refinement, some auxiliary operations corresponding to the steps of the authentication protocol are introduced. This includes encryption of messages before they are transmitted between agents.

In the second refinement, further auxiliary operations are introduced modelling the ability of the spy to initiate fake runs of the protocol. The spy is also allowed to read and replay any encrypted messages transmitted in the system. By showing that the second refinement is a refinement of the abstract service we show that the protocol does indeed achieve the desired authentication. This is the sense in which the protocol is analysed.

---

[1] Actually, the real specification treats the spy as an agent and says that the worst that the spy can do is send fake messages to itself.

A Tool for Logic Program Refinement
(Extended Abstract)

Robert Colvin, Ian Hayes, Ray Nickson and Paul Strooper

Software Verification Research Centre, School of Information Technology,
University of Queensland, Brisbane, 4072, Australia.

A refinement calculus for logic programs has been developed in [2]. It is based on

- a *wide-spectrum language* that includes both specification and executable constructs (code);
- a *refinement relation* between programs in the wide-spectrum language, which models the notion of correct implementation; and
- a collection of *refinement laws* providing the means to refine specifications to code in a stepwise fashion.

The wide-spectrum language includes assertions and general predicates (specification constructs), types and invariants, as well as a subset that corresponds to Horn clauses (code). The refinement relation is defined so that an implementation must produce the same set of solutions as the specification it refines, but it need do so only when the assertions and invariants hold. There are refinement laws for manipulating assertions and predicates, and for introducing code constructs.

Rigorous application of refinement laws can be tedious and time-consuming. Many refinement laws are conditional, so proof is needed to ensure they are correctly used. As for the imperative refinement calculus, tool support is called for.

This paper describes a tool for the refinement of logic programs, based on the refinement calculus presented in [2]. The tool, REFLP, is interactive, with the user driving the refinement process. REFLP is built on *Ergo* [3], a customisable interactive theorem prover whose basic proof paradigm is window inference [4].

So far, REFLP has been applied to two small examples. Experience with these examples has shown that although REFLP requires more steps than a typical refinement by hand, most of these steps are straightforward and in fact fewer applications of refinement laws are needed. The major advantages of REFLP are that it keeps track of most of the tedious detail involved in refinement, and that it is rigorous so that the results of the refinement are guaranteed within the framework presented. On the other hand, refinement with REFLP requires more steps because even the most trivial and obvious results must be verified.

Another advantage of REFLP is the use of window inference to manage context. In a logic program refinement by hand, assertions are used to represent assumptions about the contexts in which the program will be used. These assertions lead to cluttered program fragments and the application of many laws that distribute and manipulate assertions. REFLP maintains these assumptions automatically as window hypotheses, and allows them to be accessed when needed.

As a concrete comparison, consider the results of refining a procedure which was originally done by hand [2] requiring 13 steps, and using REFLP which required 80 steps. At first sight, the difference in the number of steps appears daunting. It must be noted though that in the 13 steps 'by hand' there are 44 applications of refinement laws. In the refinement by hand, what constitutes a step is something that is decided by the person doing the refinement and typically involves the application of multiple refinement laws, and can involve the refinement of multiple subcomponents. For the refinement with REFLP, however, many steps (31) involved opening and closing windows; these are achieved by a simple click of a mouse button. In fact, most of the REFLP commands are executed using a mouse and are thus less tedious than transcribing the refinement by hand. In addition, 18 steps are required to discharge proof obligations, which are typically straightforward and therefore not done in the refinement by hand. This leaves 31 steps that actually involve the application of a refinement law, which compares favourably with the 44 law applications in the refinement by hand. Part of the difference stems from the fact that in REFLP some of the refinement laws are encoded as window inference rules.

We have already found that the development of the REFLP tool was easier than the development of a similar tool, PRT, for the imperative refinement calculus [1]. REFLP was built on top of *Ergo* in a straightforward manner, whereas for PRT sophisticated mechanisms were needed to support state-dependent reasoning.

References

1. David Carrington, Ian Hayes, Ray Nickson, Geoffrey Watson, and Jim Welsh. A tool for developing correct programs by refinement. In He Jifeng, editor, *Proc. BCS 7th Refinement Workshop, Bath, UK*, Electronic Workshops in Computing, pages 1–17. Springer, 1996. URL http://www.springer.co.uk/eWiC/Workshops/7RW.html.

2. Ian Hayes, Ray Nickson, and Paul Strooper. Refining specifications to logic programs. In J. Gallagher, editor, *Logic Program Synthesis and Transformation. Proceedings of the 6th International Workshop, LOPSTR'96, Stockholm, Sweden, August 1996*, volume 1207 of *Lecture Notes in Computer Science*. Springer Verlag, 1997.

3. Ray Nickson, Owen Traynor, and Mark Utting. Cogito Ergo Sum: Providing structured theorem prover support for specification formalisms. In Kotagiri Ramamohanarao, editor, *Proceedings of the Nineteenth Australasian Computer Science Conference (ACSC'96)*, volume 18(1) of *Australian Computer Science Communications*, pages 149–158, 1996.

4. Peter Robinson and John Staples. Formalizing a hierarchical structure of practical mathematical reasoning. *Journal of Logic and Computation*, 3(1):47–61, 1993.

Towards a Specification of Concurrent Systems in Z Referring to a True Concurrency Semantics
(Extended Abstract)

Jean-Marie Condom and Khadir Ouriachi

Département informatique- Université de Pau et des Pays de l'Adour
Av. de l'Université - B.P. 576 - 64012 PAU UNIVERSITÉ Cedex - FRANCE
Tel : 33 5 59 92 31 54 - Fax : 33 5 59 80 83 74
(condom,ouriachi)@univ-pau.fr

The software currently used in real-time systems becomes so complex that its development must take advantage of formal methods in order to ensure its correctness and to decrease seriously the risks of failure. The success of the Z formal language used in large commercial projects [2] and the significant progress made in its application to concurrent systems [3] lead us to bring our contribution to the specification of reactive systems using Z only, that is without combining it with additionnal real-time formalisms. In particular our approach distinguishes itself from the traditionnal ones by our reference to a true concurrency semantics.

1   An execution model based on a true concurrency semantics

All the work done about the specification of concurrency in Z usually refers to the theory of interleaving. According to it, the system behaviour is viewed as a transition system whose state changes result from the execution of one operation choosen non-deterministically. If that leads to a simpler theory of specification and verification of concurrent systems, some authors notice that this approach is too restrictive compared with opposite approaches based on true concurrency which consider that more than one operations can be executed in parallel. The model of true concurrency we define and use here is a partioned transition system. This expression is borrowed from that of partioned action systems introduced by R.J.R. Back [1] as it has some analogies with it. Let P=p1,...,pn be a partioning of the actions of the systems. Each pi refer to a process and is identified with a set of atomic actions as well as a set of state variables local to the process. A transition linking two successive states is performed by the execution in parallel of one or more actions each one belonging to a different process. Shared variables can be added for the communication between processes.

2   Specifying concurrent systems referring to a true concurrency semantics

The specification of concurrent systems in Z is usually built in two steps : the first step is the independent specification of each concurrent process of the system which represents the static specification ; the second step describes the interaction between processes ; this is the dynamic specification. Considering that operations can be executed in parallel lead us to propose some solutions to the following problems : the execution of parallel actions, the management of conflictual actions, the synchronisation between the different processes and the specification of an execution model based on true concurrency.

2.1　The static specification

Each process is represented by an abstract data type. The actions of the system are specified in Z by operation schemas. In the case where two actions are in conflict, that is they are both enabled but can't be executed in parallel, these actions are merged into one schema realizing the choice between both actions. The specification of parallel actions within one abstract data type involves to split up this data type into several data types.

2.2　The dynamic specification

Considering parallel actions implies that the promotion process, which involves to take into account an overall state, is modified compared with the interleaving case. It mustn't be specified that state variables which don't belong to the promoted process, remain unchanged since these variables can be modified by other(s) process(es) running in parallel. An example of synchronisation by shared variables is specified.

2.3　The specification of the partioned transition system

The transition system is specified in Z by a sequence of states where a state change is performed by one or more enabled operations running in parallel. In particular it is specified that all the variables which are not concerned by a state change remain unchanged.

References

1. R. J. R. Back. Refinement of parallel and reactive programs. In M. Broy, editor, Lecture Notes for the Summer School on Program Design Calculi, pages 73-92, Springer-Verlag, 1993.
2. D. Craigen, S. Gerhart, T. Ralston. An International Survey of Industrial Applications of Formal Methods. NIST GCR 93/626 (Volumes 1 and 2), U.S. National Institute of Standards and Technology, March 1993.
3. A.S. Evans. An Improved Recipe for Specifying reactive Systems in Z. In Z User Workshop, Reading, 1997, April 1997

Specifications and Fault Detection
(Extended Abstract)

Martin de Groot and Ken Robinson

University of New South Wales, Sydney 2052, Australia
martindg@cse.unsw.edu.au, k.robinson@unsw.edu.au

In this paper we explore the relationship between the refinement calculus and fault detection. These may at first seem to be unrelated areas of computer science. However, we hope to demonstrate that they can be characterised in very similar ways using Dijkstra's weakest precondition semantics. Because of this common grounding it is possible to discuss relationships between them. We define some basic terms and provide proofs of their properties. Giving a precise definition of concepts such as *model, observation, confirmation* and *refutation* forms a large part of a theory of fault detection.

Models and observations

High level specifications can be reused as models. Consequently, the following is both a specification of a system which sums two positive numbers, and a model of such a system.

$$z : [x > 0 \land y > 0 \ / \ z = x + y]$$

Specifications can also be used to represent observations. The following is an observation of the system specified above:

$$z : [x = 1 \land y = 2 \ / \ z = 3]$$

Relating observations to models

We define confirmation and refutation, and introduce symbols to represent them. $\preceq$ (pronounced "confirms") will be used for consistent observations, while $\prec|$ (pronounced "refutes") will be used for inconsistent observations.

$$obs \preceq P \ \ == \ \ (\text{pre } obs \Rightarrow \text{pre } P) \land (\text{pre } obs \Rightarrow (\forall \mathbf{w} \bullet \text{post } obs \Rightarrow \text{post } P))$$

$$obs \prec| P \ \ == \ \ (\text{pre } obs \Rightarrow \text{pre } P) \land (\text{pre } obs \Rightarrow (\forall \mathbf{w} \bullet \text{post } obs \Rightarrow \neg \text{post } P))$$

The following diagram illustrates the relationship between refinement and confirmation. (The arrows represent entailment.)

<table>
<tr><td>

[pre P / post P]
$\Downarrow$ $\Uparrow$
[pre Q / post Q]
Refines ($\sqsubseteq$)

</td><td>

[pre obs / post obs]
$\Downarrow$ $\Downarrow$
[pre P / post P]
Confirms ($\preceq$)

</td></tr>
</table>

Fault detection can only be performed if it can be guaranteed that the observations are internally consistent or, that an observation does not both confirm and refute a system model. We need to develop a more robust notion of fault detection than mere refutation. The symbol $\prec\|$ will be used and is pronounced "indicates a fault in".

$$obs \prec\| P \;==\; obs \prec| P \wedge obs \not\preceq P$$

Fault detection is monotonic

While confirmation is not monotonic with respect to refinement, refutation and fault detection are.

$$obs \prec\| P \wedge P \sqsubseteq Q \Rightarrow obs \prec\| Q$$

This relationship of monotonicity for fault detection means that any fault found using an abstract model will also be detected in a more detailed model. This is an important result, as it means that fault detection can be performed with abstract models or high level specifications. As these specifications will be less complex, it should therefore be easier (less computationally expensive) to perform diagnostic tests with these models.

Probe points

The only way that we can take advantage of data refinement to give more accurate fault detection is if access points into the new data structures are added. These access points are called *probe points*. It is perfectly safe to introduce probe points at any stage provided the variable chosen does not occur free in the specification.

$$\mathbf{w} : [pre \; / \; post] \sqsubseteq \mathbf{w} : [pre' \; / \; post'] \Rightarrow \mathbf{w}, x : [pre \; / \; post] \sqsubseteq \mathbf{w}, x : [pre' \; / \; post']$$

Where x is not free in *pre* or *post*.

The issue of adding probe points is a result of the specification formalism. For example, a Z schema does not implicitly constrain variables. In this case adding probe points would represent a refinement of the specification. So there would be no need to modify the original specification to build an implementation suitable for fault monitoring.

Adding value to formal developments

Perhaps the most important aspect of our proposal is that it adds value to the process of program derivation. By interpreting specifications as models, it becomes possible to utilise specifications after the system has been implemented for purposes other than continuing to guarantee program correctness. We have shown how they can be used for fault monitoring, but there remains the possibility that other areas of system monitoring may have some use for these models. This should in turn make it more worthwhile to go through the process of formal system development.

Expressiveness and Parsimony in Z
(Extended Abstract)

David Edmond

School of Information Systems,
Queensland University of Technology,
GPO Box 2434, Brisbane, Queensland 4001, Australia
davee@icis.qut.edu.au

There are a number of commonly used Z operations that may be paired off. Such pairings include (1) the domain and range operations, (2) domain and range restriction, and (3) domain and range anti-restriction. Not only are the operations within each pair very closely related in terms of what they do, they also give rise to a number of very similar laws. Ockham's razor (or the Law of Parsimony) suggests that we should try to avoid unnecessary duplication. This may be achieved in the way shown in the paper. Furthermore, we can use this approach to make specifications more immediately readable.

Z is a specification language based on typed set theory and predicate calculus. A set, in Z terms, is a *homogeneous* collection. The notation also caters for *heterogeneous* collections, and there are two ways of constructing or specifying these – tuples and bindings (or records). A tuple is an ordered collection of possibly different things, and a record is a labelled collection. For historical reasons, tuples are the conventional mathematical expression of a heterogeneous collection. Hence, there is a bias in Z towards the use of tuples.

The concept of a *binary relation* is of central importance, allowing us to collect and state facts concerning entities of interest within the application domain. A relation is represented as a set of ordered pairs. The crucial role played by the relation is evidenced by the number of operations associated with it. These are, briefly: domain, range, composition, restriction, anti-restriction, inversion, image, and closure

This paper argues that these may be reduced and simplified. Moreover, their removal or amalgamation may lead to less cryptic specifications, and a reduction in the number of associated laws.

Heterogeneous objects: A heterogeneous object is a composite object, the components of which are either (1) of *different kinds*, or (2) of the same kind but playing *different roles*. There are two ways of defining such an object. The conventional *mathematical* approach is to use the Cartesian product operator. The conventional *computing* approach is to label the components involved. There is no simple way in Z of constructing such an object, that is, no way corresponding to the use of round brackets enclosing an ordered list to construct a tuple.

Relationships: A relationship is a set of heterogeneous objects. State observations are very commonly expressed as relationships (relations, functions, etc.). The domain and range of a relationship constitute two distinct names for what is, essentially, a single concept – that of projecting the objects that participate in the relationship. Specifications written in Z typically make extensive use of the "dom" and "ran" operators to describe the ways in which state observations, in the form of relationships, overlap, or do not overlap, one another.

There are two difficulties with the use of the domain and range operators:

1. What is, essentially, the same kind of operation is given two different names, depending upon whether it is being used on the "left-hand side" of the relationship in which case the "dom" version is used, or whether it is being used on the "right-hand side" in which case, the "ran" version is used.
2. In neither the "dom" nor the "ran" usage do we get a clear and immediate idea or picture of what is implied.

In Spivey's Z Reference Manual, ten laws are given for the domain and range operators. They are, effectively, dual laws. For each law regarding use of the domain operator, there is a corresponding law covering use of the range operator. By recognising that the two operators are simply two forms of projection, these ten laws may be reduced to five. Other operations, such as domain and range restriction and anti-restriction also give rise to laws which may be reduced in a similar manner.

In this paper, it is suggested that the two ways of defining heterogeneous collections in Z have not been fully resolved. There is a bias towards the conventional Cartesian product and tuples. This leads to an unnecessary proliferation in the number of laws concerning operations that are forms of projection. Ockham's razor (or the Law of Parsimony) suggests that we should try to avoid unnecessary duplication. This may be achieved in the way shown. Furthermore, by paying particular attention to the way that we label the constituents of a composite object, we can make specifications more immediately readable and less susceptible to errors of interpretation.

This paper should *not* be seen as criticism of Z, merely an invitation to consider some questions regarding the lack of uniformity in the treatment of composite or heterogeneous objects.

Reverse Petri Net Technology Transfer:
On the Boundary of Theory and Applications
(Extended Abstract)

Hartmut Ehrig, Maike Gajewsky, Sabine Lembke and Julia Padberg

Technical University of Berlin[**]
ehrig, magda, lembkes, padberg @cs.tu-berlin.de

The housing management system project WIS of the company LION is one of the most successful practical application of Petri net technology in the area of cooperating and distributed business process management (see also e.g. [OSS94, GGK96]). This system is based on the workflow management environment LEU and the FUNSOFT approach [Gru91]. FUNSOFT nets are very powerful in practice, but they are not well understood from a theoretical point of view.

Our work aims to improve this situation not by Petri net technology transfer from theory to practice but by reverse transfer from practice to theory. We consider the main features used explicitly in FUNSOFT or implicitly in LEU and try to abstract them in a theoretically clean and comprehensive way leading to a generic concept of high-level nets. The features are transfered to theory in order to describe, and examine them independently of any system implementation in a similar way to abstract data types. The formalization of these features is the basis for further examination of structuring, rule-based refinement, analysis etc., and the relation between them. In fact, both, the explicit data model of FUNSOFTand the explicit organizational model used in LEUare not essential for the meaning of FUNSOFT. Our new concept of parameterized FUNSOFT nets, called PARFUN, is generic with respect to the data and organizational model and includes FUNSOFT features like guarded transitions, copy arcs and job execution.

It is important for practical applications to consider other techniques for the data as well as the organizational model. However, it is difficult to exchange the modelling techniques, as they are tightly integrated into FUNSOFT and LEU. A flexible use of developing systems like LEU can be obtained, if the data and organizational models are parameterized. This is one of the main reasons to develop a parameterized abstraction of FUNSOFT. With PARFUN we want to give a framework that supports the exchange of modelling techniques and the modification of model integration. Defining a suitable data and organizational interface, the change from one modelling technique to another is reduced to the change in the implementation of the model, where the interface is unchanged. Figure 1 depicts this approach as realized in PARFUN.

Specific features of FUNSOFT nets, like place fusion and transition substitution, should not be used explicitly in the concept of a net but be superimposed

[**] This work is part of the joint research project between H. Weber (Coordinator), H. Ehrig (both from the Technical University Berlin) and W. Reisig (Humboldt University of Berlin), supported by the Deutsche Forschungsgemeinschaft (DFG).

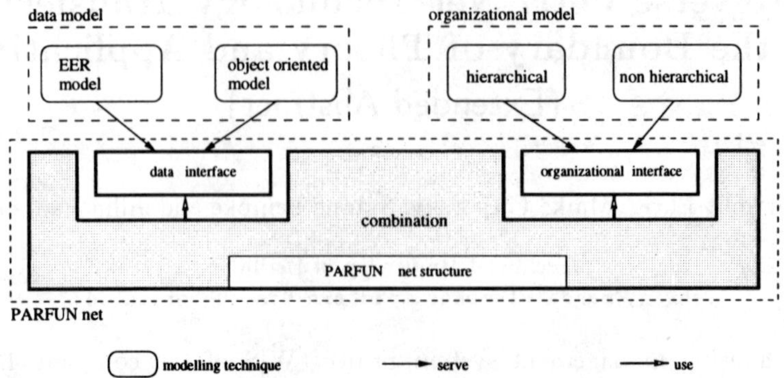

Fig. 1. ParFun: A framework for integrating different workflow models

constructions on net classes. For this reason our ParFun approach includes general notions of fusion, union, and rule-based refinement defined as structuring and refinement techniques for ParFun nets. These techniques have been studied for algebraic high-level nets [PER95] and for abstract Petri nets, a categorical approach unifying different kinds of low-level and high-level net classes [Pad96].

We define structuring and refinement concepts of ParFun nets which are essential for all kinds of practical applications. Our main technical results show under which conditions fusion and union of ParFun nets are compatible with rule-based refinement. We extend ParFun nets by access strategies and discuss briefly how the remaining concepts of FunSoft like transition invocation, time and flexible arcs might be handled by extensions of the ParFun approach. Future work comprises the formulation of these concepts in order to reach the full range of applications, that are possible for FunSoft.

On the other hand our parameterization concept and the general notions of fusion, union and rule-based refinement within the ParFun approach are going beyond the modelling power of FunSoft and our main results are important from a theoretical as well as a practical point of view. In this sense ParFun is really on the boundary of theory and applications.

References

[GGK96] G. Graw, V. Gruhn, and H. Krumm. Support of Cooperating and Distributed Business Processes. In *Proc. Int. Conf. on Parallel and Distributed Systems*, 1996. Tokyo.

[Gru91] V. Gruhn. *Validation and Verification of Software Process Models.* PhD thesis, Universität Dortmund, Abteilung Informatik, 1991.

[OSS94] A. Oberweis, G. Scherrer, and W. Stucky. INCOME/STAR: Methodology and Tools for the Development of Distributed Information Systems. *Information Systems*, 19(8):643–660, 1994.

[Pad96] J. Padberg. *Abstract Petri Nets: A Uniform Approach and Rule-Based Refinement.* PhD thesis, Technical University Berlin, 1996. Shaker Verlag,.

[PER95] J. Padberg, H. Ehrig, and L. Ribeiro. Algebraic high-level net transformation systems. *Mathematical Structures in Computer Science*, 5:217–256, 1995.

A Semantic Approach to the Solution of the Legacy Code Problem
(Extended Abstract)

Alex de V. Garcia, Edward Hermann Haeusler and Armando M. Haeberer

Laboratório de Métodos Formais
Pontifícia Universidade Católica
Marquês de São Vicente 225,Gávea, Rio de Janeiro - Brazil
{alex,hermann,armando}@lmf-di.puc-rio.br

Abstract

There are many reports on automatic translation of legacy systems to new platforms. One way of translating systems to a new platform is through program transformations. Such transformations translate a program in a **source language** — L_s — to a program written in a **target language** — L_t.

This paper reports the implementation of a semantic translating system — STS — that translates any language into a fixed object oriented language. The STS approach is based on a semantic specification of the language in which the legacy code is written. The denotation function is applied to the source code in order to produce a functional specification of the program. This specification is then translated into the object oriented language.

The figure below illustrates how the usual transformational solution translates code from L_s to L_t:

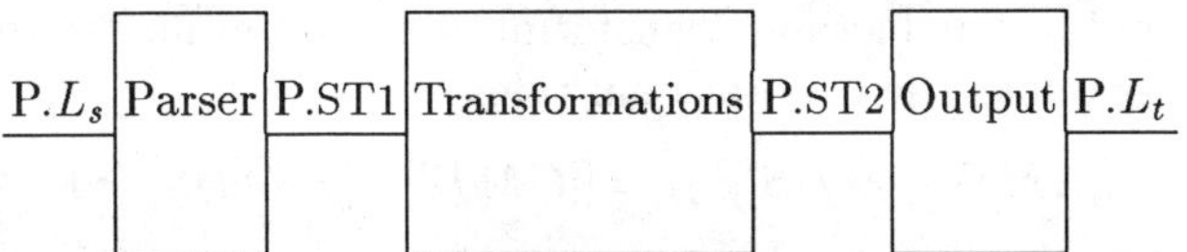

Where P.L_s is the input program written in L_s and P.ST1 its syntax tree; P.L_t is the input program written in L_t and P.ST2 is its syntax tree. In this solution the transformations provide the meaning of constructions of L_s by translating then to L_t. Hence, the semantics of the target program is given in an *ad hoc* basis, i.e., the meaning of a construction is coded in a programming language.

The approach proposed here consists of using $[[\ ]]_{sl}$, the denotational specification of L_s, to extract a functional specification of the source code. This specification is then translated to an object oriented language to be executed. In this approach, the user of the system writes no transformations, they are automatically generated from L_s denotational specification. In STS the semantics of the target program is given in a formal basis. While writing a denotational specification, one focuses on specifying the meaning of language constructions, being less concerned with flow control and data manipulation than in the case of writing transformations. The figure bellow illustrates the transformation from a specific source language to C++ with STS. All the language dependent steps (namely: the parser and the evaluation of [[ ]]) are generated automatically from L_s denotational specification. This "metatransformation" process is not illustrated in the figure.

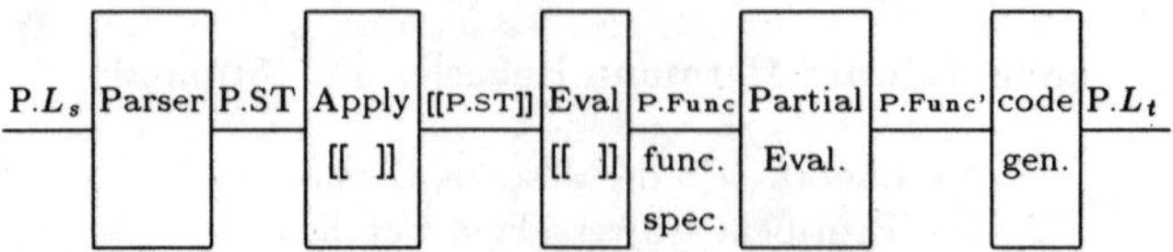

- Apply [[ ]]. Formally applies the denotation function.
- [[P.ST]]. Denotation function applied to its argument, but not yet evaluated.
- Evaluate [[ ]]. Transformations that evaluate the denotation function. These transformations are generated automatically from the semantic specification.
- P.func. Functional specification of the program.
- Partial evaluation. Applies reductions to the functional specification, compile time expressions can be solved in this phase.
- P.func'. Functional specification of the program after partial evaluation.
- Code generation. Translates the functional specification into a fixed object oriented target language. Primitive domains are implemented as classes. Class templates implement the domain constructing operators ($+$, $\times$ and $\longrightarrow$). Those templates offer operations such as pair construction and projection as built-in methods.

Each of these steps handle program code, so they are suited to the transformational paradigm. Automatic generation of the language-dependent steps is also achieved by means of transformations. Automatic generation of transformations that evaluate [[ ]] is an important step in STS. A semantic rule in L_s specification is translated to a reduction (transformation) that evaluates the denotation function. For example, the denotational rule:

$$[[CMD; CMDS]](\gamma) = [[CMD]] \circ [[CMDS]](\gamma)$$

is translated into a transformation rule:

$$[[CMD; CMDS]](\gamma) \, \triangleright \, [[CMD]] \circ [[CMDS]](\gamma)$$

There is a growing need for migrating legacy systems due to the fast development of new platforms. A shortcoming of many of the existing solutions is that they lack a formal basis to assure the correctness of the target code. STS deals with this problem in such a way that the user of the system does not specify the translation by writing imperative programs or by writing transformations, but only the denotational specification of the source language. The transformations are automatically generated by the system and are correct by construction. It has been shown that once the correctness of the metatransformations and of code generation are proven, then all translations performed by STS are correct, while in the usual transformational solution it is necessary a correctness proof for each L_s.

An Editor for Refinement Structure Diagrams
(Extended Abstract)

Lindsay Groves and Ivan Tuckwell

School of Mathematical and Computing Systems,
Victoria University of Wellington,
PO Box 600, Wellington, New Zealand
{lindsay, ivan}@mcs.vuw.ac.nz

The refinement calculus [4] is usually presented as a formalisation of the stepwise refinement approach to programming, and most discussions of program refinement assume (often implicitly) a methodology whereby programs are developed in a strictly top-down fashion. The programmer is assumed to begin with a specification and gradually turn this into a program via a sequence of transformations, discharging any proof obligations required to ensure that correctness is preserved.

While top-down design is accepted as a good ideal to strive for and a good way to record program developments for didactic purposes, it does not reflect the way in which programmers actually construct programs [6]. In practice, programmers explore many alternative design options and switch between high-level and low-level concerns, often looking a long way ahead to anticipate details of data structure implementations or to predict the impact of possible design decisions. When these predictions are imperfect, the programmer will often go back and change decisions made earlier in the design and adjust existing parts of the program accordingly. The fallacy of top-down design becomes even more apparent when we consider software maintenance and reuse, and interactions within programmer teams.

The use of formal development methods, such as the refinement calculus, might help to make the design process more systematic, for example eliminating some of the dead-ends leading to incorrect programs. But we should still expect programmers to look ahead to consider ease, efficiency and even feasibility of later implementation, and to change their minds and try different approaches when things don't work out as expected.

In formal program development, we also need to consider the interaction between the traditional decisions about program structure and those concerned with formalisation. For example, in constructing a program with a loop, we might write the loop first and then look for a loop invariant and show that it is preserved; or we might choose the loop invariant first and then derive a loop body that will preserve it. While there may be good reasons for advocating the latter approach (e.g. [1]), it is simply perverse to insist that this is the only acceptable approach to follow.

When program refinement is done with pencil and paper, the precise order in which steps are performed is relatively unimportant. In trying to decide between two design options (e.g. two possible loop invariant), we can sketch out the details of each option on a separate piece of paper and discard that one that is rejected. When we design tools to support program development, however, we need to consider the precise steps required and the order in which they can or must be performed. The decisions we make will have a profound impact on the ease with which the tool can be used (and the difficulty of implementing it). Most existing refinement tools are based on the top-down model, and thus require the programmer to work in a fairly rigid top-down manner, with limited support for the kinds of development patterns outlined above.

In this paper, we discuss the design of a refinement tool, based on Refinement Structure Diagrams [2], in which we attempt to support a mixture of formal and informal development in a flexible manner. Refinement Structure Diagrams are much like the usual structure diagrams (e.g. [3]) with formal specifications attached to the boxes. Subcomponents of a Refinement Structure Diagram can then be seen as representing the application of refinement rules in a formal development.

Most of the flexibility in the operation of this tool is due to the use of metavariables [5]. When a component of a diagram is expanded or transformed (i.e. when a refinement rule is applied), the parameters of the rule (e.g. loop invariants, guards, etc.) are left unbound, and are bound explicitly when required. By binding these metavariables as refinement steps are performed, we obtain a strictly formal top-down development. By leaving them all unbound, we obtain a completely informal development. We can also construct a complete (informal) structure diagram, and provide formal specifications and proofs later. To facilitate this, we use more general forms of refinement rules than those found, for example, in [4].

The tool allows structure diagrams to be edited in a fairly conventional way, for example, adding extra statements in a sequential decomposition or adding an extra branch to an **if** or **do** statement. These editing operations modify the refinement rules used, and modified proof obligations can be generated when required.

References

1. E. W. Dijkstra. *A Discipline of Programming*. Academic Press, 1976.
2. Lindsay Groves. Formal program derivation with structure diagrams. In *Proceedings of the Second New Zealand Formal Program Development Colloquium*. Massey University, February 1996.
3. M. A. Jackson. *Priciples of Program Design*. Academic Press, 1975.
4. Carroll Morgan. *Programming from Specifications*. Prentice-Hall, Second Edition, 1994.
5. Raymond G. Nickson and Lindsay J. Groves. "Metavariables and Conditional Refinements in the Refinement Calculus". *Proc. 6th Refinement Workshop*, Springer-Verlag, 1994, pp167–187.
6. Robert Worden. "The Process of Refinement". *Proc. 4th Refinement Workshop*, J. M. Morris and R.C. Shaw (Eds.), Springer Verlag, 1991, pp 1–5.

Transaction Decomposition:
Refinement of Timing Constraints
(Extended Abstract)

Jan Haveman

BAe Dependable Computing Systems Centre,
University of Newcastle upon Tyne

This paper describes a method to rigorously decompose end-to-end timing properties expressed in a real-time transaction model. This model aims to provide a systematic way to develop distributed real-time systems, from the first specification of the functionality and timing properties of its tasks through to an implementation by stepwise refinement of these tasks.

To enable this rigour the model, unlike existing transaction models that only allow the specification of timing constraints on transaction invocation (minimum inter-activation time) and completion (deadline or maximum response time), also allows timing constraints on individual transaction inputs and outputs. These constraints take the form of time windows relative to the transaction activation time in which a particular input must be read or an output written. We introduce a simple formalism for reasoning with time windows, and apply it to a number of examples of transaction refinement.

Real-time Transactions. A transaction definition consists of the specification of its inputs, outputs, variable access, functionality, and control (activation, guard and termination). Three types of transaction are distinguished according to the way they are activated: periodic (once every fixed period), sporadic (on arrival of a data stimulus) and conditional (when a specified condition becomes true).

Transaction timing properties are usually defined in terms of the control information only: a time T_A related to the transaction's activation (its period or minimum inter-activation time) and a time T_D related to its termination (its deadline). These properties can be formalised using a logic that deals with explicit time. Using the logic RTL the execution of a transaction Tr can be modelled as the occurrence of an action with start event $\uparrow Tr$ and end event $\downarrow Tr$. The constraint that a transaction, once activated, should terminate within T_D time units is expressed in terms of the times at which these start and end events occur.

The timing properties defined for transaction inputs and outputs can be either general time window constraints or specialised versions of them: deadline, offset, periodic or jitter constraints.

Time Window Constraints. A window constraint $W = \langle E_s, E_c, l, u \rangle$ (with $l \leq u$) prescribes that an event E_c should occur within a time interval $l \leq t \leq u$ relative to another event E_s (the 'timer start' event). Expressed in RTL this is:

$$\forall i, t \cdot \Theta(E_s, i, t) \Rightarrow \exists j, t' \cdot \Theta(E_c, j, t') \wedge l \leq t' - t \leq u$$

If the window constraint is used in the definition of a transaction Tr, the event E_s represents the transaction's activation $\uparrow Tr$, and the event E_c represents the event of reading the input or writing the output for which the window constraint was defined.

It is possible to compose time windows as $\langle E_0, E_1, l, u \rangle + \langle E_1, E_2, p, q \rangle = \langle E_0, E_2, l+p, u+q \rangle$, and windows that share their start event can be related in a number of ways (for instance to specify that the first window's constrained event may occur before the second window's constrained event, or to specify that they should occur within a specified time of each other).

Transaction Decomposition and Timing Analysis. As illustrated in Fig. 1, a transaction can be decomposed into subtransactions that operate together to satisfy its functional and timing requirements. The figure shows a conditional transaction Tc and its decomposition into a periodic transaction Tp followed sequentially by a sporadic transaction Ts. Tc's activation condition $A(in, m)$ has become Tp's execution guard, which is checked periodically every T time units. The output windows $W_{1,2}$ have to be chosen such that they satisfy Tc's timing constraints. We discuss the constraints on the timing properties of the subtransactions for a number of example decompositions.

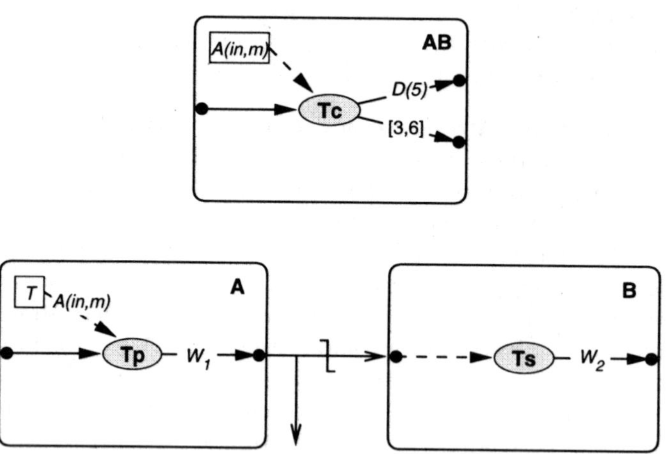

Fig. 1. Example decomposition of a transaction

As an example of how one can conduct analysis of more general timing properties in this framework, we analyse the situation where a subsystem receives data from two different sources. All three subsystems are operating periodically, but at different frequencies. The two data items that the receiving subsystem obtains in any given period can originate from a number of possible periods of the two source subsystems. We show how timing constraints on transaction inputs and outputs allow us to constrain the difference in freshness of the two inputs.

Intensional Z
(Extended Abstract)

Martin C. Henson and Steve Reeves

Department of Computer Science
University of Essex

Department of Computer Science
University of Waikato

Z as usually understood has its semantics given using classical logic and extensional set theory. What we aim to do here is to start to explore the idea of making other choices as a foundation for Z. In particular we choose to use a constructive logic and an intensional set theory.

We do this for several reasons. We believe we will get a simpler explication of Z's semantics than the standard one because, amongst other reasons, we will be returning to the original and, for a computer scientist, more natural conception of a function as an intensional object; an object that describes a computation rather than an object which is an infinite set of tuples. Sets are then intensional (informally because they are, roughly, the ranges of these intensional functions).

Our next reason is that this basis is less mathematically extravagant in the sense (related to the point above) that objects like function spaces and power sets, for example, are now also expressible at this one level, produced by functions, rather than being picked out of some large hierarchy.

Given this basis we also have open to us in a natural and simple way the notion of providing proofs about such intensional objects with computational content. Thus, as we will see in the full paper, our semantics for Z extends without much extra work to a logic for developing programs from Z specifications.

Finally, we have to say that the description of Z given in many textbooks and even in the standard is riven with obscurities and informalities; it seems a reasonable goal to describe Z in a clear and coherent fashion.

In the full paper we give the theory of types and operations, the interpretation of standard Z within that theory and the term assignment theory. Here, however, we give just a taste of the first two stages.

We introduce a constructive theory of intensional types and operations called $\mathcal{T}$. This theory is a generalisation (to permit comprehension types) of Beeson's *EON*.

The *terms* are those of combinatory algebra together with constants for the formation of natural numbers and labels for the eventual interpretation of schema types:

$$t \rightarrow x \mid l \mid 0 \mid succ \mid S \mid K \mid t\,t$$

The *types* are the natural numbers and comprehensions:

$$T \rightarrow N \mid \{x \mid \phi(x)\}$$

The *well-formed formulæ* are:

$$\phi \;\to\; t \in T \mid t = t \mid t\!\downarrow \mid \bot \mid \phi \wedge \phi \mid \phi \vee \phi \mid \phi \supset \phi \mid (\forall x)\phi \mid (\exists x)\phi$$

Negation is defined, as usual by: $\neg\phi = \phi \supset \bot$.

The *judgements* of the theory are sequents of the form:

$$\Gamma \vdash \phi$$

with Γ a *set* of assumptions $\cdots \phi_i \cdots$

The theory is presented as a system of natural deduction in sequent form. We have the standard assumption and weakening rules to allow changes to the context together with a set of rules that tell us what the well-formed formulae are. We provide a small selection of these below.

$$\frac{}{\Gamma \vdash t = t} \qquad \frac{\Gamma \vdash t_0 = t_1}{\Gamma \vdash t_1 = t_0} \qquad \frac{\Gamma \vdash t_0 = t_1 \qquad \Gamma \vdash \phi(t_0)}{\Gamma \vdash \phi(t_1)}$$

$$\frac{\Gamma \vdash (t_0\ t_1)\!\downarrow}{\Gamma \vdash t_0\!\downarrow} \qquad \frac{\Gamma \vdash (t_0\ t_1)\!\downarrow}{\Gamma \vdash t_1\!\downarrow} \qquad \frac{\Gamma \vdash t \in T}{\Gamma \vdash t\!\downarrow}$$

$$\frac{}{\Gamma \vdash 0\!\downarrow} \qquad \frac{\Gamma \vdash n\!\downarrow}{\Gamma \vdash (succ\ n)\!\downarrow} \qquad \frac{}{\Gamma \vdash 0 \in N} \qquad \frac{\Gamma \vdash n \in N}{\Gamma \vdash succ\ n \in N}$$

We also have standard rules for predicate logic (slightly amended so that we can be sure that any terms we introduce in the universal elimination rule and existential introduction rule are defined). We are now in a position to give a flavour of the translation from Z into $\mathcal{T}$.

$$[\![N]\!] = N$$
$$[\![A \to B]\!] = \{f \mid (\forall x \in [\![A]\!])(fx \in [\![B]\!])\}$$
$$[\![\cdots l_i : T_i \cdots]\!] = \{f \mid (\forall x \in \{\cdots l_i \cdots\})(fx \in B(x))\}$$
$$[\![Bool]\!] = \{0, succ\ 0\}$$

where the $B(x)$ in the schema types clause is explained thus. Let I be the indexing set, i.e. the comprehension $\{\cdots l_i \cdots\}$ then $B(x)$ will be: $\{z \mid x \in I \wedge \cdots \wedge x = l_i \supset z \in T_i \wedge \cdots\}$ where, as you can see, the variable x is free as required.

Given this we have that powerset $P(T)$ is just shorthand for $A \to Bool$. And then continuing in the same vein we have schema themselves:

$$[\![[\delta \mid \phi]]\!] = \{x \in [\![\delta]\!] \mid \phi[l_i \leftarrow x\ l_i]\}$$

where δ is a declaration.

Refining Reactive Systems Using the Refinement Calculator
(Extended Abstract)

Thomas Långbacka *

Department of Computer Science, University of Helsinki
http://www.cs.Helsinki.FI/~tlangbac/

A *reactive system* is a program that interacts with an *environment* and reacts to certain types of events. Typically reactive systems are modelled as non-terminating programs. The environment is often described in the programming notation used. An operating system is an example of a reactive system. Its job is to react to commands from its users. Although an operating system may have been terminated, it still has a state that lives on, e.g. the file system.

The Refinement Calculator [1] (references in this extended abstract are to be seen as sources of both first hand information as well as further references) is a tool that allows the user to develop formally correct programs in a stepwise manner (underlying formalism: the refinement calculus). Refinement steps are verified using the HOL theorem prover. Currently it supports refinement of sequential (and terminating) programs.

In the following we show how the Refinement Calculator can be extended to deal also with reactive systems. The formal framework supported is *action systems* [2]. One could use following (ad hoc) syntax for action systems:

$$\begin{aligned}
&var\ v \\
&ini\ A_0 \\
&act\ A_1 [\!] A_2 [\!] \ldots [\!] A_m \\
&fin\ A_F
\end{aligned}$$

where v is the list of *local variables*, A_0 is the *initialisation* predicate, A_F is the *finalisation* predicate and $A_1, \ldots, A_m$ are *actions*. Each action is a *guarded command* of the form $g \to S$ where S is a statement (as in [1]) and g is an enabling predicate (*guard*). Informally an action system behaves as follows

1. The local variables are assigned values satisfying the initialisation predicate.
2. Enabled actions are chosen for execution non-deterministically over and over.
3. If the finalisation predicate evaluates to true (at any point in the computation) then the action system *may* terminate.

Since reactive programs may run forever, action systems are given *trace* semantics. Refinement of action systems can be dealt with in this semantic framework by defining a *trace refinement* relation. However, it is more convenient to use another relation — *simulation refinement* (written $\preceq$) — for our purposes. This relation is very closely related to *data refinement* [1]. Basically this relation is

---

* **Second affiliation:** Turku Centre for Computer Science – TUCS

defined using three separate conditions that have to be proven for the relation to hold. These conditions relate the initialisation predicates, finalisation predicates and the actions to each other in a certain way. Given certain additional conditions we have a relation, which is stronger than trace refinement.

Reactive programs are executed in an *environment*. For action systems, this is modelled using parallel composition (written $\|$). Given an action system $\mathcal{A}$, executing it in the environment $\mathcal{E}$ means executing the action system $\mathcal{A}\|\mathcal{E}$. In practise the composed action system is created by combining the initialisation and finalisation predicates using conjunction and by appending together the individual action lists. The local state of the parallel composition is the combination of the two local states. The following compositionality rule can be used to refine individual action systems in a parallel context:

$$\frac{\vdash \mathcal{A} \preceq \mathcal{C}}{\vdash \mathcal{A}\|\mathcal{E} \preceq \mathcal{C}\|\mathcal{E}} \tag{1}$$

We have formalised action systems in HOL [2]. In short, this is achieved by defining three constants in HOL:

1. A parallel operator for action systems (action systems are defined only by means of their type)
2. An operator to represent non-deterministic choice of executable actions
3. The simulation relation between action systems.

Furthermore the rule in equation (1) has to be proven. In order to use the HOL formalisation in the Refinement Calculator some more things need to be done:

- Add *window rules* [1] that allow for example the compositionality rule to be used to focus on individual action systems composed in parallel with other action systems, or others that allow focusing on individual actions (and then "data refine" them). Also the reflexivity and transitivity of the simulation relation has to be proven in HOL.
- The Calculator syntax [1] has to be extended to deal with a set of added surface syntax elements (as proposed above).
- Additional convenience rules have to be added, e.g. for reordering action systems composed in parallel.

Once these things have been done one can use the Refinement Calculator also for reasoning about reactive systems. Note that — although the simulation relation is used on the top level — most of the work will actually be carried out on the level of data refinement (possibly with additional side conditions), i.e. using machinery (see [1]) that already exists in the current version of the Calculator.

References

1. M. Butler, J. Grundy, T. Långbacka, R. Rukšėnas, and J. von Wright. The refinement calculator: Proof support for program refinement. In these proceedings.
2. T. Långbacka and J. von Wright. Refining reactive systems in HOL using action systems. TUCS Technical Report 102, Turku Centre for Computer Science, 1997. Available from the following URL: `http://www.tucs.abo.fi/`.

π-Calculus Models of Synchronous and Asynchronous Message Passing
(Extended Abstract)

J. M. Potter[1], J. J. Zic[2], and A. Yun-Wen Liu[2]

[1] Microsoft Research Institute, MPCE,
Macquarie University, Sydney
`potter@mri.mq.edu.au`
[2] Department of Software Engineering,
University of New South Wales
`johnz,annaliu@cse.unsw.edu.au`

We consider extensibility and re-use of π-calculus models [MPW92], and the extent to which existing behavioural equivalences for the π-calculus [San93] are adequate. We discuss extending specifications so that models conform to lower level protocols. This permits more flexible composition of the models, with switchable protocols, all of which can be modeled in the π-calculus.

In this paper, we model a channel as a simple one place buffer, then show how the definition of the buffer may be extended to allow channel connection and disconnection.

```
agent C(in,out) = in(x).'out<x>.C<in,out>
```

A switchable channel accepts a connection request, and hands back the input and output port names which may then be used for communication until a disconnection request is received.

```
agent SWC(conn) = conn(reply).(^in,out,disc)
                    'reply <in,out,disc>.C1<in,out,conn,disc>
agent C1(in,out,conn,disc) = in (x).'out <x>.C1<in,out,conn,disc>
                    + disc().SWC<conn>
```

The agent `SWC` represents a switchable channel which on a request to the `conn` port offers its communication port names `in`, `out`, `disc`, fresh for each connection, on the `reply` port given to it in the request. Communication occurs via the `in` and `'out` ports, until a disconnect (`disc`). The switchable channel then disables further use of the communication port names used for the existing connection, and is ready for another connection request.

It is clear that `C1` is in some sense derived from `C`. In fact, `C1` is just `C` extended with a choice allowing for disconnection. We could not re-use `C` in the definition of `C1` because the recursion behaviour needed to be based on the new process. This suggests two options to get reusable specification components: either replace recursive occurrences of the process by a variable which is bound in the defining context, or adopt a convention of self-reference as in object-oriented models, which allows the recursive call to adapt to the new defining context.

Within the context of the definition of SWC, we want the self-reference to carry through to its recursive use in the definition of C1 without causing clashes with the self-reference of C1. The notion of defining context should be formalised.

Combining existing agents with standard components is another aspect of extensibility. For example, the way the communication channel was made switchable is generic. We could substitute other channel specifications where C occurred and achieve the same switching behaviour. Furthermore we would like to identify a standard testing for which the communication behaviour should be the same as if there were no switching at all.

The motivation for the next examples came from attempting to model prioritised reception of data from two different channels, either of which may be synchronous (SC) or asynchronous (ASC). The lack of data on the higher priority channel needs to be ascertained before accepting data from the lower priority channel. We elected to capture this through a demand driven protocol. The point of the exercise is to demonstrate the same protocol being applied to two different types of channels, so that they remain interchangeable, at least as far as the demand protocol goes. The receiver issues **ready** requests then receives a null or a datum in response.

```
agent SC  = readyout.'readyin.SC1
agent SC1 = nullin.'nullout.SC + in(x).'out<x>.SC

agent ASC  = 'readyin.(^r)(readyout.'r.0 | ASC1)
agent ASC1 = nullin.r.'nullout.ASC + in(x).r.'out<x>.ASC
```

When we compose either of these channels with appropriate controllers, they can both be shown to be weakly bisimilar to the simple C above. In other words we have two different types of channel component which have identical behaviours in a given context. That is, they satisfy the same specification, but also conform to a demand driven protocol. This notion of satisfying a simple higher level specification and conforming to a lower level protocol is also worthy of a precise formulation.

References

[MPW92]   R. Milner, J. Parrow, and D. Walker. A Calculus of Mobile Processes, Parts 1–2. *Information and Computation*, 100(1):1–77, 1992.

[San93]   D. Sangiorgi. A theory of bisimulation for the π-calculus. *Proceedings CONCUR'93*. Lecture Notes in Computer Science, 715:127–142, 1993.

Specifying a Collaborative Editor
(Extended Abstract)

Steve Reeves

Department of Computer Science
University of Waikato
Private Bag 3105
Hamilton
NEW ZEALAND
stever@waikato.ac.nz

The aim of this paper is to illustrate how formal specifications for collaborative interactive systems might be written. Specifically we choose to specify a collaborative editor. The full specification of the system follows Hussey and Carrington's adaptation of Foley and van Dam's work, and has three parts: the semantics of the system; the syntax of the system; the semantics of the awareness aspects of the system. The third aspect of the specification is required in order to say what awareness users should have of each other and of the system. This is a new requirement that comes from the fact that we are dealing with collaborative systems in which users need to know about one another.

We first have to specify the basis for our editor, which is the paragraph. This is done in the full paper, though here we omit its specification. It fairly standardly models that state of a paragraph as a sequence of characters and allow operations like insertion and deletion of text. The specification is then used in the main specification for Editor, some of which we give here:

$$
\begin{array}{|l}
\hline\ Editor \\\hline
\quad
\begin{array}{l}
users : \mathbb{F}\ People \\
paras : \mathrm{seq}\ Paragraphs \\
locked\_by : Paragraph \nrightarrow People
\end{array} \\\hline
\quad
\begin{array}{l}
\mathrm{dom}\ locked\_by \subseteq \mathrm{ran}\ paras \\
\mathrm{ran}\ locked\_by \subseteq users
\end{array} \\\hline
\quad
\begin{array}{|l}
\hline\ INIT \\\hline
\quad locked\_by = \varnothing \\\hline
\end{array} \\\hline
\end{array}
$$

$$
\begin{array}{|l}
\hline\ NewUser \\\hline
\quad
\begin{array}{l}
\Delta(users) \\
user? : People
\end{array} \\\hline
\quad
\begin{array}{l}
user? \notin users \\
users' = users \cup \{user?\}
\end{array} \\\hline
\end{array}
\qquad
\begin{array}{|l}
\hline\ Leave \\\hline
\quad
\begin{array}{l}
\Delta(users) \\
user? : People
\end{array} \\\hline
\quad users' = users - \{user?\} \\\hline
\end{array}
$$

$$
\begin{array}{|l}
\hline
\text{\_\_} Take \text{\_\_\_\_\_\_\_\_\_\_\_\_\_\_\_\_\_\_\_\_\_\_\_\_} \\
\Delta(locked\_by) \\
user? : People \\
para? : Paragraph \\
\hline
para? \notin (\mathrm{ran}\, paras) - \\
\qquad\qquad (\mathrm{dom}\, locked\_by) \\
locked\_by' = \\
\quad locked\_by \cup \{para? \mapsto user?\} \\
\hline
\end{array}
\qquad
\begin{array}{|l}
\hline
\text{\_\_} Drop \text{\_\_\_\_\_\_\_\_\_\_\_\_\_\_\_\_\_\_\_\_\_} \\
\Delta(locked\_by) \\
user? : People \\
para? : Paragraph \\
\hline
\{para? \mapsto user?\} \in locked\_by \\
locked\_by' = \\
\quad locked\_by - \{para? \mapsto user?\} \\
\hline
\end{array}
$$

$$
\ldots\ldots UserLeave \mathrel{\widehat{=}} AnyUser \bullet Leave
$$
$$
CursorLeft \mathrel{\widehat{=}} AssociatedUser \bullet selected\_para?.cursor\_left
$$
$$
\ldots\ldots
$$

We next have to show how the syntax of the system works, i.e. how the
user concretely interacts with the system. This is done by specifying the allowed
interactions using CSP. For example we have

$$
\begin{aligned}
EDITPARA = \; & x : \{CursorLeft, CursorRight\} \rightarrow EDITPARA \\
& \mid EDITTEXT \\
& \mid Drop \rightarrow EDIT
\end{aligned}
$$

Here there is a choice: either the user can do one of $CursorLeft$ or $CursorRight$
or they can proceed directly with the process $EDITTEXT$ or they can do $Drop$
and the system behaves as the process $EDIT$.

$$
\begin{aligned}
EDITTEXT = \; & x : \{DeleteLeft, DeleteRight, \\
& \quad InsertLeft, InsertRight\} \rightarrow EDITPARA
\end{aligned}
$$

We finally introduce the specification style for the awareness aspects of such
a system, using the following presentation:

$$
\begin{array}{|ll}
\hline
\text{\_\_\_\_\_\_\_\_\_\_\_\_\_\_\_\_\_\_} & \\
x, y : People & \qquad\qquad \text{signature} \\
p : Paragraph & \\
\hline
present(x) & \qquad\qquad \text{precondition} \\
\hline
Take(x, p) \quad \mid \quad x & \qquad\qquad \text{action} \mid \text{agent} \\
\hline
aware(y, locked\_by(p) = x) & \qquad\qquad \text{postcondition} \\
\hline
\end{array}
$$

Specification and Implementation of a Random Access Data Type: A Case Study Using the B-Method
(Extended Abstract)

Ken Robinson

School of Computer Science and Engineering
University of New South Wales
Sydney NSW 2052, Australia

In this tutorial paper we explore the specification and implementation of a random access date type. The development is carried out within the discipline of the B-Method[1] and using the B-Toolkit[2]. The treatment presented here is intended to be the first part of a unified treatment of data types, in which the basic functionality of the data types is emphasized. This contrasts with the treatment given in conventional intermediate data structure treatments in which data types are differentiated on the basis of implementation specific functionality. The paper is a companion to the paper by Martin Schwenke[3].

A random access data type (RADT) allows the storage and retrieval of data (of type DATA) that is associated with a key (of type KEY). The following algebraic specification captures the behaviour of an *empty* RADTelement, and functions for updating a (key, data) binding and retrieving and removing a data element from a RADTobject. Note that the syntax used here is consistent with B.

$$empty \in \mathrm{RADT}$$

$$update \in (\mathrm{KEY} \times \mathrm{DATA}) \longrightarrow (\mathrm{RADT} \longrightarrow \mathrm{RADT})$$

$$retrieve \in \mathrm{KEY} \longrightarrow (\mathrm{RADT} \longrightarrow (\mathsf{BOOL} \times \mathrm{DATA}))$$

$$remove \in \mathrm{KEY} \longrightarrow (\mathrm{RADT} \longrightarrow \mathrm{RADT})$$

$$\forall\, key \bullet (key \in \mathrm{KEY} \Rightarrow ($$
$$\exists\, dd \bullet (dd \in \mathrm{DATA} \wedge retrieve(key)(empty) = \mathsf{FALSE} \mapsto dd)))$$

$$\forall(xx, kk, dd, rr) \bullet ((xx \in \mathrm{KEY} \wedge kk \in \mathrm{KEY} \wedge dd \in \mathrm{DATA} \wedge rr \in \mathrm{RADT}) \Rightarrow ($$
$$xx = kk \Rightarrow retrieve(xx)(update(kk \mapsto dd)(rr)) = \mathsf{TRUE} \mapsto dd) \wedge ($$
$$xx \neq kk \Rightarrow retrieve(xx)(update(kk \mapsto dd)(rr)) = retrieve(xx)(rr)))$$

$$\forall(xx, kk, rr) \bullet ((xx \in \mathrm{KEY} \wedge kk \in \mathrm{KEY} \wedge rr \in \mathrm{RADT}) \Rightarrow ($$
$$xx = kk \Rightarrow \exists\, dd \bullet (dd \in \mathrm{DATA} \wedge ($$
$$retrieve(xx)(remove(kk)(rr)) = \mathsf{FALSE} \mapsto dd))) \wedge ($$
$$xx \neq kk \Rightarrow retrieve(xx)(remove(kk)(rr)) = retrieve(xx)(rr)))$$

The above specification is captured in B by a stateless machine containing only constants.

We then define a machine with a deferred set RADT, a state consisting of a single instance and operations **UPDATE**, **RETRIEVE**, and **REMOVE** whose semantics are given entirely in terms of the functions defined in the algebraic specification.

The next step is to refine the previous machine to one in which the set RADT is represented by a partial function, but the operations are still defined abstractly. The subsequent step is to refine the operations using standard relational constructs to replace the abstract functions from the algebraic specification.

In the final steps we give a number of possible implementations: one using arrays, and another using hashing. The latter implementation is interesting as it involves two RADTs, each of which might be implemented differently.

The abstract framework behind this development is as old as abstract data types, but it is interesting to perform all of the development under a single formal development method and toolkit. We are also interested in this as a basis for the development of other data types.

References

1. J.-R. Abrial. *The B-Book*. Cambridge University Press, 1996.
2. B-Core(UK) Ltd, Oxford Science Park, Oxford UK. *B-Toolkit*, release 2.0 edition, 1994.
3. Martin Schwenke. High-level refinement of random access data structures. Abstract in this proceedings.

Modular Refinement Diagrams
(Extended Abstract)

Martin de Groot and Ken Robinson
martindg@cse.unsw.edu.au, k.robinson@unsw.edu.au

University of New South Wales, Sydney 2052, Australia

Once a system has been implemented, descriptions can be used to monitor the behaviour of that system. When used for monitoring purposes such descriptions are often called *models*. Using a model for systems monitoring is generally referred to as *model-based reasoning* (MBR), which includes the important area of model-based diagnosis (MBD). MBR employs models in many different forms. The most important feature of these models is that they can be automatically manipulated, (i.e. reasoned about), by a monitoring system. A typical solution to this requirement is to express the model in the Horn clause subset of predicate calculus and then write a Prolog program to implement and also manipulate that model.

We propose that formal program derivation effectively produces hierarchical models of the system being derived. A specification can be seen as representing any one level, while refinement ($\sqsubseteq$) defines the ordering of the models within the hierarchy. Refinement diagrams, first proposed by Ralph Back, can be used to represent this development process. This paper discusses the issue of how the components of a system can be identified from the program derivation. We show that *modular* refinement diagrams represent the decomposition process and, that modular refinement diagrams can be produced from any program derivation.

Automating the extraction of modular refinement sequences should be relatively easy. Given that a system has been developed using a formal development tool, there will be a record of every refinement step taken to produce the final implementation. This is sufficient to enable another tool to be produced which identifies every stage of the system refinement and constructs a corresponding modular refinement diagram. Thus, every formally developed system can be delivered with a completely accurate hierarchical model containing the actual breakdown into its constituent components, sub-components, sub-sub-components, etc.

Modular refinement diagram extraction example

In Figure 1 a program derivation is described that includes all the different kinds of refinement links defined for refinement diagrams. ¿From this development history a modular refinement diagram has been extracted (Figure 2).

This example shows that the extracted diagram includes the final version of all components and composite systems. All functional partitions found in Figure 2 persist through to the actual implementation. Also, components such as

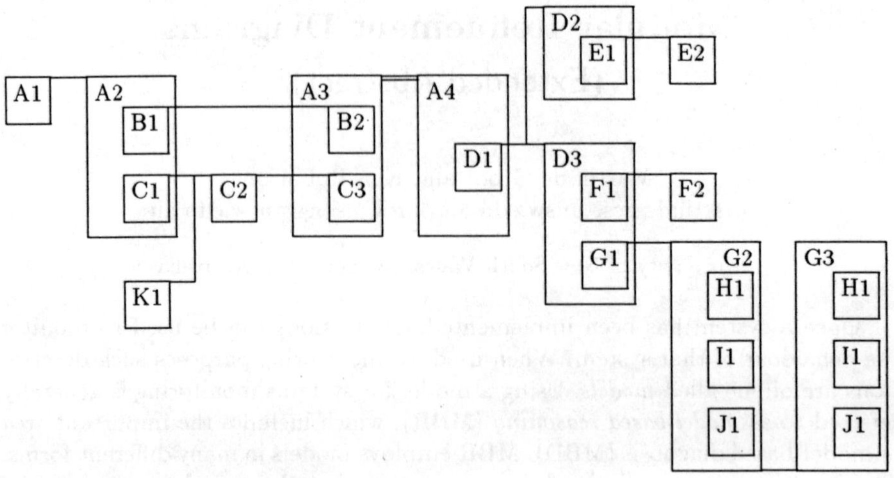

Fig. 1. Original refinement diagram

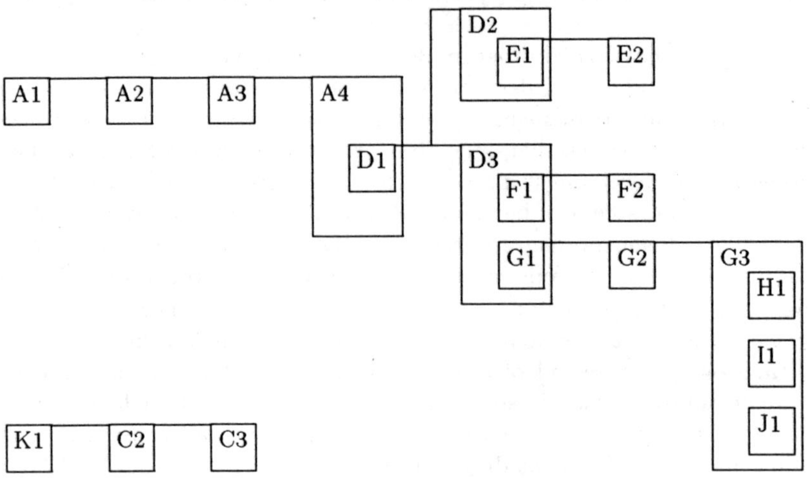

Fig. 2. Extracted modular refinement diagram

$B1$ and $C1$ are removed from the model because they are not part of the final implementation. The extended version of this paper includes the specification of a function that collapses composite systems into a single component. This function is used to produce the new version of systems such as $A2$ in the modular refinement diagram that has exactly the same functionality as the original three systems ($A2$, $B1$ and $C1$) combined. Thus modular refinement gives us the decomposition of a system into its actual components.

High-level Refinement of
Random Access Data Structures
(Extended Abstract)

Martin Schwenke

Department of Computer Science
The Australian National University
Canberra, ACT 0200, Australia
Martin.Schwenke@cs.anu.edu.au

Data refinement [3, 4] is a method for replacing abstract data types by concrete ones. This paper explores data refinement in the context of data abstraction and software reuse. There are two parts of the exploration:

- The construction of reusable data structures.
- The use of abstract data structures in a design process beginning with mathematical specifications.

These two notions are closely related. Many existing 'type classes' are based too strongly on implementation requirements. It is more fruitful to attempt to find common ground between the mathematical abstractions found in specification notations and related concrete implementations. A helpful suggestion is to distinguish kernel features from nonkernel ones (see Rüping *et al*[5]) so the that size of the structures being matched is smaller. This means that abstract and concrete types can be reconciled at a core level, rather than trying to reconcile large numbers of complex operations. Gries's transform [2] solves the problem of allowing abstract types to be added to programs at a syntactic level and uses data refinement as its implementation mechanism. Although Gries seems to advocate the use of mathematical abstractions, the problem of finding good reusable abstractions is still difficult.

The data types that are the primary vehicle for this paper are bulk types, which are used for storing homogeneous collections of data. Bulk types are pervasive in large software systems, at both the specification and implementation stages. They can be separated into two main groups:

- iterative data structures; and
- random access data structures.

Iterative data structures are those that are constructed and used in some particular order. The most common example of an iterative structure is a list. Lists, and related structures, have been subjected to enormous amounts of study in the field of program construction [1]. There are many high-level operations that manipulate lists and combinations of these operations are subject to equally high-level transformations. Derivation of programs that use variants of lists, including stacks, queues and sequential files, can benefit from a study of program transformations on lists.

There are similar benefits to be gained by studying *random access data structures* (RADs). In the area of mathematical specification, RADs correspond to limited forms of partial functions or mappings. At the implementation level, RADs correspond to things like arrays and hash-tables. Interestingly, they also correspond to things like association lists, which are also iterative structures. A data structure can be both an iterative structure and a random access data structure. However, it is important to remember that such hybrid structures belong to two families instead of a single, combined family that appears to be more common! There are many structures that allow random access operations to be performed but are not amenable to iterative operations.

So, what is a RAD? A RAD has the following kernel operations:

empty constructs an empty RAD.

update adds a key/data pair to a RAD, cancelling any previous association for the given key.

remove deletes a key and its associated data from a RAD.

retrieve returns the data associated with a key, if one exists.

Note that there are no operations that can be applied to all 'elements" of a RAD, because such operations belong to iterative structures.

Some implementations of RADs are discussed, as are the associated implementation techniques. One nice implementation is hash-tables using two nested RADs. The retrieve operation of the outer RAD has constant time complexity, while the inner RADs are buckets for resolving hashing clashes. Some of the implementations contain limitations such that their use entails proofs obligations. Some examples are given to illustrate how RADs can be carried through the design process, simplifying the process of producing an implementation from a mathematical specification.

References

1. R. Bird. An introduction to the theory of lists. Technical Monograph PRG-56, Oxford University Computing Laboratory, Programming Research Group, Oxford University, Oct. 1986.
2. D. Gries. Data refinement and the transform. In M. Broy, editor, *Deductive Program Design*, volume 152 of *Series F: Computer and Systems Sciences*, pages 205–232, Berlin Heidelberg, 1996. NATO Scientific Affairs Division, Springer-Verlag. Proceedings of the NATO Advanced Study Institute on Deductive Program Design, held in Marktoberdorf, Germany, July 26 – August 7, 1994.
3. C. A. R. Hoare. Proof of correctness of data representations. *Acta Informatica*, 1(4):271–281, 1972.
4. C. Morgan and P. H. B. Gardiner. Data refinement by calculation. In C. Morgan and T. Vickers, editors, *On the Refinement Calculus*, Formal Approaches to Computing and Information Technology, pages 85–110. Springer-Verlag, London, 1994.
5. A. Rüping, F. Weber, and W. Zimmer. Demonstrating coherent design — a data structure catalogue. In R. Ege, M. Singh, and B. Meyer, editors, *TOOLS 11 — Technology of Object-Oriented Languages and Systems*. Prentice Hall, 1993.

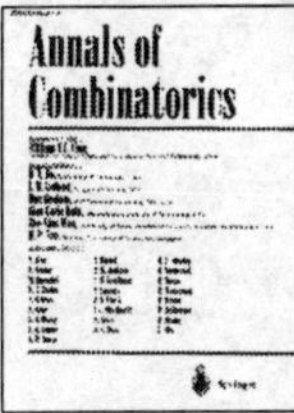

Annals of Combinatorics

Annals of Combinatorics will publish outstanding contributions to combinatorial mathematics in all its aspects. Special regard will be given to new developments which have yet to be given proper recognition, but which in the opinion of the editors show promise of eventual mathematical breakthroughs. Papers published in *Annals of Combinatorics* will not be limited to the field of combinatorics in the strict sense. They will range over problems and theories that have arisen, or will arise, in applications to computer science, biology, statistics, probability, physics and chemistry, as well as over work of a combinatorial nature in representation theory, number theory, topology, algebraic geometry and the theory of special functions.

Annals of Combinatorics will also publish research announcements and book reviews and will at all times ensure the efficient and speedy processing of submitted manuscripts.

Subscription Information
ISSN 0218-0006 Title no. 798
1997 Volume 1   4 issues per volume